Fundamentals of Combustion Engineering

Fundamentals of Combustion Engineering

Achintya Mukhopadhyay
Swarnendu Sen

CRC Press
Taylor & Francis Group
Boca Raton London New York

CRC Press is an imprint of the
Taylor & Francis Group, an **informa** business

CRC Press
Taylor & Francis Group
6000 Broken Sound Parkway NW, Suite 300
Boca Raton, FL 33487-2742

First issued in paperback 2020

ISBN-13: 978-1-4822-3330-8 (hbk)
ISBN-13: 978-0-367-73154-0 (pbk)

Visit the Taylor & Francis Web site at
http://www.taylorandfrancis.com

and the CRC Press Web site at
http://www.crcpress.com

To

Our Parents

Late Sambhu Prasad Mukherjee and Late Lila Mukherjee

and

Late Sukhendu Sen and Mrs. Lina Sen,

and to Our Wives and Daughters

Sumita, Susmita, Supurna and Riti

and to Our Respected Teachers and Beloved Students

Contents

Preface

Combustion of fuels has been the principal source of energy for humans since the dawn of civilisation. Combustion of fuels generates many products which are harmful to the environment. Consequently, much work has been done over the centuries to improve our understanding of the processes involved and thus to design more efficient and more environment-friendly energy conversion systems. Increasing concerns about adverse effects of combustion on the environment and dwindling fuel reserves have led to exploration of other sources of energy such as solar, wind, hydel and geothermal, which have the advantages of being environmentally benign and renewable. With the present level of technological development, however, we are still not in a position to completely depend on renewables alone as energy sources in the immediate future, and combustion is expected to remain an important source of energy for a considerable time. Thus, combustion remains an important subject of study, albeit with simultaneous adoption of new approaches. For example, in the immediate future, the world is expected to see a growth of hybrid systems where combustion systems would be integrated with renewable energy systems like solar or wind. Similarly, combustion itself is expected to see many changes, for example, use of newer grades of fuels such as biofuels, hydrogen and synthetic gas. The potential of combustion systems as high-power and high-energy systems makes them attractive for portable mechatronic and electronic systems, provided the challenges of stable combustion in micro-combustors can be adequately addressed. All these call for introduction of emerging concepts at an early stage to equip students today to become professional leaders of the future. The need for a textbook to address these emerging issues in continuity with the fundamental concepts of combustion has been the motivation behind *Fundamentals of Combustion Engineering*.

The book is primarily intended to serve as a textbook for senior-level undergraduate and postgraduate students in the disciplines of mechanical engineering, power engineering and aerospace engineering. But it can also be used by researchers and practising engineers in the field of combustion. The treatment of the topics assumes a basic undergraduate background in the subjects of thermodynamics, fluid mechanics, heat transfer and mathematics. The focus of the book is on fundamentals of combustion. But equal emphasis is given on the underlying physics and potential applications of the topics covered without sacrificing the essential mathematical rigour. Some of the topics included in the book are beyond the usual coverage of introductory textbooks on combustion. But we feel that exposure to these topics is essential for scientists and engineers to meet future challenges. The worked-out examples and the end-of-chapter problems also bring out the importance of the various topics in practical applications.

Fundamentals of Combustion Engineering is the outcome of several years of teaching and research on the subject by both of us. Most of the topics covered in this book have been taught by both of us at both senior-level undergraduate and postgraduate-level courses at Jadavpur University and Indian Institute of Technology Madras. The response of our students has given us confidence in the suitability of the topics at these levels. Combustion is a subject that has been developed over hundreds of years by different groups of people. Hence, it is not possible to give equal emphasis to all aspects of the subject between the two covers of a single textbook. Our personal preference and expertise through our teaching and research influence the choice of topics and the depth of coverage. We also feel that the contents of the book are more than what can be covered in a one-semester course. We leave it to the course instructors to adapt the contents of the book according to the objectives of the course and the composition of the class.

We are indebted to a number of persons without whom we would not have been able to complete this project. First, we would like to express our gratitude to our teachers who gave us our first exposure to the fundamental subjects and inspired us to pursue a career in teaching and research. We are especially indebted to Dipankar Sanyal and Ishwar K. Puri, who first introduced us to the world of combustion. We also owe a lot to our students, whose inquisitive questions and brilliant ideas, both in the classroom and in our research group, helped us improve our own understanding of the subject. The congenial atmosphere at Jadavpur University gave us the peace and tranquillity of mind for our academic pursuits in the form of both teaching and research. The culture of mutual cooperation among colleagues at Jadavpur enabled us to pursue our research and also improve our knowledge. A special mention must be made of our colleagues at Jadavpur: Dipankar Sanyal, Amitava Datta, Koushik Ghosh, Ranjan Ganguly, Saikat Mookherjee and Rana Saha, with whom we collaborated on a number of projects. The knowledge gained there has contributed in a big way towards writing this book. We are also deeply indebted to our collaborators, Ishwar K. Puri (McMaster University, Canada), Suresh K. Aggarwal (University of Illinois at Chicago, USA), Wolfgang Polifke (Technical University Munich, Germany) and Nilanjan Chakraborty (University of Newcastle, UK). We express our gratitude to our students who offered many perspectives on the topics covered in this book. It is difficult to mention all their names, but Aranyak Chakravarty, Sourav Sarkar, Somnath De, and Arijit Bhattacharya must be mentioned for their active cooperation.

Finally, we would like to pay homage to our parents, who braved many odds to bring us up. Without their sacrifice, we would have never reached this level. We are also grateful to our wives, Sumita and Susmita, and to our daughters, Supurna and Riti, for supporting us in all our professional activities.

We hope that *Fundamentals of Combustion Engineering* will be useful to both new learners of the subject and to experienced researchers and practising engineers. We look forward to feedback and suggestions from all.

Achintya Mukhopadhyay
Swarnendu Sen
Kolkata, India

Authors

Achintya Mukhopadhyay is a professor of Mechanical Engineering at Jadavpur University, Kolkata (Calcutta), India. He also served as professor of Mechanical Engineering at Indian Institute of Technology, Madras. Dr. Mukhopadhyay also held visiting positions at Technical University of Munich, where he was an Alexander von Humboldt Fellow, and at the University of Illinois at Chicago. Dr. Mukhopadhyay's teaching and research interests include thermodynamics, heat transfer, combustion, multiphase flows and design and analysis of thermal systems. Dr. Mukhopadhyay has more than 275 research publications to his credit, including more than 100 international journal publications. Dr. Mukhopadhyay is a fellow of the West Bengal Academy of Science and Technology and International Society for Energy, Environment and Sustainability and life member of Indian Society of Heat and Mass Transfer and Indian section of the Combustion Institute.

Swarnendu Sen is a professor of Mechanical Engineering at Jadavpur University, Kolkata (Calcutta), India. Dr. Sen held a visiting position at Technical University of Munich, where he was a DAAD fellow. He also held visiting positions at the University of Illinois at Chicago and Virginia Tech, Blacksburg, in the United States. He worked in HCL Ltd. and Development Consultants Ltd. as a graduate engineer in the area of design and analysis. His areas of interest include reacting and multiphase flow, magnetic fluid and nano-fluid transport, heat transfer augmentation and combustion synthesis of carbon nano-structures. Dr. Sen has approximately 300 research publications to his credit, including more than 100 international journal publications. Dr. Sen is a fellow of the West Bengal Academy of Science and Technology and International Society for Energy, Environment and Sustainability. He is a life member of the Indian section of the Combustion Institute; the Indian Society of Heating, Refrigerating and Air Conditioning Engineers; and the Indian Society of Heat and Mass Transfer.

Chapter 1

Introduction

1.1 INTRODUCTION TO COMBUSTION

What is combustion? The Oxford dictionary has provided an elegant answer – 'the process of burning something'. Then elaborated – 'Rapid chemical combination of a substance with oxygen, involving the production of heat and light'. The Webster's dictionary says, 'Rapid oxidation generating heat or both light and heat; also, slow oxidation accompanied by relatively little heat and no light'. These definitions provide some idea about the subject. The obvious next question is – why combustion? It is difficult to answer this in a word or in a sentence. So we look forward.

The global demand of energy is increasing every day. US Energy Information Administration (EIA) projects a 28% increase in energy use by 2040 [1]. Figure 1.1 shows the increase, by energy source, since 1990.

The major shareholders of energy resources are petro-product, coal and natural gas at present. Even after 20 years, this trend will not alter. Use of natural gas may cross the demand of coal at that time. Renewable energy generation took a higher gradient before a decade. It is still far behind the fossil fuels. Nuclear energy is not at all taking off because of the safety aspect mainly. In electricity generation also, fossil fuel contributes around 70% share. Figure 1.2 shows a chart for world electricity generation. We need to burn fossil fuel to generate energy. There is a need for efficient burning: one factor is cost, and the other one is not to overspend the fuel reserve for the sake of future use.

In the power sector, about 40% of power is generated by using coal. For the sake of efficient burning, various methods have been developed. With the new methodologies, not only we are burning coal efficiently but also low-grade coals are being burnt nowadays. The transport sector, on the other hand, is greatly dependent on oil firing. Day by day, the demand is growing for higher car mileage. This also requires a better technology of burning.

The increasing demand of energy is associated with increased emission. The different pollutants released by combustion devices already reached an

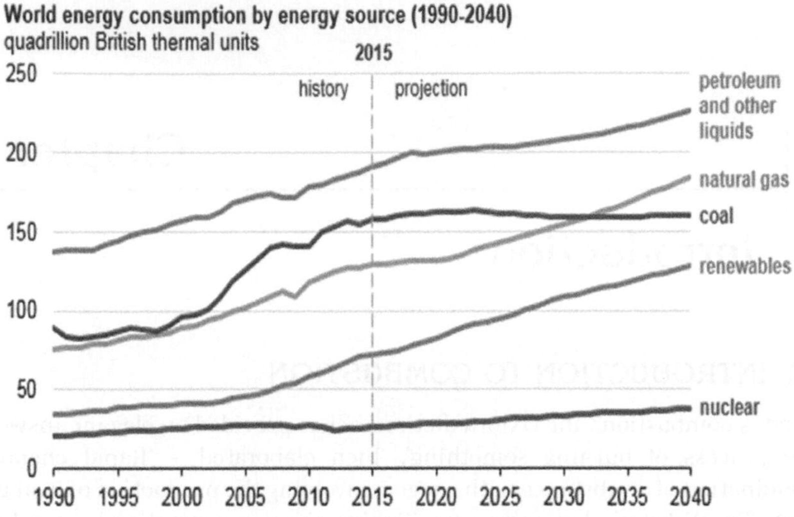

Figure 1.1 Predicting Energy Usage by Energy Source. (From Today in Energy, EIA document, 14 September 2015, https://www.eia.gov/todayinenergy/detail.php?id=32912.)

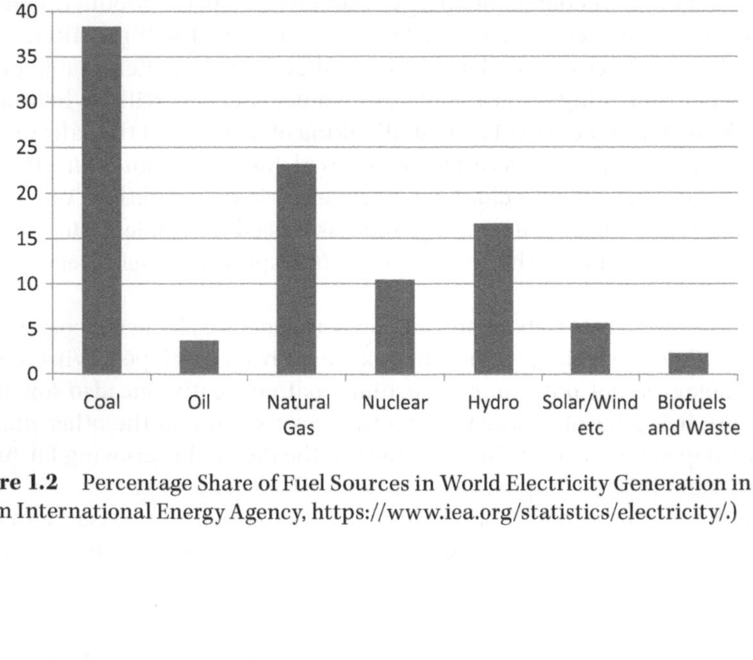

Figure 1.2 Percentage Share of Fuel Sources in World Electricity Generation in 2016. (From International Energy Agency, https://www.iea.org/statistics/electricity/.)

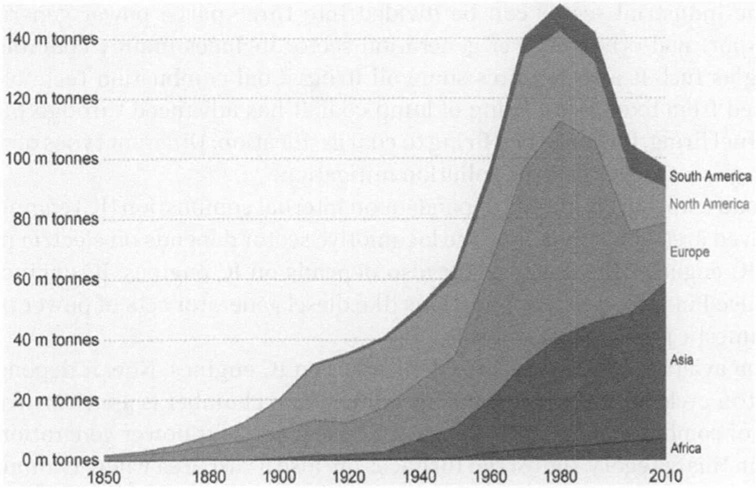

Figure 1.3 Worldwide SO_2 Emissions by Region. (From Ritchie, H., Roser, M., Air Pollution, Our World in Data, 2017, https://ourworldindata.org/air-pollution.)

alarming level. Figure 1.3 depicts the world SO_2 emission [2]. The figure shows that, for last two decades, emissions were reduced in some parts of our globe. In some places, however, emissions are still showing upward trends. Thus, awareness of the problem has been transformed into action in some parts of the world and some parts are yet to come up to the mark. However, the awareness and thus development of technology has brought a good effect on overall emissions around the world. The pollution norms are also becoming stricter day by day. To cope with these stringent criteria while maintaining the ever-growing energy demand calls for new technologies.

The above facts motivated people to develop the knowledge base of combustion. Study of combustion started on the day when humanity first lighted fire. Today, we understand many things about combustion. But still there is more darkness than light. We need an answer to the questions, Why combustion? We need torch bearers to tear away the darkness.

1.2 APPLICATIONS OF COMBUSTION

Applications of combustion can be classified broadly into two parts – domestic and industrial. The domestic applications cover cigarette lighters, kitchen burners, room-heating fireplaces and so on. Though the number of applications is not high, they do claim a good amount of understanding.

The industrial sector can be divided into three parts: power generation, transport and others. Power generation sector includes mainly coal and gas firing as fuel. It also requires some oil firing. Coal combustion technologies started from fixed grate firing of lump coal. It has advanced through pulverized fuel firing, fluidized bed firing to coal gasification. Different types of methodologies are developed for pollution mitigation.

Road transport is mainly dependent on internal combustion (IC) engines, an involved area of combustion. The locomotive sector depends on electric power and IC engines. The naval sector also depends on IC engines. IC engines are also used in land-based applications like diesel generator sets of power plants or domestic power generator sets.

The aviation sector was once dependent on IC engines. Now it depends on Brayton cycle based propulsion. The combustion chamber is another involved area of combustion. Land-based gas turbine plants, for power generation, are also in this category. Industrial furnaces are also a vast area where combustion technology has a very important role to play.

Rockets, missiles, gun shells are also propelled by combustion. But this is a different type of combustion. Fossil fuel is not the fuel here. We may call it solid propellant. Here, two chemicals start reacting in solid phase. Fire research is another area where experience in combustion is solicited.

1.3 APPROACHES TO COMBUSTION STUDY

The study approach of *Fundamentals of Combustion Engineering* starts with the thermodynamics of combustion chemistry, thermochemistry in short. The next chapter is on chemical kinetics. Chapter 4 deals with different simple reactor models, where spatial variation of properties is neglected. To know the detailed evaluation of the properties, we need to know the conservation equations. These are discussed in the next chapter. Chapters 6 and 7 develop an understanding about premixed and non-premixed flame, respectively. A brief idea about partially premixed flame is also covered. In the next chapter, fundamentals of liquid combustion are treated. This chapter deals with droplet and spray combustion. Chapter 9 contributes on understanding and analysis of turbulent combustion. Combustion of coal is covered in the next chapter. This chapter deals also with the preliminaries of solid propellants. Chapter 11 briefly discusses about combustion emissions and mitigation. The last chapter introduces some diagnostic techniques used in combustion.

The goal of this book is to impart an introductory but necessary exposure of the subject to beginners. The main focus is undergraduate/postgraduate students. The fundamental aspects of combustion along with some application-oriented basic features are discussed here. Those who want additional details may consult advanced books or other related materials.

1.4 TYPES OF COMBUSTION

Combustion can be divided into two groups. In some applications, fuel and oxidizers are mixed before going to a combustion chamber. When they burn, a flame is visible. The flame is acting as an interface. On one side of the flame, burnt gas is present. On the other side, there remains the unburned mixture. This type of combustion is called premixed combustion. In the other group, fuel and oxidizers enter into the combustion chamber as separate streams. They are mixed with each other and burn inside the combustion chamber. A flame is visible here also, and the flame is demarcating fuel at one side and oxidizer on the other. This type of flame is called a non-premixed flame.

In SI engine, fuel and air are supplied in the form of a homogeneous mixture. A spark then occurs to initiate a flame. The flame is normally initiated at the spark plug. The flame propagates through the mixture from one end to the other. This type of combustion is called a propagation mode. In a CI engine, in the contrary, fuel and air are supplied as separate streams. The combustion chamber temperature is then increased through compression by piston movement. After reaching a certain temperature (and pressure), the fuel ignites automatically. This is called explosion mode combustion.

Another classification can be based on the consideration of phases of the reactants. If two reactants (fuel and air) are in the same phase, it is called homogeneous reaction. Gaseous and liquid hydrocarbon fuels burn in this mode with air or oxygen. When coal and air react, as in thermal power stations, one is in solid and the other in gas phase. This is an example of heterogeneous reaction.

Combustion can be divided according to flame and flameless conversion. We have already talked about the combustion that includes flames, like premixed, non-premixed and so on. In some combustion the reaction is very slow and thus the heat release. In this combustion, the reaction front is not visible and hence it is called flameless or mild combustion. One can think of glucose burning in our cells, which is an example of mild combustion.

REFERENCES

1. Today in Energy, EIA document, 14 September, 2015 [https://www.eia.gov/todayinenergy/detail.php?id=32912].
2. Ritchie, H. and Roser, M., Air Pollution, Our World in Data, 2017 [https://ourworldindata.org/air-pollution].

Chapter 2

Thermodynamics of Reacting Systems

2.1 REVIEW OF THERMODYNAMICS

2.1.1 Thermodynamic Properties

Thermodynamic properties are of two types: intensive and extensive. **Intensive properties** are those whose values are independent of the mass of the system, while **extensive properties** are those whose values are proportional to the mass of the system. Examples of the former are pressure and temperature, while examples of the latter are volume, energy, enthalpy and entropy. Since extensive properties are proportional to the mass of the system, they can be converted to corresponding intensive properties by defining them on a unit mass or unit molar basis. These properties are referred to as **specific** properties. Thus, if the enthalpy of the system of mass m is denoted by *H*, specific enthalpy (on a unit mass basis) can be defined as:

$$h = H/m \tag{2.1}$$

We can also define specific property on a molar basis as:

$$h^* = H/N \tag{2.2}$$

N is the number of moles in the system. For a single component system, the mass and the number of moles are related through the molecular weight M of the substance as:

$$m = NM \tag{2.3}$$

The specific properties on mass and molar basis (for example, enthalpy) are related as:

$$h^* = Mh \tag{2.4}$$

In this book, specific properties will be defined both on the mass and the molar basis depending on convenience of analysis. We shall also follow the convention that extensive properties will be denoted by capital letters while the corresponding specific properties will be defined by the respective lowercase letters. The specific molar properties will be denoted throughout the text by the asterisk (*) superscript. The overbar often used to denote molar quantities will be reserved for mean quantities.

For a single-component, simple compressible system (i.e., a system whose only work mode is the compression work), the state of the system can be defined by two independent intensive (or specific) properties. For a multicomponent system, on the other hand, in addition to the two independent properties required to define the state of a single component system (e.g., any combination of pressure P, temperature T, specific volume v, specific internal energy u, specific enthalpy h and specific entropy s, provided they are independent), composition of the system in terms of the quantity of each constituent present in the multicomponent mixture needs to be specified.

Apart from property, another important concept from thermodynamics that will be used frequently in the coming chapters is the concept of equilibrium. A system is said to be in **equilibrium** with its surroundings if the state of the system (and hence the surroundings) does not change. This will be achieved when all the properties defining the state of the system have identical values with their counterparts in the surroundings. Subsets of complete equilibrium can be achieved if one of the properties of the system has identical value with that of the surroundings. When the system and the surroundings have the same **pressure**, they are said to be in **mechanical** equilibrium. **Thermal** equilibrium is attained through a common value of **temperature**. In addition, absence of changes in phase and/or composition of a multicomponent multiphase system implies equality of a property called **chemical potential**, which will be formally introduced later. Equality of chemical potential leads to **chemical** equilibrium. When a system is in mechanical, thermal and chemical equilibrium, it is said to be in **thermodynamic** equilibrium. Thermodynamic equilibrium indicates complete equilibrium between system and surroundings. A system in thermodynamic equilibrium cannot undergo any further change of state. Thus, the state of a system can be completely specified by a combination of any two of pressure, temperature, specific volume, specific enthalpy, specific internal energy and specific entropy and a set of chemical potentials. This will be used in later sections.

2.1.2 Property Relations

The relation between different equations of state is called **equation of state.** Most commonly, equation of state implies a relation between pressure, temperature and volume. The simplest form of equation of state is of the following form:

$$pV = NR^{*}T \tag{2.5}$$

In Equation 2.5, R^{*} has the value 8.314 kJ/kmolK for all gases and is known as the **universal gas constant.** Any gas which obeys the above equation of state is known as **ideal gas.** The basic premise of the ideal gas equation of state is the absence of intermolecular interaction. In reality, no gas fully satisfies Equation 2.5, known as the ideal gas equation of state. Equations of state for real gases are more complicated in form. However, most of the gases encountered in combustion approximately satisfy the ideal gas equation of state at pressures close to atmospheric pressure. Moreover, the high temperature encountered during combustion generally results in sufficiently low density. Hence in this book, only the ideal gas equation of state will be followed. Though such treatment is adequate for most applications, it is not valid at high (near critical) ambient pressures. This limit is often reached in connection with liquid hydrocarbon fuels in combustion engines and gas turbine combustors.

The above equation of state can also be expressed on a mass basis as follows:

$$PV = \frac{m}{M}R^{*}T = m\frac{R^{*}}{M}T = mRT \tag{2.6}$$

The quantity $R = \left(R^{*}/M\right)$ is called the **characteristic gas constant.** Unlike the universal gas constant, the value of the constant changes with gases. The ideal gas equation of state is expressed more conveniently in terms of specific properties on mass and molar basis as follows:

$$Pv = RT \tag{2.7}$$

or

$$P = \rho RT \tag{2.8}$$

and

$$Pv^{*} = R^{*}T \tag{2.9}$$

or

$$P = \rho^{*}R^{*}T \tag{2.10}$$

In the above equations, ρ and ρ^* denote mass density and molar density, respectively, which are reciprocals of the respective specific volumes. For simple compressible single-component systems, the state of a system is defined by two independent properties. Consequently, any thermodynamic property can be expressed as a function of two other independent thermodynamic properties. The relation between internal energy (or enthalpy) with temperature and specific volume (or pressure) is known as **calorific equation of state.** Mathematically this can be expressed as:

$$u = u(T, v) \tag{2.11}$$

and

$$h = h(T, P) \tag{2.12}$$

Thus, differential changes in these properties can be written as:

$$du = \left(\frac{\partial u}{\partial T}\right)_v dT + \left(\frac{\partial u}{\partial v}\right)_T dv \tag{2.13}$$

$$dh = \left(\frac{\partial h}{\partial T}\right)_P dT + \left(\frac{\partial h}{\partial P}\right)_T dP \tag{2.14}$$

The specific heats at constant pressure and constant volume are defined as:

$$C_v = \left(\frac{\partial u}{\partial T}\right)_v \tag{2.15}$$

and

$$C_p = \left(\frac{\partial h}{\partial T}\right)_P \tag{2.16}$$

The ideal gas equation of state is a necessary and sufficient condition for u and h to be functions of T only [1]. Thus, for ideal gases:

$$C_v = \frac{du}{dT} \tag{2.17}$$

$$C_p = \frac{dh}{dT} \tag{2.18}$$

The specific heats for ideal gases are therefore functions of temperature only. This temperature dependence can sometimes be neglected for small changes in temperature. However, in combustion, large changes in temperature are encountered. Hence in this book, this variation has to be considered unless explicitly omitted for simplicity of calculations. Integrating Equations 2.17 and 2.18, we obtain:

$$u(T) = u_{ref} + \int_{T_{ref}}^{T} C_v(T)dT \tag{2.19}$$

$$h(T) = h_{ref} + \int_{T_{ref}}^{T} C_p(T)dT \tag{2.20}$$

In an analogous manner, one can also define specific internal energy, specific enthalpy and specific heats on a molar basis also as follows:

$$C_v^* = \frac{du^*}{dT} \tag{2.21}$$

$$C_p^* = \frac{dh^*}{dT} \tag{2.22}$$

Integrating Equations 2.21 and 2.22, we obtain:

$$u^*(T) = u_{ref}^* + \int_{T_{ref}}^{T} C_v^*(T)dT \tag{2.23}$$

$$h^*(T) = h_{ref}^* + \int_{T_{ref}}^{T} C_p^*(T)dT \tag{2.24}$$

2.1.3 Mixture of Gases

For a mixture of gases, in addition to the properties required to define the state of a single component system, one needs to specify the composition of the mixture. Thus, for an N-component mixture, composition of each of the components is essential. Consistent with the use of intensive properties for definition of thermodynamic states, one can specify the composition of the mixture in terms of **mass fraction** (y_i), defined as the ratio of mass of ith component (m_i) to total mass of the mixture (m). Thus, we have:

$$y_i = \frac{m_i}{m} \tag{2.25}$$

It should be noted that in a mixture with N components, one needs to define only N–1 mass fractions since the Nth mass fraction can be derived from the following constraint:

$$\sum_{i=1}^{N} y_i = 1 \tag{2.26}$$

One can also specify the mixture composition in terms of mole fraction (x_i), defined as the ratio of the number of moles of ith component (N_i) to the total number of moles in the mixture (N). Thus the expression of mole fraction is given by:

$$x_i = \frac{N_i}{N} \tag{2.27}$$

Obviously, mole fractions also satisfy the constraint:

$$\sum_{i=1}^{N} x_i = 1 \tag{2.28}$$

One also needs to introduce the concepts of partial volume and partial pressure in the context of multicomponent gas mixtures. **Partial pressure (p_i)** is defined in the context of mixtures occupying a fixed volume V as the pressure exerted by the ith component if it alone occupies the entire volume at the same temperature T as that of the mixture. **Partial volume (V_i)** is defined as the volume occupied by the ith component alone at the same pressure P and temperature T of the mixture. The absence of intermolecular interactions in an ideal gas mixture leads to the following law:

1. ***Dalton's Law of Partial Pressure:*** In an ideal gas mixture, total pressure of the mixture is equal to the sum of the partial pressures of its constituent ideal gases.
2. ***Amagat's Law of Partial Volume:*** In an ideal gas mixture, total volume of the mixture is equal to the sum of the partial volumes of its constituent ideal gases.

From the definition of partial pressures, one can write:

$$p_i V = N_i R^* T \tag{2.29}$$

For the whole mixture one can similarly write:

$$PV = NR^* T \tag{2.30}$$

From the definition of mole fraction, it therefore follows that:

$$p_i = x_i P \tag{2.31}$$

It is easy to verify that this definition of partial pressure satisfies Dalton's Law of Partial Pressure. Likewise, one can write for partial volume:

$$V_i = x_i V \tag{2.32}$$

It is also convenient to derive the relation between mole fraction and mass fraction since conversion between the two will often be needed in the subsequent sections:

$$y_i = \frac{m_i}{\sum_{i=1}^{N} m_i} = \frac{N_i M_i}{\sum_{i=1}^{N} N_i M_i} = \frac{\frac{N_i}{N} M_i}{\sum_{i=1}^{N} \frac{N_i}{N} M_i} = \frac{x_i M_i}{\sum_{i=1}^{N} x_i M_i} \tag{2.33}$$

Likewise, one can also define:

$$x_i = \frac{N_i}{\sum_{i=1}^{N} N_i} = \frac{\frac{m_i}{M_i}}{\sum_{i=1}^{N} \frac{m_i}{M_i}} = \frac{\frac{m_i / M}{M_i}}{\sum_{i=1}^{N} \frac{m_i / M}{M_i}} = \frac{y_i / M_i}{\sum_{i=1}^{N} y_i / M_i} \tag{2.34}$$

For real gases, due to intermolecular interactions, mixture properties like internal energy depend in a complex manner on the properties of the constituent gases. However, in ideal gases, such intermolecular interactions are absent. Hence the extensive properties of the mixture (like volume V, internal energy U, enthalpy H and entropy S) are simply the sum of the corresponding extensive properties of the constituent gases. Thus, for enthalpy of a mixture of N_{comp} ideal gases, one can write:

$$H_{mix} = \sum_{i=1}^{N_{comp}} H_i \tag{2.35}$$

One can also define specific properties (e.g., specific enthalpy) for a mixture as:

$$h_{mix} = \frac{H_{mix}}{m} = \frac{\sum_{i=1}^{N_{comp}} H_i}{m} = \frac{\sum_{i=1}^{N_{comp}} m_i h_i}{m} = \sum_{i=1}^{N_{comp}} \frac{m_i}{m} h_i = \sum_{i=1}^{N_{comp}} y_i h_i \tag{2.36}$$

Similarly, for a molar basis, it is possible to write:

$$h_{mix}^{*} = \frac{H_{mix}}{N} = \frac{\sum_{i=1}^{N_{comp}} H_i}{N} = \frac{\sum_{i=1}^{N_{comp}} N_i h_i^{*}}{N} = \sum_{i=1}^{N_{comp}} \frac{N_i}{N} h_i^{*} = \sum_{i=1}^{N_{comp}} x_i h_i^{*} \tag{2.37}$$

2.1.4 First Law of Thermodynamics

The First Law of Thermodynamics is a statement of conservation of energy. For a **closed** system (i.e., a system with only energy transfer but no mass transfer with the surroundings across the system boundary), the First Law of Thermodynamics can be expressed as:

$$\delta Q = dE + \delta W \tag{2.38}$$

Here δQ and δW denote the small heat and work transfers across the system boundary during an infinitesimal change of state of the system identified by an infinitesimal change in thermodynamic property dE, where E denotes the total energy of the sum and is expressed as the sum of internal energy U, kinetic energy, KE and potential energy PE. The symbols δ and d have been used (and in the following sections also) to denote *inexact* (corresponding to quantities which are not thermodynamic properties) and *exact* (corresponding to thermodynamic properties) differentials, respectively. Thus, one can write:

$$E = U + KE + PE \tag{2.39}$$

In most of the cases relevant to combustion, with the exception of supersonic combustion, changes in kinetic and potential energies are negligible compared to the change in internal energy. Thus, one can approximate Equation (2.38) as:

$$\delta Q = dU + \delta W \tag{2.40}$$

In Equations (2.38) and (2.40), all heat entering a system and all work done by the system are considered positive. This sign convention will be followed consistently throughout the text. However, many of the combustion devices like gas turbine combustors and fuel-fired furnaces involve inflow and outflow of reactants and products; hence it is not convenient to express the energy balance in terms of First Law of Thermodynamics for closed systems.

An alternative and more convenient form is the First Law of Thermodynamics for control volumes, which can be written in the following form:

$$\frac{dU_{CV}}{dt} = \sum_i \dot{m}_i h_i - \sum_e \dot{m}_e h_e + \dot{Q} - \dot{W}_{CV} \tag{2.41}$$

The symbol $\dot{m}$ denotes mass flow rate. The above equation accounts for multiple inlets and multiple exits. It should be noted here that use of enthalpy to denote the energy associated with incoming and outgoing streams is due to inclusion of work done by or against the surrounding pressure during inflow and outflow. Thus, these works, often referred to as flow work, are not included in $\dot{W}_{CV}$.

2.1.5 Second Law of Thermodynamics

The Second Law of Thermodynamics can be expressed mathematically as:

$$dS \geq \frac{\delta Q}{T} \tag{2.42}$$

In the above equation, S denotes entropy, which is a thermodynamic property, and T is the absolute temperature. The inequality in Equation (2.42) corresponds to irreversible processes, while the equality is valid for reversible processes. Equation (2.42) can be converted to an equality by adding a non-negative term, δS_{gen} as follows:

$$dS = \frac{\delta Q}{T} + \delta S_{gen} \tag{2.43}$$

Here, δS_{gen} is zero for reversible processes and positive for irreversible processes. Thus, the term represents entropy generation in a system due to irreversibilities in the process and is referred to as **entropy generation**. For a control volume, one can express the Second Law of Thermodynamics as:

$$\frac{dS_{CV}}{dt} = \sum_i \dot{m}_i s_i - \sum_e \dot{m}_e s_e + \sum_j \frac{\dot{Q}_j}{T_j} + \dot{S}_{gen} \tag{2.44}$$

One can also write:

$$\delta Q_{rev} = TdS \tag{2.45}$$

Also, for a simple compressible system in absence of frictional irreversibilities one can write:

$$\delta W_{rev} = PdV \tag{2.46}$$

Substituting Equations (2.45) and (2.46) in Equation (2.40), one can write:

$$TdS = dU + PdV \tag{2.47}$$

Using the definition of enthalpy, one can also express Equation (2.47) as:

$$TdS = d(H - PV) + PdV = dH - VdP \tag{2.48}$$

Equations (2.47) and (2.48) are two fundamental equations in thermodynamics. It may be noted that, although the equations were derived for reversible processes, they involve only changes in thermodynamic properties and are therefore valid for all processes (i.e., they are path independent). However, the physical significances implied in Equations (2.45 and 2.46) are valid only for reversible processes.

Equations (2.47 and 2.48) can be used for evaluating entropy changes in a process. For unit mass of a system, one can write:

$$ds = \frac{1}{T}du + \frac{P}{T}dv \tag{2.49}$$

$$ds = \frac{1}{T}dh - \frac{v}{T}dP \tag{2.50}$$

For an ideal gas using Equations 2.17, 2.18 and 2.7, the entropy change can be expressed as:

$$ds = \frac{1}{T}C_v dT + \frac{R}{v}dv \tag{2.51}$$

$$ds = \frac{1}{T}C_p dT - \frac{R}{P}dP \tag{2.52}$$

Integrating Equations (2.51) and (2.52), one obtains:

$$s(T,P) = s_{ref} + \int_{T_{ref}}^{T} C_v(T)\frac{dT}{T} + R\ln\frac{v}{v_{ref}} \tag{2.53}$$

$$s(T,P) = s_{ref} + \int_{T_{ref}}^{T} C_p(T)\frac{dT}{T} - R\ln\frac{P}{P_{ref}} \tag{2.54}$$

It is important to note that, unlike internal energy and enthalpy, entropy is not a function of temperature alone, even for an ideal gas. For a mixture of ideal gases at a given pressure, the specific entropy of each constituent is given by:

$$s_i(T,P) = s_{i,ref} + \int_{T_{ref}}^{T} C_{p_i}(T)\frac{dT}{T} - R_i\ln\frac{p_i}{P_{ref}} \tag{2.55}$$

In the above equation, p_i denotes partial pressure. The mixture specific entropy is obtained by calculating the following:

$$s(T,P)=\sum_{i=1}^{N} y_i s_i(T,p_i) \tag{2.56}$$

2.2 FUELS

Most of the common fuels used today are hydrocarbons with the generic molecular formula C_xH_y. Common fuels having this generic molecular formula belong to the class of aliphatic hydrocarbons, for example, alkanes (C_xH_{2x+2}), alkenes (C_xH_{2x}) and alkynes (C_xH_{2x-2}). The same generic formula is also applicable to aromatic hydrocarbons (e.g., benzene, C_6H_6).

The structures of alkanes (e.g., methane, CH_4; ethane, C_2H_6,; propane, C_3H_8 etc.), alkenes (e.g., ethene or ethylene, C_2H_4; propylene or propene, C_3H_6 etc.) and alkynes (e.g., ethyne or acetylene, C_2H_2 etc.) are shown in Figure 2.1. The compounds containing only single bonds between carbon atoms are known as saturated hydrocarbons; those with double or triple bonds are known as unsaturated hydrocarbons. Generally, double or triple bonds in the unsaturated hydrocarbons make these compounds more reactive. Aromatic hydrocarbons are generally formed during soot formation. Consequently, many of these aromatic hydrocarbons are considered soot precursors. The formation of these aromatic hydrocarbons is an important component in the study of soot formation. Alcohols, in which one or more of the H atoms of the hydrocarbons

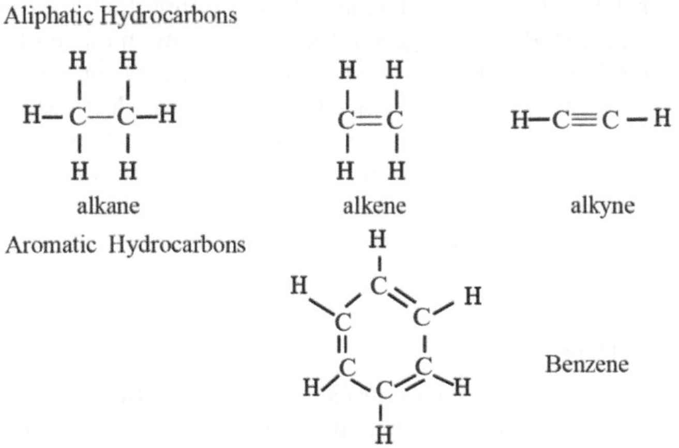

Figure 2.1 General structure of common hydrocarbons.

TABLE 2.1 SOME COMMON FUELS AND THEIR COMPOSITIONS

Substance	Density (kg/m^3)	Composition (Volumetric %)
LPG (Liquid)	540–560	$C_3H_8 = 18$, $C_4H_{10} = 80$, Higher Hydrocarbons (HHC) = rest
Biogas	1.1–1.2	$CH_4 = 60–80$, $CO_2 = 40–20$
Producer gas (wood)	0.9–1.2	$CO = 16–20$, $H_2 = 16–18$, $CO_2 = 8–10$, N_2 = rest, HHC = trace
Blue water gas	0.65	$H_2 = 50$, $CO = 40$, $CO_2 = 6$, N_2 and others = rest
Coke oven gas	0.40	$H_2 = 54$, $CH_4 = 24$, $CO = 8$, $CO_2 = 6$, N_2 and HHC = rest.

Source: Mukunda, H.S., "Understanding combustion," *Universities Press (India) Private Limited Publication.*

is replaced by an OH radical, are also important as fuels. In particular, alcohols like methanol and ethanol are obtainable from biological sources and are considered as promising alternative fuels. The general molecular formula for alcohols is $C_xH_yO_z$. Most of the common fuels are, however, blends of several hydrocarbons. Common liquid fuels like gasoline, diesel and kerosene contain hundreds of hydrocarbons. Even common gaseous fuels like natural gas (predominantly CH_4 with a small amount of C_3H_8) and liquefied petroleum gas (LPG; primarily a mixture of C_4H_{10} and C_3H_8) are hydrocarbon blends. Though a majority of the fuels are hydrocarbons or their derivatives, there are a number of important inorganic fuels. Possibly, the most important non-hydrocarbon fuel is hydrogen. Hydrogen is gaining importance in the face of dwindling reserves of fossil fuels and also because of its high energy content. Other than hydrogen, inorganic compounds like hydrogen peroxide (H_2O_2), ammonia (NH_3) and hydrazine (N_2H_4) are also considered fuels. Carbon monoxide (CO) is also an important carbon-containing inorganic fuel. Synthetic gas obtained from coal gasification primarily contains CO and H_2 as the combustible components. Compositions of some common fuels are listed in Table 2.1.

2.3 STOICHIOMETRY

Combustion reactions are essentially oxidation of fuels. The most common oxidiser used is air, though in recent years, combustion in pure oxygen (oxyfuel combustion) or combustion in environments with enhanced oxygen concentration (oxygen-enhanced combustion) is being explored as these techniques offer benefits like higher flame temperature and lower NO_x emission. The overall or global chemical reaction for a fuel with molecular formula C_xH_y can be expressed as:

$$C_xH_y + \left(x + \frac{y}{4}\right)O_2 = xCO_2 + \frac{y}{2}H_2O \tag{2.57}$$

The above equation implies that 1 mole of the fuel requires $\left(x + \frac{y}{4}\right)$ moles of O_2 for complete combustion to produce x moles of CO_2 and $y/2$ moles of H_2O. For fuels containing oxygen, like alcohols, part of the oxygen requirement is supplied by the fuel itself. Thus, the global reaction can be expressed as:

$$C_xH_yO_z + \left(x + \frac{y}{4} - \frac{z}{2}\right)O_2 = xCO_2 + \frac{y}{2}H_2O \tag{2.58}$$

The quantity of oxygen needed for combustion of unit quantity of fuel is known as the **oxygen-fuel ratio.** This ratio can be defined both on a mass basis (mass of oxygen needed for combustion of unit mass of fuel) and a molar basis (moles of oxygen needed for combustion of 1 mole of fuel). On a molar basis, the oxygen-fuel ratio for the chemical reaction represented by Equation (2.58) is $\left(x + \frac{y}{4} - \frac{z}{2}\right):1$. The corresponding value on a mass basis is given by $\left(x + \frac{y}{4} - \frac{z}{2}\right)M_{O_2} : M_F$ where M_F and M_{O_2} are the molecular weights of the fuel and oxygen, respectively. The reciprocal of the oxygen-fuel ratio is known as the **fuel-oxygen ratio.** As mentioned previously, air is the most common source of oxygen. The constituents of air, other than oxygen, mostly do not take part in the oxidation. Collectively they are denoted as nitrogen and often referred to as **atmospheric nitrogen.** In air, oxygen constitutes 21% on a molar (or equivalently volume) basis. Thus, with every mole of O_2, $79/21 \approx 3.76$ moles of N_2 are present. Although this nitrogen does not take part in the chemical reaction during oxidation of fuel, this additional mass alters the physical properties like temperature and mole fraction and needs be included in the chemical reaction. Thus, Equation (2.58) can be modified to:

$$C_xH_yO_z + \left(x + \frac{y}{4} - \frac{z}{2}\right)(O_2 + 3.76N_2) = xCO_2 + \frac{y}{2}H_2O + 3.76\left(x + \frac{y}{4} - \frac{z}{2}\right)N_2 \tag{2.59}$$

Analogous to the oxygen-fuel ratio, one can also define an **air-fuel ratio** as the amount of air needed for complete combustion of a unit quantity of fuel. The air-fuel ratio can also be defined on both a mass and a molar basis. On a mass basis, the air-fuel ratio for the reaction represented by Equation (2.59) is given by $4.76(x+\frac{y}{4}-\frac{z}{2})M_{air} : M_F$, where M_{air} is the molecular weight of air and is generally taken as 28.84kg/kmol. The reciprocal of the air-fuel ratio is known as **fuel-air ratio.** When fuel and air are present in chemically correct proportion (that is, at the end of the reaction there is neither any unburnt fuel nor any unused air), they are said to be present in **stoichiometric proportion**, and the corresponding air-fuel ratio is known as the **stoichiometric air-fuel ratio.** A fuel-air mixture containing excess air is known as a **lean** mixture; that containing excess fuel is known as a **rich** mixture. Thus, the air-fuel ratio for a lean mixture is higher and that for a rich mixture is lower than the stoichiometric air-fuel ratio. Since the stoichiometric air-fuel ratio is different for different fuels, it is difficult to understand whether a mixture is rich or lean from knowing the air-fuel ratio alone without reference to the stoichiometric value. A more convenient way of expression is the **equivalence ratio (φ)**, which is defined as the ratio of stoichiometric air-fuel ratio to the actual air-fuel ratio. Thus, for a lean mixture, the equivalence ratio is less than 1, while for a rich mixture it is greater than 1. One can also define the following:

$$\%\ \text{theoretical air} = \frac{\text{Actual air-fuel ratio}}{\text{Stoichiometric air-fuel ratio}} \times 100 = \frac{100}{\varphi}$$

One can also define the percentage of excess air as:

$$\begin{aligned}\%\ \text{excess air} &= \frac{(\text{Actual air-fuel ratio}) - (\text{Stoichiometric air-fuel ratio})}{(\text{Stoichiometric air-fuel ratio})} \times 100 \\ &= 100\left(\frac{1}{\varphi} - 1\right)\end{aligned}$$

Thus, a mixture with an equivalence ratio of 0.8 can also be expressed as a mixture with 120% theoretical air or 20% excess air.

For lean mixtures, the product will contain some oxygen due to the presence of excess air. For a lean mixture with an equivalence ratio $\varphi(<1)$ the chemical reaction can be expressed as:

$$\begin{aligned}C_xH_yO_z + \frac{x+\frac{y}{4}-\frac{z}{2}}{\varphi}(O_2+3.76N_2) = xCO_2 + \frac{y}{2}H_2O + \left(\frac{1}{\varphi}-1\right)\left(x+\frac{y}{4}-\frac{z}{2}\right)O_2 \\ + \frac{3.76}{\varphi}\left(x+\frac{y}{4}-\frac{z}{2}\right)N_2\end{aligned} \tag{2.60}$$

For rich mixtures ($\varphi > 1$), on the other hand, the oxygen (or air) supplied is inadequate for complete oxidation of the fuel to CO_2 and H_2O. Instead, partially oxidised products like CO, H_2 or even C may form. The chemical reaction may be expressed as:

$$C_xH_yO_z + \frac{x+\frac{y}{4}-\frac{z}{2}}{\varphi}(O_2+3.76N_2) = aCO_2 + bH_2O + cCO + dH_2 + \frac{3.76}{\varphi}\left(x+\frac{y}{4}-\frac{z}{2}\right)N_2 \tag{2.61}$$

Here *a, b, c* and *d* are the unknown molar compositions for the four species. The equations needed to solve for these unknowns come from the elemental mass balances. However, only three such conditions are available from the mass balance of C, H and O atoms. Thus, additional relations are needed. These relations are obtained from considerations of chemical equilibrium discussed in Section 2.1.3 and involve rather complicated calculations. Sometimes, an easier, though less accurate, alternative approach is adopted for quick calculations. This is done by assigning priority to the different oxidation reactions. The reactions in order of decreasing priority are as follows:

$$2H_2 + O_2 = 2H_2O \tag{2.62}$$

$$C + \frac{1}{2}O_2 = CO \tag{2.63}$$

$$CO + \frac{1}{2}O_2 = CO_2 \tag{2.64}$$

The implication of the above sequence is that all the hydrogen would be first oxidised to H_2O. Then the remaining oxygen will be used for oxidizing C to CO, and finally the remaining O_2 will be used to oxidise CO to CO_2. If sufficient O_2 is not available for oxidation of all C atoms to CO, the combustion products will contain C particles. Thus, it is important to determine this threshold limit of oxygen content for a rich mixture.

Sometimes the exact fuel composition is unknown. In such cases, the fuel composition and the mixture equivalence ratio can be determined from the knowledge of the composition of the combustion products, often referred to as **exhaust gas** or **flue gas**. The composition can be specified in terms of mass fraction (**gravimetric**) or mole fraction (**volumetric**). In addition, one can specify the composition considering the water vapour in the exhaust gas or after condensing the water vapour. The former is known as **wet basis**; the latter is known as **dry basis**.

Example 2.1

Liquefied petroleum gas (LPG) can be considered as a mixture of 40% propane and 60% butane on a volumetric basis. Calculate the mass of air required per kg of fuel if the gas is to be burned at an equivalence ratio of 0.8. Also calculate the composition of the products of combustion on a dry volumetric basis.

Solution

The stoichiometric equations for propane and butane are:

$$C_3H_8 + 5(O_2 + 3.76N_2) = 3CO_2 + 4H_2O + 18.8N_2$$

$$C_4H_{10} + 6.5(O_2 + 3.76N_2) = 4CO_2 + 5H_2O + 24.44N_2$$

Hence, the reaction for LPG-air mixture at an equivalence ratio of 0.8 is given by:

$$(0.4C_3H_8 + 0.6C_4H_{10}) + \frac{0.4\times5 + 0.6\times6.5}{0.8}(O_2 + 3.76N_2)$$
$$= 3.6CO_2 + 4.6H_2O + 27.73N_2 + 1.475O_2$$

Air-fuel ratio (by mass) $= [1.25\times5.9\times(32 + 3.76\times28)] : [0.4\times44 + 0.6\times58] = 19.605:1$

Product composition on a dry volumetric (molar) basis:

$$n_{CO_2} = 3.6 \ (11\%)$$

$$n_{O_2} = 1.475 \ (4.5\%)$$

$$n_{N_2} = 27.73 \ (84.5\%)$$

Example 2.2

The products of combustion of a saturated aliphatic hydrocarbon fuel of unknown composition have 8% CO_2, 0.9% CO, 8.8% O_2 and 82.3% N_2 measured on volumetric dry air basis. Calculate the actual air-fuel ratio, composition of the fuel and the % theoretical air.

Solution

Let us do the calculations on a basis of 100 kmol of dry products. The chemical reaction can be written as:

$$aC_xH_y + b(O_2 + 3.76N_2) = 8CO_2 + 0.9CO + 8.8O_2 + 82.3N_2 + cH_2O$$

Since we are considering a saturated aliphatic hydrocarbon, $y = 2x + 2$. From elemental mass balance:

C balance: $ax = 8.9$
H balance: $ay = 2(x+1)a = 2c$

N balance: $3.76b = 82.3$
O balance: $2b = 16 + 0.9 + 17.6 + c$

Solving, we obtain $a = 0.36, b = 21.88, c = 9.26, x = 24.72$. Thus, the chemical formula of the fuel is given by $C_{24.72}H_{51.44}$.

On a volume basis, stoichiometric oxygen-fuel ratio $= 24.72 + \dfrac{51.44}{4} = 37.58:1$
Actual oxygen-fuel ratio $= \dfrac{21.88}{0.36} = 60.78:1$
Hence, % theoretical air $= \dfrac{60.78}{37.58} \times 100\% = 161.73\%$

2.4 FIRST LAW FOR REACTING SYSTEMS

The starting point of thermodynamic analysis of chemically reacting systems is the application of the laws of thermodynamics to chemically reacting systems. Although the basic premise remains the same as that of non-reacting systems, the multicomponent nature of the system and change in chemical composition introduces certain complexities to the analysis, as will be discussed in the following sections.

Let us first consider a non-flow reactor. For a non-flow reactor with negligible changes in kinetic and potential energies, the First Law of Thermodynamics for a non-cyclic process states (Figure 2.2):

$$Q_{12} = \Delta U + W_{12} \tag{2.65}$$

For a constant volume reactor, no work is done by the system. We have $W_{12} = 0$. So, Equation (2.65) becomes:

$$\Delta U = U_P - U_R = Q_{12} \tag{2.66}$$

For a constant pressure non-flow reactor, the work done is given by:

$$W_{12} = P\left(V_P - V_R\right) \tag{2.67}$$

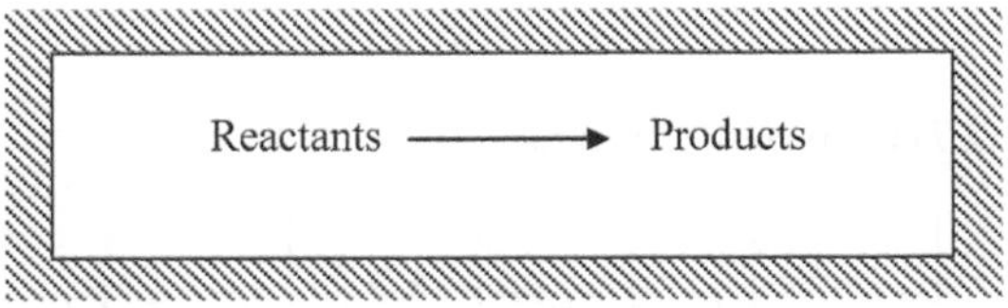

Figure 2.2 Reacting system.

The energy balance on the reactor gives:

$$Q_{12} = U_P - U_R + P(V_P - V_R) = H_R - H_P = \Delta H \tag{2.68}$$

Although non-flow reactors are encountered in practice, one encounters steady flow constant pressure reactors more commonly. For such reactors, neglecting changes in kinetic and potential energies, one can write:

$$\dot{m}_R h_R + \dot{Q} = \dot{m}_P h_P + \dot{W} \tag{2.69}$$

Very often in these reactors, no or negligible work transfer occurs. Thus the heat transfer is given by:

$$\dot{Q} = \dot{m}_P h_P - \dot{m}_R h_R \tag{2.70}$$

For steady flow reactors, one can write $\dot{m}_P = \dot{m}_R = \dot{m}$ (say). Thus the above equation can be written as:

$$\frac{\dot{Q}}{\dot{m}} = h_P - h_R \tag{2.71}$$

For most situations, it is adequate to assume that the reactants and the products are mixtures of ideal gases. This assumption becomes questionable, however, at near critical conditions. For ideal gases, the mixture enthalpy is given by:

$$h_R = \sum_i y_{R_i} h_{R_i} \tag{2.72}$$

$$h_P = \sum_i y_{P_i} h_{P_i} \tag{2.73}$$

For individual species, the enthalpy is given by:

$$h_i(T) = h_{ref_i} + \int_{T_{ref}}^{T} C_{p_i}(T)\,dT \tag{2.74}$$

Thus Equation (2.71) can be written as:

$$\frac{\dot{Q}}{\dot{m}} = \sum_i y_{P_i}\left[h_{P_i,ref} + \int_{T_{ref}}^{T} C_{P\,P_i}(T)\,dT\right] - \sum_i y_{R_i}\left[h_{R_i,ref} + \int_{T_{ref}}^{T} C_{P\,R_i}(T)\,dT\right] \tag{2.75}$$

In single-component and non-reacting systems, the reference enthalpy is the same at inlet and exit states and cancels out. So the choice of the reference state is not important as long as the same reference state is chosen for all states. On the other hand, for multicomponent, particularly reacting, systems, the composition of the inlet and the exit states are different. So one needs to carefully define a consistent reference state for all the species. This is achieved through definition of enthalpy of formation.

2.4.1 Enthalpy of Formation

Enthalpy of formation $\left(\Delta h_f\right)$ is the energy required when 1 mole of the species is formed from its constituent elements in their naturally occurring state at standard conditions (1 bar, 298.15 K). Physically this represents the energy required to break the bonds of the standard state elements from the bonds of the compound. For gases, the standard state is defined as 1 bar pressure and 298.15 K temperature. For solids and liquids, the standard state is 298.15 K. Some examples of elements in their naturally occurring states are H_2, O_2, N_2, C(s, graphite), and so on. A natural consequence of the definition of the enthalpy of formation is that the enthalpy of formation of all elements in their naturally occurring state is zero at standard conditions.

Let us consider the following reaction:

$$H_2 + 0.5O_2 \rightarrow H_2O + 241{,}845 \text{ kJ/kmol} \tag{2.76}$$

This reaction liberates 241,845 kJ of energy per kmol of H_2O formed. Since enthalpy of formation, by definition, is the energy required to form the compound from its elemental constituents, the enthalpy of formation for H_2O is –241,845 kJ/kmol. For all compounds formed by exothermic reactions from their elemental constituents, enthalpy of formation is negative.

Let us now consider the following reaction:

$$0.5H_2 + 217{,}997 \text{ kJ/kmol} \rightarrow H \tag{2.77}$$

This reaction requires 217,997 kJ of energy supply per kmol of H. Thus, enthalpy of formation for H is 217,997 kJ/kmol. For all compounds formed endothermically, enthalpy of formation is positive. It may be noted that most of the stable compounds like H_2O have negative enthalpies of formation, while for unstable atoms and radicals like H, enthalpy of formation is positive. There are a few exceptions like hydrogen peroxide, H_2O_2, which has a positive enthalpy of formation.

Two important statements (laws) which are useful for calculating energy transfer in a chemical reaction are the following:

1. ***Lavoisier and Laplace Law*:** Heat change accompanying a chemical reaction in one direction is exactly equal and opposite in sign to that associated with the same reaction in the reverse direction.

$$CO + 0.5O_2 \rightarrow CO_2 - 283{,}005 \text{kJ/kmol} \tag{2.78}$$

$$CO_2 \rightarrow CO + 0.5O_2 + 283{,}005 \text{kJ/kmol} \tag{2.79}$$

2. ***Hess Law*:** Heat change at constant pressure or constant volume in a given chemical reaction is the same whether the reaction takes place in one or several stages.

$$C(s) + 0.5O_2 \rightarrow CO - 110{,}541 \text{kJ/kmol} \tag{2.80}$$

$$CO + 0.5O_2 \rightarrow CO_2 - 283{,}005 \text{kJ/kmol} \tag{2.81}$$

$$C(s) + O_2 \rightarrow CO_2 - 393{,}546 \text{kJ/kmol} \tag{2.82}$$

Let us again consider the chemical reaction (2.65) at the reference state. At this stage, let us define the reference state as 1 bar, 298.15 K. Thus, the corresponding enthalpy of formation at the reference state is taken as the reference enthalpy. In other words, the common datum used in the definition of enthalpies of reactants and products is that all elements in their naturally occurring state at the reference state have zero enthalpy. It may be noted that, for ideal gases, the enthalpy is a function of temperature only.

2.4.2 Calculation of Enthalpy of Formation at Elevated Temperatures

Let us consider a chemical reaction in which the reactants are at reference temperature, T_{ref}, and the products are at temperature T. Thus the change in enthalpy is given by:

$$\sum_P H_P(T) - \sum_R H_R(T) \tag{2.83}$$

Since enthalpy is a thermodynamic property, its changes are functions of the initial and the final states only and are independent of the path followed. This can be used to determine the enthalpy of formation at any arbitrary temperature. Let us consider two alternative paths (see Figure 2.3) A (1-3′-2) and B (1-3″-2). Along path A, the reactants undergo chemical changes to form products at T_{ref}.

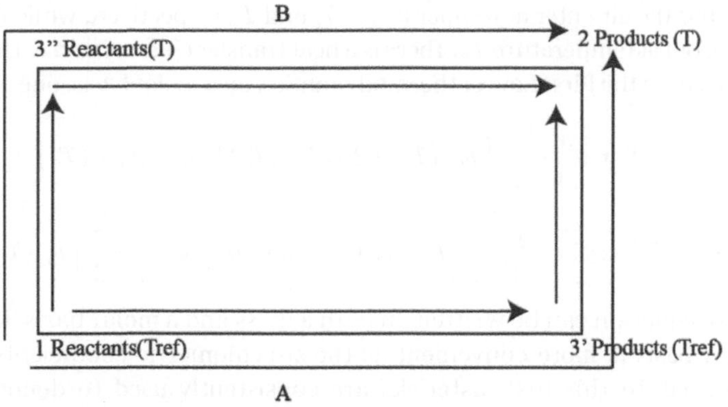

Figure 2.3 Formation of products from reactants following A and B path.

The products then undergo a change in sensible enthalpy to temperature T. Thus, the enthalpy change is given by $\Delta h_f(T_{ref})(1\rightarrow 3')+\int_{T_{ref}}^{T} C_{p_P}dT(3'\rightarrow 2)$. For this integral, we shall consistently use the shorthand $\Delta h_P(T)$.

Along path B, the reactants first undergo a change in sensible enthalpy to temperature T, and then the reactants at temperature T are converted to products at temperature T. Thus, the enthalpy change is given by $\int_{T_{ref}}^{T} C_{p_R}dT(1\rightarrow 3'')+\Delta h_f(T)(3''\rightarrow 2)$.

Since the total enthalpy change is same along thc two paths, one can write:

$$\Delta h_f\left(T_{ref}\right)+\Delta h_P\left(T\right)=\Delta h_f\left(T\right)+\Delta h_R\left(T\right) \tag{2.84}$$

Thus, the enthalpy of formation at any temperature is given by:

$$\Delta h_f\left(T\right)=\Delta h_f\left(T_{ref}\right)+\Delta h_P\left(T\right)-\Delta h_R\left(T\right) \tag{2.85}$$

2.4.3 Enthalpy of Combustion and Adiabatic Flame Temperature

Let us consider the combustion of a generic hydrocarbon $C_xH_yO_z$ in excess air where the fraction of excess air is given by α. The chemical reaction is given by:

$$C_xH_yO_z+(1+\alpha)\left(x+\frac{y}{4}-\frac{z}{2}\right)[O_2+3.76N_2]=xCO_2+\frac{y}{2}H_2O+\alpha\left(x+\frac{y}{4}-\frac{z}{2}\right)O_2$$
$$+3.76\left(1+\alpha\right)\left(x+\frac{y}{4}-\frac{z}{2}\right)N_2 \tag{2.86}$$

The fuel and the air enter at temperatures T_F and T_A, respectively, while the products come out at temperature T_P. There is a heat transfer Q associated with the process. Thus, from the First Law of thermodynamics, on a molar basis, one obtains:

$$h_F^*(T_F)+(1+\alpha)\left(x+\frac{y}{4}-\frac{z}{2}\right)\left[h_{O_2}^*(T_A)+3.76h_{N_2}^*(T_A)\right]+Q=xh_{CO_2}^*(T_P)$$
$$+\frac{y}{2}h_{H_2O}^*(T_P)+\alpha\left(x+\frac{y}{4}-\frac{z}{2}\right)h_{O_2}^*(T_P)+3.76(1+\alpha)\left(x+\frac{y}{4}-\frac{z}{2}\right)h_{N_2}^*(T_P) \quad (2.87)$$

The above equation can be written on both a mass and a molar basis. However, the molar basis is more convenient as the stoichiometric coefficients can be directly used. In this text, asterisks are consistently used to denote molar quantities. Using enthalpy of formation at T_{ref} as the reference enthalpy, the above equation can be written as:

$$\left\{\Delta h_{fF}^*+\Delta h_F^*(T_F)\right\}+(1+\alpha)\left(x+\frac{y}{4}-\frac{z}{2}\right)\left[\left\{\Delta h_{fO_2}^*+\Delta h_{O_2}^*(T_A)\right\}\right.$$
$$\left.+3.76\left\{\Delta h_{fN_2}^*+\Delta h_{ON_2}^*(T_A)\right\}\right]+Q=x\left\{\Delta h_{fCO_2}^*+\Delta h_{CO_2}^*(T_P)\right\}$$
$$+\frac{y}{2}\left\{\Delta h_{fH_2O}^*+\Delta h_{H_2O}^*(T_P)\right\}+\alpha\left(x+\frac{y}{4}-\frac{z}{2}\right)\left\{\Delta h_{fO_2}^*+\Delta h_{O_2}^*(T_P)\right\}$$
$$+3.76(1+\alpha)\left(x+\frac{y}{4}-\frac{z}{2}\right)\left\{\Delta h_{fN_2}^*+\Delta \mathrm{h}_{N_2}^*(T_P)\right\} \quad (2.88)$$

The sensible enthalpy change for each species is generally tabulated as a function of temperature. Also, the enthalpies of formation are tabulated for different compounds. Thus, knowing the temperatures of the incoming and outgoing fluids enables one to determine the enthalpies. Generally, the inlet temperatures, T_F and T_A are specified. However, the product temperature T_P depends on the amount of heat transfer, Q. Thus, there are two unknowns, T_P and Q. One needs to specify one of these to solve for the other. Here we shall consider two special cases. First, we simplify the above equation by assuming $T_F=T_A=T_{ref}=298$ K and $\alpha=0$. With these assumptions and using the definition that elements in their naturally occurring states have zero enthalpy of formation, the above equation simplifies to:

$$\Delta h_{fF}^*+Q=x\left\{\Delta h_{fCO_2}^*+\Delta h_{CO_2}^*(T_P)\right\}+\frac{y}{2}\left\{\Delta h_{fH_2O}^*+\Delta h_{H_2O}^*(T_P)\right\}$$
$$+3.76\left(x+\frac{y}{4}-\frac{z}{2}\right)\Delta h_{N_2}^*(T_P) \quad (2.89)$$

Case I: $T_P = T_{ref} = 298$ K
The above equation reduces to:

$$Q = x\,\Delta h^*_{fCO_2} + \frac{y}{2}\Delta h^*_{fH_2O} - \Delta h^*_{fF} \tag{2.90}$$

The heat transfer Q determined as above is known as the enthalpy of combustion, Δh^*_c of the fuel. It is the heat transferred to the combustor if fuel and air at reference temperature undergo complete combustion and the products are again cooled to the reference temperature. Since combustion reactions are exothermic in nature, heat has to be removed from the combustor. Hence enthalpy of combustion is usually a negative quantity. The negative of the enthalpy of combustion is known as **calorific value** or **heating value** of the fuel. It represents the energy released from complete combustion of unit mass or unit mole of the fuel. Since the reference temperature (298 K) is lower than the normal boiling point of water, the state of water in the cooled products need to be carefully considered. If the water is allowed to condense, additional heat will be released in the form of latent heat. Thus the total heat released will be equal to the sum of the heat released due to combustion and condensation. This value is referred to as the **higher calorific value** (**HCV**) or **higher heating value** (**HHV**) of the fuel. If the heat release due to condensation is neglected, the heat release is referred to as **lower calorific value** (**LCV**) or **lower heating value** (**LHV**) of the fuel. Heating value is often written as Q_{CV}.

Case II: Q = 0

The energy equation can be written as:

$$x\left\{\Delta h^*_{fCO_2} + \Delta h^*_{CO_2}(T_P)\right\} + \frac{y}{2}\left\{\Delta h^*_{fH_2O} + \Delta h^*_{H_2O}(T_P)\right\} + 3.76\left(x + \frac{y}{4} - \frac{z}{2}\right)\Delta h^*_{N2}(T_P) - \Delta h^*_{fF} = 0 \tag{2.91}$$

The value of T_P which satisfies the above equation is known as the **adiabatic flame temperature** (***T_{ad}***) of the fuel. This is generally the maximum product temperature that can be achieved.

2.4.4 Constant Volume Combustion

For constant volume adiabatic combustion, the energy equation can be written as:

$$\sum_P U_P(T_P, P_f) = \sum_R U_R(T_R, P_i) \qquad (2.92)$$

The internal energy of the reactants and the products can be written as:

$$u_R = \sum_i y_{R_i} u_{R_i} = \sum_i y_{R_i}(h_{R_i} - R_i T_R) = \sum_i y_{R_i}\left[\Delta h_{f_{R_i}} + \int_{T_{ref}}^{T} C_{P_{R_i}}(T)dT - R_i T_R\right] \quad (2.93)$$

$$u_P = \sum_i y_{P_i} u_{P_i} = \sum_i y_{P_i}(h_{P_i} - R_i T_P) = \sum_i y_{P_i}\left[\Delta h_{f_{P_i}} + \int_{T_{ref}}^{T} C_{P_{P_i}}(T)dT - R_i T_P\right] \quad (2.94)$$

The final pressure in the combustor is calculated from:

$$\frac{P_i}{N_R T_R} = \frac{P_f}{N_P T_P} \qquad (2.95)$$

Example 2.3

LPG can be considered as a mixture of 40% propane and 60% butane on a volumetric basis. Determine the adiabatic flame temperature for combustion of stoichiometric pressure at constant pressure if the reactants enter at 298 K.

Solution

The chemical reaction for stoichiometric LPG-air mixture is:

$$(0.4C_3H_8 + 0.6C_4H_{10}) + 5.9(O_2 + 3.76N_2) = 3.6CO_2 + 4.6H_2O + 22.184N_2$$

From energy balance we can write:

$$0.4 \times \Delta h^*_{f_{C_3H_8}} + 0.6 \times \Delta h^*_{f_{C_4H_{10}}} = 3.6 \times \left[\Delta h^*_{f_{CO_2}} + \Delta h^*_{CO_2}(T_{ad})\right] + 4.6 \times \left[\Delta h^*_{f_{H_2O}} + \Delta h^*_{H_2O}(T_{ad})\right]$$
$$+ 22.184 \Delta h^*_{N_2}(T_{ad})$$

Substituting the values of enthalpies of formation from the table, we have:

$$3.6 \times \Delta h^*_{CO_2}(T_{ad}) + 4.6 \times \Delta h^*_{H_2O}(T_{ad}) + 22.184 \Delta h^*_{N_2}(T_{ad})$$
$$= 0.4 \times (-103847) + 0.6 \times (-146440) - 3.6 \times (-393546) - 4.6 \times (-241845)$$
$$= 2399849.8$$

The adiabatic flame temperature T_{ad} has to be suitably chosen such that the sensible enthalpy values from the corresponding tables satisfy the above equation. This requires iterative solution. There is no standard iterative technique. However, some informed initial guess can quicken the convergence. One such procedure is outlined below.

Assumption of constant specific heat makes the left-hand side an explicit function of T_{ad} and iteration is not necessary. However, depending on the choice of the reference temperature at which the specific heat is evaluated, we can overestimate or underestimate the adiabatic flame temperature.

As a first guess, we evaluate the specific heat at 298 K. Since this gives very low values of specific heats, T_{ad} is overpredicted.

Guess 1: T_{ad} determined using C_p at 298 K. This gives $T_{ad} = 2871.06K$.

We use a guessed value of $T_{ad} = 2800K$ to calculate LHS, which comes out to be 2955017.79. This value is greater than the RHS value, which shows that the adiabatic flame temperature is overestimated, as expected.

Guess 2: T_{ad} is determined using C_p at 2800 K. This gives $T_{ad} = 2135.35K$.

We use a guessed value of $T_{ad} = 2100K$ to calculate LHS, which comes out as 2052645.592. This value is less than the RHS, showing that the adiabatic flame temperature is underestimated. This is expected as we have used C_p corresponding to a very high temperature. However, these values of T_{ad} can be used to interpolate the correct T_{ad}. On interpolation, we obtain: $T_{ad} = 2369.34K$.

A more refined iterative process using computational techniques can be used for more accurate determination of adiabatic flame temperature.

2.5 CHEMICAL EQUILIBRIUM

The evaluation of adiabatic flame temperature requires knowledge of composition of the combustion products. For lean mixtures, as mentioned before, the composition of the products of complete oxidation of fuel can be evaluated from elemental mass balance alone. On the other hand, for rich mixtures, partial oxidation of fuels gives rise to products like CO and H_2 in addition to CO_2 and H_2O. This increase in the number of products makes it impossible for us to determine the product composition from conservation of elemental mass alone. This additional relation comes from consideration of thermodynamic equilibrium, which involves both the First and Second Laws of Thermodynamics. Even for complete oxidation, for temperatures below 1250 K [3], the stable species do not dissociate appreciably. So the actual product composition is similar to that predicted by stoichiometry and elemental mass balance. However, under most circumstances, as the products reach a much higher temperature, dissociation of species like CO_2

and H_2O occurs. Since dissociation reactions are endothermic, even a small fraction of the species dissociation can lead to a significant reduction in flame temperature.

The starting point of calculation of thermodynamic equilibrium is the well-known relation (recalling Equations 2.47 and 2.48):

$$TdS = dU + PdV$$

or equivalently:

$$TdS = dH - VdP$$

From the First Law of Thermodynamics, we have for a simple compressible system:

$$\delta Q = dU + PdV = dH - VdP \tag{2.96}$$

From the Second Law of Thermodynamics, we have (Equation 2.43):

$$dS = \frac{\delta Q}{T} + \delta\left(S_{gen}\right)$$

Combining the above two equations, we get:

$$TdS = dH - VdP + T\delta\left(S_{gen}\right) \tag{2.97}$$

Let us define a new thermodynamic property, the **Gibbs function**, given by:

$$G = H - TS \tag{2.98}$$

The corresponding differential relation (from Equations 2.98 and 2.97) is given by:

$$dG = dH - TdS - SdT = -SdT + VdP - T\delta\left(S_{gen}\right) \tag{2.99}$$

Since $T\delta\left(S_{gen}\right) \geq 0$,

$$dG\big|_{T,P} = -T\delta\left(S_{gen}\right) \leq 0 \tag{2.100}$$

Now the equality in the above is valid only for reversible processes (and hence not for spontaneous processes). For spontaneous processes, $dG\big|_{T,P} < 0$. The process will cease when equilibrium is reached and the properties will not change any further. Thus, the condition for equilibrium is $dG\big|_{T,P} = 0$, which coupled with the condition that $dG\big|_{T,P} < 0$ for a spontaneous process implies that at equilibrium G reaches a minimum. For a multicomponent system, in addition to the two independent variables needed to define a simple system, we need to specify the mass or number of moles of each of the constituents. Thus, on a molar basis one can write:

$$G^* = G^*(T, P, n_1, n_2, ..., n_N) \tag{2.101}$$

Therefore:

$$dG^* = \left.\frac{\partial G^*}{\partial T}\right|_{P,n_i} dT + \left.\frac{\partial G^*}{\partial P}\right|_{T,n_i} dP + \sum_{j=1}^{N} \left.\frac{\partial G^*}{\partial n_j}\right|_{T,P,n_{i\neq j}} dn_j \tag{2.102}$$

From the relation for single-component systems, we write:

$$\left.\frac{\partial G^*}{\partial T}\right|_{P,n_i} = -S \tag{2.103}$$

$$\left.\frac{\partial G^*}{\partial P}\right|_{T,n_i} = V \tag{2.104}$$

We define a new property called chemical potential μ_j^* as:

$$\left.\frac{\partial G^*}{\partial n_j}\right|_{T,P,n_{i\neq j}} = \mu_j^* \tag{2.105}$$

For a mixture of ideal gases, the Gibbs function is given by:

$$g_i^*(T, p_i) = h_i^*(T) - Ts_i^*(T, p_i) \tag{2.106}$$

It should be noted that, for ideal gases, mixture properties are simply the sum of the corresponding properties of the constituents. Since there is no interaction between species, g_i^* should be evaluated for the condition that ith species alone is present. For this reason, the use of partial pressure is appropriate.

Now, for a pure substance, $h_i^*(T) = \Delta h_{f_i}^* + \Delta h_i^*(T)$. Similarly for entropy for an ideal gas:

$$s_i^*(T, p_i) = s_{i,ref}^* + \int_{T_{ref}}^{T} C_{p_i}^*(T)\frac{dT}{T} - R^* \ln\frac{p_i}{P_{ref}} \tag{2.107}$$

As in the case of enthalpy, the reference entropy is taken as the entropy of formation (Δs^*_{fi}) at T_{ref} = 298 15 K and $P_{ref} = P_{atm}$ = 1 bar. Thus, one can write:

$$s^*_i(T, p_i) = \Delta s^*_{fi} + \int_{T_{ref}}^{T} C^*_{p_i}(T)\frac{dT}{T} - R^* \ln \frac{p_i}{P_{ref}} \tag{2.108}$$

The first two terms on RHS depend on temperature only and are together tabulated in an ideal gas table. It is denoted as $s^{o*}_i(T)$ and physically signifies entropy at temperature T and atmospheric pressure P_{atm}. Substituting in the expression (Equation 2.106) for molar specific Gibbs energy, we get:

$$g^*_i(T, p_i) = h^*_i(T) - T\left[s^{o*}_i(T) - R^* \ln \frac{p_i}{P_{ref}}\right] = \left[h^*_i(T) - Ts^{o*}_i(T)\right] + R^* T \ln \frac{p_i}{P_{ref}} \tag{2.109}$$

The first two terms are again functions of temperature and denoted as $g^{o*}_i(T)$. Since $p_i = x_i P$:

$$g^*_i(T, p_i) = g^{o*}_i(T) + R^* T \ln \frac{p_i}{P_{ref}} \tag{2.110}$$

The **Gibbs function of formation** is defined as follows:

$$\Delta g^{o*}_{fi}(T) = g^{o*}_i(T) - \sum_{j\ elements} \nu_j g^{o*}_j(T) \tag{2.111}$$

where ν_j are the stoichiometric coefficients of elements in their natural state to form one mole of species i. For example, in the reaction $H_2 + 0.5O_2 = H_2O$:

$$\Delta g^*_{fH_2O}(T) = g^{o*}_{H_2O}(T) - g^{o*}_{H_2}(T) - 0.5 g^{o*}_{O_2}(T) \tag{2.112}$$

For all elements in their naturally occurring states, $\Delta g^*_{fi} = 0$ at reference temperature. For equilibrium at constant temperature and pressure:

$$dG(T, P) = d\left(\sum_{i=1}^{N} n_i g^*_i(T, p_i)\right) = 0 \tag{2.113}$$

Substituting the expression for $g_i^*(T, p_i)$ (Equation. 2.110) in the above equation, one obtains:

$$d\left[\sum_{i=1}^{N} n_i \left\{ g_i^{o^*}(T) + R^*T \ln \frac{p_i}{P_{ref}} \right\}\right] = 0 \tag{2.114}$$

Expanding the above, one can write:

$$\sum_{i=1}^{N} \left\{ g_i^{o^*}(T) + R^*T \ln \frac{p_i}{P_{ref}} \right\} dn_i + \sum_{i=1}^{N} n_i d\left\{ g_i^{o^*}(T) + R^*T \ln \frac{p_i}{P_{ref}} \right\} = 0 \tag{2.115}$$

At constant T and P, the second term can be written as zero. Hence:

$$\sum_{i=1}^{N} \left\{ g_i^{o^*}(T) + R^*T \ln \frac{p_i}{P_{ref}} \right\} dn_i = 0 \tag{2.116}$$

Let us consider the generic reaction:

$$aA + bB \Leftrightarrow cC + dD \tag{2.117}$$

where a, b, c and d are the stoichiometric coefficients. Since change in each species is proportional to its stoichiometric coefficient, we have:

$$dn_A = -\nu a \tag{2.118}$$

$$dn_B = -\nu b \tag{2.119}$$

$$dn_C = \nu c \tag{2.120}$$

$$dn_D = \nu d \tag{2.121}$$

The signs in the above equations indicate that production (consumption) of C and D is accompanied by consumption (production) of A and B. The conditions for equilibrium are given by the following:

$$\begin{aligned} &\left\{ g_A^{o^*}(T) + R^*T \ln \frac{p_A}{P_{ref}} \right\} dn_A + \left\{ g_B^{o^*}(T) + R^*T \ln \frac{p_B}{P_{ref}} \right\} dn_B \\ &+ \left\{ g_C^{o^*}(T) + R^*T \ln \frac{p_C}{P_{ref}} \right\} dn_C + \left\{ g_D^{o^*}(T) + R^*T \ln \frac{p_D}{P_{ref}} \right\} dn_D = 0 \end{aligned} \tag{2.122}$$

This implies that:

$$\left[-ag_A^{o^*}(T)-bg_B^{o^*}(T)+cg_C^{o^*}(T)+dg_D^{o^*}(T)\right]$$
$$+R^*T\left[-a\ln\frac{p_A}{P_{ref}}-b\ln\frac{p_B}{P_{ref}}+c\ln\frac{p_C}{P_{ref}}+d\ln\frac{p_D}{P_{ref}}\right]=0 \tag{2.123}$$

Substituting the expression for Gibbs function of formation, one can write:

$$\sum_{i=1}^{N} g_i^{o^*}(T)dn_i = c\sum_{i=1}^{N}\nu_i g_i^{o^*}(T) = c\sum_{i=1}^{N}\nu_i\left[\Delta g_{fi}^{o^*}(T)+\sum_{j\text{ elements}}\nu_j g_j^{o^*}(T)\right]$$
$$= c\sum_{i=1}^{N}\nu_i\Delta g_{fi}^{o^*}(T) = \sum_{i=1}^{N}\Delta g_{fi}^{o^*}(T)dn_i \tag{2.124}$$

This is because the term under double summation is equal to zero from elemental mass balance. Hence, one can write:

$$\ln\left[\frac{p_C^c p_D^d}{p_A^a p_B^b}P_{ref}^{a+b-c-d}\right] = -\frac{\left[-ag_A^{o^*}(T)-bg_B^{o^*}(T)+cg_C^{o^*}(T)+dg_D^{o^*}(T)\right]}{R^*T} \tag{2.125}$$

The numerator of the RHS is a function of temperature T only and is called the **Standard State Gibbs Function**, $\Delta G^*(T)$. The argument of the logarithmic function is called the **equilibrium constant** and is denoted by $K_p(T)$. Using these notations, one can write:

$$K_p(T) = \exp\left[-\frac{\Delta G^*(T)}{R^*T}\right] \tag{2.126}$$

Expressing partial pressures in terms of mole fractions, the equilibrium constant can be expressed as follows:

$$K_p = \frac{x_C^c x_D^d}{x_A^a x_B^b}\left(\frac{P}{P_{ref}}\right)^{c+d-a-b} \tag{2.127}$$

One can define the mole fraction of species A as the ratio of the number of moles of A to the total number of moles in the same volume. Denoting the number of moles of A per unit volume (i.e., molar concentration of A) as $[X_A]$ and noting that the total number of moles per unit volume is given by $\frac{P}{R^*T}$, the mole fraction can be written as:

$$x_A = \frac{[X_A]}{\left(P / R^*T\right)} \tag{2.128}$$

Thus, the equilibrium constant $K_p(T)$ can be written in terms of molar concentration as:

$$K_p(T) = \frac{[X_C]^c [X_D]^d}{[X_A]^a [X_B]^b} \left(\frac{R^*T}{P_{ref}}\right)^{c+d-a-b} \tag{2.129}$$

The term $\frac{[X_C]^c [X_D]^d}{[X_A]^a [X_B]^b}$ is denoted as $K_C(T)$. Thus, one can write:

$$K_C(T) = K_P(T) \left(\frac{R^*T}{P_{ref}}\right)^{a+b-c-d} \tag{2.130}$$

Example 2.4

Calculate K_p for the water gas shift reaction $H_2O + CO \Leftrightarrow H_2 + CO_2$ at 2000 K. What will be the value of K_c?

Solution

The reaction is $H_2O + CO \Leftrightarrow H_2 + CO_2$.

$$\Delta G^*(T) = \Delta g^*_{f_{CO_2}}(T) + \Delta g^*_{f_{H_2}}(T) - \Delta g^*_{f_{CO}}(T) - \Delta g^*_{f_{H_2O}}(T)$$

At $T = 2000$ K, $\Delta G^*(T) = -396410 + 0 - (-285,948) - (-135643) = 25181$ kJ/kmol

Hence, $K_p(T) = \exp\left[-\Delta G^*(T) / R^*T\right] = 0.2199$

$$K_C(T) = K_P(T)\left(\frac{R^*T}{P_{ref}}\right)^0 = 0.2199$$

2.5.1 Effects of Pressure and Temperature on Equilibrium Composition

Since $K_p(T)$ for a given reaction is a function of temperature only, at a given temperature, $K_p(T)$ is constant for a given reaction. Hence, at a given temperature, one can write:

$$\left[\frac{x_C^c x_D^d}{x_A^a x_B^b}\right]\left(\frac{P}{P_{ref}}\right)^{c+d-a-b} = \text{constant} \tag{2.131}$$

As pressure P increases, the square bracketed term decreases; that is, product concentration decreases or reactant concentration increases if $c+d-a-b>0$. This implies that the backward reaction is favoured. The opposite is true if $c+d-a-b<0$.

The influence of temperature on the equilibrium composition is less straightforward. To derive that, let us first start with the derivative:

$$\frac{d}{dT}\left(\frac{g^*}{T}\right)=\frac{T\dfrac{dg^*}{dT}-g^*}{T^2} \tag{2.132}$$

From the definition of Gibbs function, $g_i^* = h_i^* - Ts_i^*$, one can write:

$$\frac{dg_i^*}{dT}=\frac{dh_i^*}{dT}-T\frac{ds_i^*}{dT}-s_i^* \tag{2.133}$$

At constant pressure, $dh_i^* = Tds_i^*$. Under isobaric conditions one can write:

$$\frac{dg_i^*}{dT}=-s_i^* \tag{2.134}$$

Hence, one can write:

$$\frac{d}{dT}\left(\frac{g_i^*}{T}\right)=\frac{Ts_i^*-\left(h_i^*-Ts_i^*\right)}{T^2}=-\frac{h_i^*}{T^2} \tag{2.135}$$

Considering the thermodynamic state at P_{ref} and T, g_i^* and h_i^* become Δg_{fi}^* and Δh_{fi}^*, respectively. Hence, the derivative $\frac{d}{dT}(\frac{\Delta G^*}{T})$ can be written as:

$$\frac{d}{dT}\left(\frac{\Delta G^*}{T}\right)=\frac{d}{dT}\left[\sum_i\frac{\nu_i\Delta g_{fi}^*}{T}\right]=-\sum_i\frac{\nu_i\Delta h_{fi}^*}{T^2} \tag{2.136}$$

Using the definition of enthalpy of combustion, $\sum_i \nu_i \Delta h^*_{f_i} = \Delta h^*_c$, one can write:

$$\frac{d}{dT}(\ln K_p) = \frac{d}{dT}\left(-\frac{\Delta G^*}{R^* T}\right) = \frac{\Delta h^*_c}{R^* T^2} \tag{2.137}$$

For exothermic reactions, Δh^*_c is negative. Hence $\ln K_p$ (and hence K_p) decreases with an increase in temperature. At a given pressure, this implies a decrease in product concentration; that is, the reverse reaction is favoured. The opposite is true for endothermic reactions. Since dissociation of CO_2 and H_2O is endothermic in nature, these reactions are favoured at high temperatures for a given pressure.

2.5.2 Equilibrium Constants in the Presence of Condensed Phase

The constant of equilibrium constant is slightly modified if one of the species is in a solid or a liquid state. To illustrate the calculation of equilibrium constant, let us consider the following reaction:

$$C(s) + O_2 \Leftrightarrow CO_2 \tag{2.138}$$

The condition for thermodynamic equilibrium at constant T and P is given by:

$$g^*_{C(s)}(T,P) + g^*_{O_2}(T,P) - g^*_{CO_2}(T,P) = 0 \tag{2.139}$$

For solids and liquids, properties depend very weakly on pressure. Hence:

$$g^*_{C(s)}(T,P) \approx g^*_{C(s)}(T) \tag{2.140}$$

Thus, the equilibrium constant can be written as:

$$g^*_{C(s)}(T) + g^*_{O_2}(T,P) - g^*_{CO_2}(T,P) = 0 \tag{2.141}$$

In terms of standard state properties, one can write:

$$g^{o*}_{C(s)}(T) + g^{o*}_{O_2}(T) - g^{o*}_{CO_2}(T) = R^* T \ln \frac{p_{CO_2}}{P_{ref}} - R^* T \ln \frac{p_{O_2}}{P_{ref}} \tag{2.142}$$

As shown earlier, the LHS can be written as follows:

$$g^{o*}_{C(s)}(T) + g^{o*}_{O_2}(T) - g^{o*}_{CO_2}(T) = \Delta g_{f_{C(s)}}(T) + \Delta g_{f_{O_2}}(T) - \Delta g_{f_{CO_2}}(T) = \Delta G^*(T) \tag{2.143}$$

Hence, the equilibrium constant can be expressed as:

$$\ln K_p(T) = -\frac{\Delta G^*(T)}{R^*T} = \ln \frac{p_{CO_2}}{p_{O_2}} \tag{2.144}$$

Thus, the expression for the equilibrium constant does not contain any contribution from the condensed phase.

2.5.3 Determination of Equilibrium Composition

Let us consider the following reaction:

$$CO_2 \Leftrightarrow CO + \frac{1}{2}O_2 \tag{2.145}$$

To determine the composition, one needs the values for the three unknowns: $x_{CO_2}, x_{CO}, x_{O_2}$ The equilibrium composition at pressure P and temperature T can be obtained as follows:

$$\frac{x_{CO}x_{O_2}^{1/2}}{x_{CO_2}}\left(\frac{P}{P_{ref}}\right)^{1+0.5-1} = K_p(T) \tag{2.146}$$

Since initially only CO_2 was present, from atomic mass balance one can write:

$$\frac{\text{number of C atoms}}{\text{number of O atoms}} = \frac{x_{CO} + x_{CO_2}}{x_{CO} + 2x_{CO_2} + 2x_{O_2}} = \frac{1}{2} \tag{2.147}$$

Finally, from the definition of mass fraction, one can write:

$$x_{CO} + x_{CO_2} + x_{O_2} = 1 \tag{2.148}$$

The three equations above can be solved simultaneously to obtain three unknowns: x_{CO_2}, x_{CO}, x_{O_2}. However, the presence of highly non-linear equation(s) makes numerical solutions imperative. In a chemical reaction, hundreds of stable and unstable species are formed. To determine the equilibrium composition, one has to specify the species that needs to be taken into account. Consideration of a larger number of species obviously improves the accuracy of calculation but only at the cost of significant additional computational effort. This requires careful optimisation of the species to be included in the analysis based on requirement. For example, for calculation of energy release, it is generally acceptable to neglect any species with a concentration less than 0.1% [4]. But for determining the quantity of some trace species or radical, one needs to consider these minor species also. Some such species may be oxides of nitrogen, NO_x or radicals like OH, which radiate energy of specific wavelengths and are often used in experiments for quantification of chemical reaction rate.

2.5.4 Equilibrium Composition for Hydrocarbon Combustion in Air

Since the reactant-product mixture for hydrocarbon combustion in air consists of the elements C, H, O and N, the system is often called C-H-O-N system [5]. Generally, two equilibrium models are considered in literature. The first one is called the **Full Equilibrium Model** and is more detailed. The other one is simpler, involves fewer computations and is called the **Water Gas Equilibrium Model.**

2.5.4.1 Full Equilibrium Model

Let us consider the combustion of a generic hydrocarbon C_xH_y in air. The chemical reactions containing detailed composition of products may be given as:

$$
\begin{aligned}
C_xH_y + \frac{x+\frac{y}{4}}{\varphi}(O_2 + 3.76N_2) = {} & \nu_{CO_2}CO_2 + \nu_{CO}CO + \nu_{C(s)}C(s) + \nu_{C(g)}C(g) \\
& + \nu_{H_2O}H_2O + \nu_{OH}OH + \nu_{H}H + \nu_{H_2}H_2 \\
& + \nu_{O}O + \nu_{O_2}O_2 + \nu_{N}N + \nu_{N_2}N_2 + \nu_{NO}NO
\end{aligned} \tag{2.149}
$$

In the above equation, there are 13 species. To find their stoichiometric coefficients (unknowns), we need 13 equations. Four of them come from the elemental mass balance of C, H, O and N as follows:

C balance: $$\nu_{CO_2} + \nu_{CO} + \nu_{C(s)} + \nu_{C(g)} = x \tag{2.150}$$

H balance: $$2\nu_{H_2O} + \nu_{OH} + \nu_{H} + 2\nu_{H_2} = y \tag{2.151}$$

O balance: $$2\nu_{CO_2} + \nu_{CO} + \nu_{H_2O} + \nu_{OH} + \nu_{O} + 2\nu_{O_2} + \nu_{NO} = \frac{2\left(x+\frac{y}{4}\right)}{\phi} \tag{2.152}$$

N balance: $$\nu_{N} + 2\nu_{N_2}\nu_{NO} = \frac{7.52\left(x+\frac{y}{4}\right)}{\phi} \tag{2.153}$$

It is important to note here that the elemental mass balance of nitrogen is also necessary here as nitrogen is not an inert species but takes part in reactions. The remaining nine equations come from the equilibrium relations of the following chemical equations:

$$C(s) + O_2 \Leftrightarrow CO_2 \qquad K_{p_1} = \frac{p_{CO_2}}{p_{O_2}} \tag{2.154}$$

$$C(s) + \frac{1}{2}O_2 \Leftrightarrow CO \qquad K_{p_2} = \frac{p_{CO}}{p_{O_2}} - \tag{2.155}$$

$$C(s) \Leftrightarrow C(g) \qquad K_{p_3} = p_{C(g)} \tag{2.156}$$

$$H_2 + \frac{1}{2}O_2 \Leftrightarrow H_2O \qquad K_{p_4} = \frac{p_{H_2O}}{p_{H_2}\, p_{O_2}{}^{1/2}} \tag{2.157}$$

$$\frac{1}{2}H_2 \Leftrightarrow H \qquad K_{p_4} = \frac{p_H}{p_{H_2}{}^{1/2}} \tag{2.158}$$

$$\frac{1}{2}H_2 + \frac{1}{2}O_2 \Leftrightarrow OH \qquad K_{p_6} = \frac{p_{OH}}{p_{H_2}{}^{1/2}\, p_{O_2}{}^{1/2}} \tag{2.159}$$

$$\frac{1}{2}O_2 \Leftrightarrow O \qquad K_{p_7} = \frac{p_O}{p_{O_2}{}^{1/2}} \tag{2.160}$$

$$\frac{1}{2}N_2 + \frac{1}{2}O_2 \Leftrightarrow OH \qquad K_{p_8} = \frac{p_{NO}}{p_{N_2}{}^{1/2}\, p_{O_2}{}^{1/2}} \tag{2.161}$$

$$\frac{1}{2}N_2 \Leftrightarrow N \qquad K_{p_9} = \frac{p_N}{p_{N_2}{}^{1/2}} \tag{2.162}$$

Equations (2.154) – (2.162) take a long time for computation as a large number of nine nonlinear equations need to be solved. An approximate but simpler approach is explained in the next section.

2.5.4.2 Water Gas Equilibrium

In this approach, production of minor species is neglected for both lean and rich mixtures. For lean mixtures, the composition is the same as that obtained from elemental mass balance. For rich mixtures, it may be recalled that the product composition cannot be determined by elemental mass balance alone as the number of species in the product (unknowns) exceeds the number of elemental mass balance relations (equations) available. A single equilibrium reaction, called the water gas shift reaction, is used as an additional relation for the simultaneous presence of both CO and H_2. The details are as follows:

$$C_xH_y + \frac{x+\frac{y}{4}}{\varphi}(O_2+3.76N_2) = aCO_2 + bH_2O + cH_2 + dCO + eO_2 + \frac{3.76\left(x+\frac{y}{4}\right)}{\varphi}N_2 \quad (2.163)$$

For lean mixtures ($\varphi < 1$), it is assumed that no CO or H_2 is formed. Here the number of unknowns (a, b, e) is equal to the number of elements (C, H, O) present. Thus, the composition is completely determined by elemental mass balance.

For rich mixtures ($\varphi > 1$), it is assumed that both CO and H_2 are formed but all the available oxygen is used up. Thus the chemical reaction becomes:

$$C_xH_y + \frac{x+\frac{y}{4}}{\varphi}(O_2+3.76N_2) = aCO_2 + bH_2O + cH_2 + dCO + \frac{3.76\left(x+\frac{y}{4}\right)}{\varphi}N_2 \quad (2.164)$$

Here, the number of unknowns (a, b, c, d) exceeds the number of elements present (C, H, O) by 1. The additional relation is obtained from the following equilibrium reaction:

$$CO_2 + H_2 \Leftrightarrow CO + H_2O \quad (2.165)$$

The corresponding equilibrium relation is:

$$K_p(T) = \frac{p_{CO}\, p_{H_2O}}{p_{CO_2}\, p_{H_2}} = \frac{b.d}{c.e} \quad (2.166)$$

The equilibrium calculations given by the above equation give accurate results if ϕ is not very close to unity and the temperature is not too high. When $\varphi \approx 1$ or the temperature is above 2000K, there is considerable dissociation of both CO_2 and H_2O. The corresponding equilibrium relations are:

$$CO_2 \Leftrightarrow CO + \frac{1}{2}O_2 \qquad K_{p_1} = \frac{p_{CO}\, p_{O_2}^{1/2}}{p_{CO_2}} \quad (2.167)$$

$$H_2O \Leftrightarrow H_2 + \frac{1}{2}O_2 \qquad K_{p_2} = \frac{p_{H_2}\, p_{O_2}^{1/2}}{p_{H_2O}} \quad (2.168)$$

Now there are five unknowns (*a, b, c, d, e*) and three elemental mass balance (C, H, O) equations. The remaining two equations come from the two equilibrium relations. It is important to note that the above two equations and the water gas shift reaction are not independent. The water gas shift equation can be obtained as a linear combination of the above two equilibrium relations.

Unlike the full equilibrium approach, in this approach, the equilibrium concentrations of trace species and radicals do not come directly from the calculations. So determining the concentrations of minor species in this approach calls for some additional calculations. In this approach, the procedure for determining the concentration of minor species is explained below.

It is assumed that the production of minor species does not affect the concentrations of the major species. In this context, any species not included in the equilibrium calculations is a minor species. Thus, for rich mixtures, O_2 is a minor species, while CO is a minor species for lean mixtures. Other examples are NO, OH, O, H, N and so on. As an example, let us consider the production of NO. For lean mixtures, the concentration of NO can be described by the following reaction:

$$\frac{1}{2}N_2 + \frac{1}{2}O_2 \Leftrightarrow NO \qquad K_{p_1} = \frac{p_{NO}}{p_{N_2}^{1/2} p_{O_2}^{1/2}} \tag{2.169}$$

The variables p_{N_2} and p_{O_2} are obtained from the equilibrium composition of the major species.

For rich mixtures, O_2 is not obtained from the equilibrium calculations as it is a minor species. An alternate reaction which can be used is:

$$H_2O + \frac{1}{2}N_2 \Leftrightarrow H_2 + NO \qquad K_{p_1} = \frac{p_{NO} p_{H_2}}{p_{N_2}^{1/2} p_{H_2O}} \tag{2.170}$$

The variables p_{N_2}, p_{H_2} and p_{H_2O} are obtained from equilibrium calculations of major species.

Example 2.5

Consider the combustion of propane in air with an equivalence ratio of 1.25. If the composition products exit at 2000 K, what is the composition of the products if the only dissociation reaction involved is the water gas shift reaction?

Solution

Since we are considering a rich mixture and want to use the water shift equilibrium model, the chemical reaction is given by:

$$C_3H_8 + \frac{1}{1.25} \times 5(O_2 + 3.76N_2) = aCO_2 + bH_2O + cCO + dH_2 + 15.04N_2$$

From elemental mass balance, we obtain:

C balance: $a + c = 3$
H balance: $b + d = 4$
O balance: $2a + b + c = 8$

Expressing all the unknowns in terms of d, we get $a = d + 1; b = 4 - d; c = 2 - d$. Thus:

$$K_p = \frac{x_{CO_2} x_{H_2}}{x_{CO} x_{H_2O}} = \frac{a.d}{b.c} = \frac{d^2 + d}{d^2 - 6d + 8}$$

For the water gas shift reaction, at $T = 2000$ K:
$\Delta G^*(T) = -396410 + 0 - (-285,948) - (-135643) = 25181$ kJ/kmol
Thus:

$$K_p(T) = \exp\left[-\Delta G^*(T)/R^*T\right] = 0.2199$$

Solving for d, we obtain $d = 0.6267$. Thus, the mole fractions are obtained as follows:

$$x_{H_2} = 0.0284$$

$$x_{H_2O} = 0.1530$$

$$x_{CO} = 0.0623$$

$$x_{CO_2} = 0.0738$$

$$x_{N_2} = 0.6825$$

2.5.5 Determination of the Equilibrium Flame Temperature

In determining the adiabatic flame temperature in Section 2.4.3, the product composition was known a priori from stoichiometry. On the other hand, all equilibrium calculations in this section have been carried out at a known temperature. A more realistic requirement would be to determine both the equilibrium composition and the temperature. This would require simultaneous solution of the composition and energy equations. A possible iteration sequence is the following:

1. Determine flame temperature assuming stoichiometry-based composition (neglecting dissociation).
2. Use the temperature obtained in Step 1 to determine the equilibrium composition.
3. Determine the flame temperature using the composition from Step 2.
4. Go to Step 2 and repeat till convergence.

2.6 APPLICATIONS/CASE STUDIES

2.6.1 Oxyfuel Combustion

Combustion of fuels is one of the major sources of environmental pollution. Since most of the fuels are fossil fuel by nature, combustion leads to release of carbon dioxide in the atmosphere. Release of CO_2 in the atmosphere, being a major cause of global warming, is an area of great concern. At the same time, since air is used as the oxidant, the nitrogen present in the air undergoes oxidation at the high temperatures encountered in combustors to produce oxides of nitrogen like NO and NO_2 (generically represented as NO_x), which are some of the prime sources of air pollution. Since combustion of fossil fuels is expected to remain the dominant source of energy in the foreseeable future, new technologies have been developed to mitigate these hazards. One such promising technology is oxyfuel combustion. This technology involves use of pure oxygen as the oxidiser instead of air. This completely eliminates nitrogen from the combustion environment and thus no NO_x is formed. To reduce CO_2 emission from combustion, different pre- and post-combustion technologies have been developed for carbon dioxide capture and sequestration (CCS). Oxyfuel combustion provides one of the cheapest post-combustion CCS techniques [6]. In this method, for a hydrocarbon fuel, the combustion products are water vapour and carbon dioxide. Once the water vapour is removed (condensed), the combustion product is nearly pure CO_2, which can be easily captured and sequestered without further processing.

Compared to combustion in air, combustion in oxygen gives a much higher temperature as the thermal inertia of the nitrogen is absent. This gives higher efficiency and better utilisation of the chemical energy of the fuel. However, the resulting temperature may be too high for the combustor material to withstand. High temperature also implies a very high chemical reaction rate, which can lead to loss of flame stability due to a phenomenon known as *flashback* (which will be discussed in detail later). To overcome these problems, oxygen is often diluted with an inert gas like CO_2, which increases the total mass, thereby reducing the temperature of the combustion products. It may be noted that since pure oxygen is not naturally available like air, it is expensive. Hence, in oxyfuel combustion generally fuel and oxygen are mixed in stoichiometric proportions rather than forming lean mixture with excess oxygen. Example 2.5 illustrates the role of CO_2 dilution in determining the adiabatic flame temperature.

Example 2.6

Hydrogen gas is burned in oxygen diluted with carbon dioxide. Fuel and oxygen are present in stoichiometric proportions. If the combustion takes place adiabatically at constant pressure, what should be the proportion of oxygen and carbon dioxide on a volumetric basis so that the temperature of the products is 2400K? Neglect dissociation effects and assume that the reactants enter at 298 K.

Solution

Let α:1 be the volumetric proportion of CO_2:O_2 required to achieve this flame temperature. For this dilution, the chemical reaction can be written as:

$$H_2 + 0.5(O_2 + \alpha CO_2) = H_2O + \frac{\alpha}{2} CO_2$$

Using energy balance, we can write:

$$\frac{\alpha}{2} \Delta h^*_{fCO_2} = \left(\Delta h^*_{fH_2O} + \Delta h^*_{H_2O}\Big|_{2400} \right) + \frac{\alpha}{2} \left(\Delta h^*_{fCO_2} + \Delta h^*_{CO_2}\Big|_{2400} \right)$$

On substituting the enthalpy values from the table, we obtain $\alpha = 2.55$. Thus, 2.55 moles of CO_2 need to be added for every mole of O_2.

2.6.2 Combustion of Synthetic Gas (Syngas)

Coal is the most abundant fossil fuel in many parts of the world like India, China and the United States and is used as the principal fuel in many applications, like thermal power plants and process industries. However, combustion of coal leads to serious levels of environmental pollution due to the presence of impurities like sulphur in addition to NO_x production. Consequently, a number of technologies have been attempted for reducing emissions from coal combustion. These technologies are often grouped under a common name, *clean coal technology*.

One of the clean coal technologies developed is coal gasification. In this technique, coal is partially oxidised in an environment of insufficient oxygen in a process called *gasification*. Gasification of coal leads to production of a gas mixture called *synthetic gas*, or *syngas* [6]. Syngas may also be produced by gasification of biomass [7]. The composition of syngas varies according to the process of production but the principal combustible constituents are hydrogen and carbon monoxide. Example 2.7 illustrates the thermochemical calculations for syngas combustion.

Example 2.7

Synthetic gas (syngas) has CO and H_2 as its principal constituents. A syngas mixture has equal proportions of CO and H_2 on a molar basis. If the mixture equivalence ratio is 0.7, calculate the heat loss from the combustor if the combustion products exit at 1000 K. The reacting mixture enters at 298 K.

Solution

The stoichiometric equations for CO and H_2 are:

$$CO + 0.5(O_2 + 3.76N_2) = CO_2 + 1.88N_2$$

$$H_2 + 0.5(O_2 + 3.76N_2) = H_2O + 1.88N_2$$

Thus, the stoichiometric equation for syngas with 50%CO and 50%H_2 is:

$$CO + 0.5(O_2 + 3.76N_2) = CO_2 + 1.88N_2$$

The chemical reaction for $\phi = 0.7$ mixture is:

$$0.5CO + 0.5H_2 + \frac{(0.5\times 0.5 + 0.5\times 0.5)}{0.7}(O_2 + 3.76N_2)$$

$$= 0.5CO + 0.5H_2O + 2.6856N_2 + 0.214250_2$$

From energy balance, we can write:

$$0.5\Delta h^*_{f_{CO}} + Q^* = 0.5\left[\Delta h^*_{f_{CO_2}} + \Delta h^*_{CO_2}\Big|_{1000K}\right] + 0.5\left[\Delta h^*_{f_{H_2O}} + \Delta h^*_{H_2O}\Big|_{1000K}\right] + 2.6856\,\Delta h^*_{N_2}\Big|_{1000K}$$

$$+ 0.21425\Delta h^*_{O_2}\Big|_{1000K}$$

Substituting the enthalpy values from the table:

$$Q^* = 0.5[-393546 + 33425] + 0.5[-241845 + 25993]$$

$$+ 2.6856\times 21468 + 0.21425\times 22721 - 0.5\times(-110541)$$

$$= -340387.13 \text{ kJ/kmol fuel}$$

Thus, 340387.13 kJ of heat has to be removed per kmol of fuel from the combustor.

2.6.3 Flue Gas/Exhaust Gas Recirculation

One of the NO_x reduction strategies involves mixing a part of the combustion products, known as flue or exhaust gas, after partial cooling, with the reactants. This technique is known as flue gas or exhaust gas recirculation. Due to the presence of a large mass of non-combustible gas in the exhaust gas, the addition of the exhaust in the reactant stream reduces the flame temperature and thus brings down NO_x production. This technique is widely applied in internal combustion engines and furnaces. Example 2.8 shows how the product temperature can be brought down by exhaust gas recirculation.

Example 2.8

Methane at 298 K is burned with 20% excess air at 400 K in a constant pressure combustor. Part of the flue gas is cooled to 500 K and recycled with the reactants. Determine the fraction of the flue gas (by volume) that has to be recirculated such that the flue gas temperature is 1700 K.

Solution

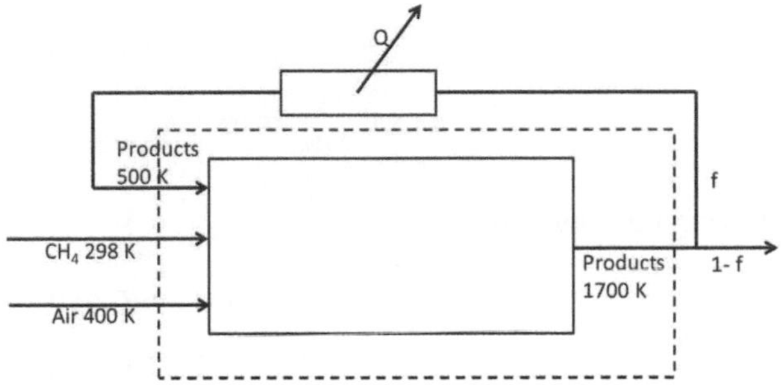

Let f be the fraction of the flue gas recirculated. Considering the control volume shown by dashed lines, the chemical reaction becomes:

$$CH_4 + 1.2 \times 2(O_2 + 3.76N_2) + f\left[aCO_2 + bH_2O + cO_2 + dN_2\right]$$
$$= aCO_2 + bH_2O + cO_2 + dN_2$$

From elemental mass balance, we get:

$$\text{C balance: } 1 + af = a \Rightarrow a = \frac{1}{1-f}$$

$$\text{H balance: } 4 + 2bf = 2b \Rightarrow b = \frac{2}{1-f}$$

$$\text{N balance: } 3.76 \times 1.2 \times 2 \times 2 + 2df = 2d \Rightarrow d = \frac{9.024}{1-f}$$

$$\text{O balance: } 2.4 \times 2 + 2af + bf + 2cf = 2a + b + 2c \Rightarrow c = \frac{0.4}{1-f}$$

Thus, the chemical equation becomes:

$$CH_4 + 2.4(O_2 + 3.76N_2) + \frac{f}{1-f}\left[CO_2 + 2H_2O + 0.4O_2 + 9.024N_2\right]$$
$$= \frac{1}{1-f}\left[CO_2 + 2H_2O + 0.4O_2 + 9.024N_2\right]$$

From energy balance:

$$\Delta h^*_{fCH_4} + 2.4\left[\Delta h^*_{O_2}\Big|_{400K} + 3.76\,\Delta h^*_{N_2}\Big|_{400K}\right] + \frac{f}{1-f}\left[\begin{array}{l}\left(\Delta h^*_{fCO_2} + \Delta h^*_{CO_2}\Big|_{500K}\right) \\ +2\left(\Delta h^*_{fH_2O} + \Delta h^*_{H_2O}\Big|_{500K}\right) \\ +0.4\,\Delta h^*_{O_2}\Big|_{500K} + 9.024\,\Delta h^*_{N_2}\Big|_{500K}\end{array}\right]$$

$$= \frac{1}{1-f}\left[\begin{array}{l}\left(\Delta h^*_{fCO_2} + \Delta h^*_{CO_2}\Big|_{1700K}\right) + 2\left(\Delta h^*_{fH_2O} + \Delta h^*_{H_2O}\Big|_{1700K}\right) \\ +0.4\,\Delta h^*_{O_2}\Big|_{1700K} + 9.024\,\Delta h^*_{N_2}\Big|_{1700K}\end{array}\right]$$

Substituting the enthalpy values from the table:

$$-74831 + 2.4\times[3031 + 3.76\times 2973] + \frac{f}{1-f}\left[\begin{array}{l}(-393546 + 8301) + 2\times(-241845 + 6947) \\ +0.4\times 6097 + 9.024\times 5920\end{array}\right]$$

$$= \frac{1}{1-f}\left[\begin{array}{l}(-393546 + 73446) + 2\times(-241845 + 57786) \\ +0.4\times 47943 + 9.024\times 45243\end{array}\right]$$

Solving, we get $f = 0.288$. Thus, 28.8% of the flue gas has to be cooled and recirculated to keep the exhaust temperature at 1700 K.

2.6.4 Fire Suppression Using Water Sprays

One of the ways of suppressing or controlling fires, particularly in enclosed spaces, is by use of water sprays. Water droplets evaporate, extracting the necessary latent heat from the flame. This lowers the flame temperature and helps in controlling and eventually suppressing the flame. Although the actual interaction between the flame and the water droplets and the water vapour resulting from their evaporation is quite complex, it is possible to get some rough estimate of the final state from thermodynamic calculations alone [8]. For combustion in an enclosed space, an important parameter is the final pressure. The introduction of water spray affects this parameter in two opposite ways. On one hand, the droplets evaporate, taking the latent heat from the combustion products and leading to a drop in temperature and pressure inside the volume. However, evaporation of water produces steam. The introduction of a large volume of steam in the confined space tends to increase the pressure. These effects are important in mitigating fire hazards in process industries and during severe accidents in nuclear reactors. Example 2.9 illustrates the effect of water addition on the temperature of combustion products.

Example 2.9

Stoichiometric H_2–O_2 mixture at 298 K, 1 bar is burned at constant pressure. Water is added as spray and evaporates completely. If the mass of water added at 25°C is 1% of the mass of hydrogen, find the final temperature.

Solution

From energy balance, one can write:

$$m_{H_2}h_{H_2}(298)+m_{O_2}h_{O_2}(298)+m_w h_w(298)=m_{H_2O}h_{H_2O}(T)+m_w h_{H_2O}(T)$$

Here, m_w and h_w denote mass of water spray added and enthalpy of liquid water, respectively. In the final state, water vapour is the only product, though it is produced from two different sources:

$$m_{H_2}h_{H_2}(298)+m_{O_2}h_{O_2}(298)+m_w h_w(298)=m_{H_2O}h_{H_2O}(T)+m_w h_{H_2O}(T)$$

$$\frac{m_{H_2O}}{m_{H_2}}\left[\Delta h_{f_{H_2O}}+\Delta h_{H_2O}(T)\right]+m_w\left[\left\{h_{H_2O}(T)-h_{H_2O}(373)\right\}\right.$$

$$\left.+\,h_{fg}+\left\{h_w(373)-h_w(298)\right\}\right]=0$$

Note that:

$$h_w(373)-h_w(298)=C_W(373-298))$$

$$h_{H_2O}(T)-h_{H_2O}(373)=\Delta h_{H_2O}(T)-\Delta h_{H_2O}(373)$$

Rearranging, we have:

$$\left(\frac{m_{H_2O}}{m_{H_2}}+\frac{m_w}{m_{H_2}}\right)\Delta h_{H_2O}(T)+\frac{m_{H_2O}}{m_{H_2}}\Delta h_{f_{H_2O}}$$

$$+\frac{m_w}{m_{H_2}}\left[C_W(373-298))+h_{fg}-\Delta h_{H_2O}(373)\right]=0$$

Note that for a stoichiometric mixture where $\frac{m_{H_2O}}{m_{H_2}}=9$ and taking the enthalpy and property values from the table:

$$\Delta h_{H_2O}(T)=-\frac{\frac{m_{H_2O}}{m_{H_2}}\Delta h_{f_{H_2O}}+\frac{m_w}{m_{H_2}}\left[C_W(373-298))+h_{fg}-\Delta h_{H_2O}(373)\right]}{\left(\frac{m_{H_2O}}{m_{H_2}}+\frac{m_w}{m_{H_2}}\right)}$$

$$=-\frac{9\times(-13428.37)+0.01\times[4.18\times75+2443.64-141.045]}{9+0.01}$$

$$=13410.56\ \text{kJ/kg}$$

$$=\ 241524.53\text{kJ/kmol}$$

Interpolating, we have $T=4862K$.

EXERCISES

2.1 Methanol is burned with 20% excess air. Calculate the air-fuel ratio on a mass and a molar basis and gravimetric and volumetric compositions of combustion products on a dry and a wet basis.

2.2 Syngas, a promising alternate fuel, has CO and H_2 as its principal constituents. If a syngas mixture has equal proportions of CO and H_2 on a molar basis, calculate (a) mass of air per kg of fuel required for a mixture equivalence ratio of 0.7 and (b) the composition of the products on a wet volumetric basis.

2.3 In oxyfuel combustion, fuel is burned in pure oxygen instead of air. However, to reduce the temperature of the products, oxygen is often diluted with CO_2. Calculate the total reactant mixture volume rate if LPG (composition as in Exercise 2.2) is burned in a 65% O_2-35% CO_2 (by volume). Calculate the number of moles of combustion products per mole of fuel. Consider stoichiometric mixture.

2.4 A four-stroke, four-cylinder spark ignition engine has cylinder bore and stroke equal to 10 cm and 15 cm, respectively. The engine runs at 3500 rpm. In the cruising range, the engine uses a fuel-air mixture with equivalence ratio of 0.9. For a volumetric efficiency of 85% calculate (a) the volume of air drawn in per minute, and (b) the mass of fuel supplied per minute. Consider the fuel to be pure octane and supply of mixture takes place at 1 bar and 25°C. (*Hint:* Volumetric efficiency is defined as the ratio of the volume of the charge sucked in to the stroke volume).

2.5 Consider the data in Exercise 2.2. To what temperature should the exhaust gas be cooled for water vapour to start condensing? Assume the total pressure to be 1 bar.

2.6 Methane is burned with excess air. The dry products contain 9.17% CO_2, 4.59% O_2 and the rest N_2. Calculate the amount of excess air supplied.

2.7 Bituminous coal has the ultimate ash free dry analysis given by 80% C, 9% H, 11% O. If the coal is burnt with 20% excess air, calculate the dew point of the mixture. How much moisture will be present in the flue gas if it is cooled 5°C below the dew point?

2.8 Calculate the enthalpy of combustion of methanol using enthalpies of formation of methanol and the product species. Will the value change if the fuel is burned in pure oxygen instead of air?

2.9 LPG can be considered as a mixture of 40% propane and 60% butane on a volumetric basis. Determine the adiabatic flame temperature for combustion at a constant volume. Comment on the relative magnitudes of the final temperature in this case of Example 2.3. Calculate the final pressure if the initial pressure is 1 bar.

2.10 Repeat Exercise 2.9 but assume constant specific heat evaluated at (a) 298 K and (b) 1200 K, and compare the results with that of Exercise 2.9.

2.11 In a constant pressure oxyfuel combustor, methane is burned in pure oxygen with the fuel and oxygen present in stoichiometric proportions. Determine the volume of carbon dioxide that has to be added per unit volume of oxygen to achieve the same adiabatic flame temperature as that of constant pressure combustion of stoichiometric methane-air mixture. Assume all reactants are at 298 K.

2.12 A spherical vessel of diameter 10 cm is filled with stoichiometric H_2–O_2 mixture at 298 K, 1 bar. The mixture is ignited and water is added in the form of liquid spray. If the mass of water added is 0.1% of the mass of hydrogen, find the final temperature and pressure in the vessel if the added water evaporates completely. Assume that the evaporation takes place at 100°C.

2.13 Consider the reaction $CO_2 \Leftrightarrow CO + 0.5O_2$. Calculate K_p and K_c at 2000 K, 1.5 bar.

2.14 Calculate K_p for the water gas shift reaction $H_2O+CO \Leftrightarrow H_2+CO_2$ at 2000 K. What will be the value of K_c?

2.15 Consider the reaction $H_2 \Leftrightarrow 2H$ in a closed vessel. Calculate the composition of the vessel at (a) 1000K, 1 bar; (b) 2000 K, 1 bar; and (c) 2000 K, 5 bar.

2.16 Consider the combustion of methane in 25% excess air. If the reactants enter at 298.15 K and the final temperature of the products is 2000 K, calculate the heat transferred from the combustor considering and neglecting the dissociation of CO_2.

REFERENCES

1. Kestin, J. *A Course in Thermodynamics*. Vol. 1. New York: CRC Press, 1979.
2. Mukunda, H. S. "Understanding combustion." *Universities Press (India) Private Limited Publication.*
3. Glassman, I., Richard A. Y. and Nick G. G. *Combustion*. New York: Academic Press, 2014.
4. Law, C. K. *Combustion Physics*. Cambridge, CA: Cambridge University Press, 2010.
5. Turns, S. R. *An Introduction to Combustion*. Vol. 287. New York: McGraw-Hill, 1996.
6. Williams, T. C., Shaddix, T. R. and Schefer, R. W., Effect of syngas composition and CO_2-diluted oxygen on performance of a premixed swirl-stabilised combustor, *Combustion Science and Technology*, **180**, 64–88, 2008.

7. Bhattacharya, A., Bhattacharya, A. and Datta, A., Modeling of hydrogen production process from biomass using oxygen blown gasification, *International Journal of Hydrogen Energy*, **37**, 18782–18790, 2012.
8. Joseph-Auguste, C., Cheikhravat, H., Djebaili-Chaumeix, N. and Deri, E., On the use of spray systems: An example of R&D work in hydrogen safety for nuclear applications, *International Journal of Hydrogen Energy*, **34**, 5970–5975, 2009.

Chapter 3

Chemical Kinetics

3.1 INTRODUCTION

Chemical equilibrium calculations give us the final equilibrium state of a chemically reacting system, but they do not give any information about the rate at which the chemical reactions take place. The equilibrium composition may be achieved if a long time is available for the chemical reactor. An analogous situation arises in the study of equilibrium thermodynamics which gives information about the total heat transfer and/or work transfer but gives no idea about the rate of heat or work transfer. For this information, one needs to study the laws of heat transfer, which deal with the finite rate of heat transfer. Similarly, the finite rate of a chemical reaction is attained from the study of chemical kinetics.

3.2 GLOBAL AND ELEMENTARY REACTIONS

Let us consider the chemical reaction described by an overall or global reaction mechanism:

$$H_2 + \frac{1}{2}O_2 \rightarrow H_2O \tag{3.1}$$

From experimental measurements, the rate of consumption of H_2 is given by:

$$\frac{d[X_{H_2}]}{dt} = -k_G(T)[X_{H_2}]^m[X_{O_2}]^n \tag{3.2}$$

$[X_{H_2}]$ denotes number of moles of H_2 per unit volume and is denoted as the molar concentration of H_2. The rate of reaction is proportional to the molar concentration of H_2 and O_2. The exponents m and n are related to the reaction order. The above expression implies that the reaction is of order m with respect to H_2 and of order n with respect to O_2. The overall order of reaction is $(m + n)$. The exact meaning of the reaction order will be discussed later. However, for global reaction mechanisms, the values of m and n are obtained from a curve fitting of experimental data and are not necessarily integers. In some cases, exponents can be negative. The proportionality factor $k_G(T)$ is a function of temperature and is known as *pre-exponential factor*.

The global reaction mechanism is an over-simplified description, however, and is good for some gross predictions only. The actual picture is much more complicated and is not represented by the global mechanism. For example, simultaneous collision of m molecules H_2 with n molecules of O_2 is unrealistic as it requires destruction and creation of several bonds at a time. In reality, the reaction proceeds through multiple (often hundreds) of intermediate steps. In these steps, several intermediate species, which can also include radicals, may be formed and destroyed. The reaction mechanisms describing these intermediate reactions are known as *elementary reaction mechanisms*. For example, for the global reaction in Equation (3.1), some of the important elementary reactions are:

$$H_2 + O_2 \rightarrow HO_2 + H \tag{3.3}$$

$$H + O_2 \rightarrow OH + O \tag{3.4}$$

$$H_2 + OH \rightarrow H_2O + H \tag{3.5}$$

$$H + O_2 + H \rightarrow HO_2 + H \tag{3.6}$$

The collection of elementary reactants is called the *reaction mechanism*. For elementary reactions, order or molecularity of reactions refers to the molecules of reactants that interact with each other. This reaction order refers to the number of molecules that collide with each other. For example, for the reaction in Equation (3.4), two molecules (one each of H and O_2) collide to form products. The order of reaction m and n denotes the influence of H_2 and O_2 concentrations on the reaction rate. Thus, the empirically determined exponents refer to the reaction order and not reaction molecularity. The reaction orders denote reaction molecularities only for elementary reactions. However, for global reactions where the reactants and the products are the starting and the final products of a reaction mechanism, the reaction orders determine the integrated effects of the molecularities of the individual elementary reactions and thus can have fractional values. The most common elementary reactions are bimolecular in nature, though unimolecular and termolecular reactions are also encountered.

3.3 BIMOLECULAR REACTIONS

Most elementary reactions are bimolecular in nature. In these reactions, two molecules collide and chemically react to form two other molecules. For the general reaction:

$$A + B \rightarrow C + D \tag{3.7}$$

the rate at which the species A is consumed is given by:

$$\frac{d[X_A]}{dt} = -k_{bimolecular}(T)[X_A][X_B] \tag{3.8}$$

However, unlike the global reactions, here the pre-exponential factor has a theoretical basis and can be estimated from kinetic molecular theory. The unit of $k_{bimolecular}$ is m^3/kmol-s.

3.4 COLLISION THEORY OF REACTION RATES

The collision theory of reaction rates states that the reaction rate is equal to the number of collisions per unit time between molecules having collision energy above a threshold value (activation energy). The major assumptions are as follows:

1. Collision energy needed for molecular change is derived from the relative translation energy of colliding molecules.
2. Gas is sufficiently dilute that only binary collisions are important.
3. Equilibrium Maxwell velocity distribution can be used even though the chemical reaction is highly transient.

The chemical reaction rate is determined by summing all possible collisions with energies above the threshold value.

It has been shown from detailed computations that the effective collision energy is the component of relative translation energy along the line of centre instead of the total relative translational energy. Physically this component represents the head-on collision. The other two components normal to the line of centre only affect the dynamics of the centre of the mass and hence cannot affect chemical changes.

Here, we derive a simple 1D model of a head-on collision. Let us consider n_i molecules, each of mass m_i and diameter D_i of species i and n_j molecules each of mass m_j and diameter D_j within unit volume. Let $\overline{V_{ij}}$ be the relative velocity. The maximum separation between molecules for collisions to take place is (D_{ij}) $\frac{D_i+D_j}{2}$. For such colliding pairs of molecules, the volume swept by the molecules in unit time is a cylinder volume given by $\Pi D_{ij}^2 \overline{V_{ij}}$. Each molecule of species i would thus collide with $\Pi D_{ij}^2 \overline{V_{ij}} n_j$ molecules of species j in unit time.

Therefore with n_i molecules of species i per unit volume, number of collisions per unit volume per unit time:

$$Z_{ij} = \Pi D_{ij}^2 \overline{V_{ij}} n_i n_j \tag{3.9}$$

For a gas at temperature T for the molecules having a Maxwellian velocity distribution:

$$\overline{V_{ij}} = (V_i^2 + V_j^2)^{\frac{1}{2}} = \left(\frac{8k_B T}{m_{ij}} \right)^{\frac{1}{2}} \tag{3.10}$$

Here, k_B is the Boltzmann constant and defined by:

$$k_B = \frac{R^*}{N_{av}} \tag{3.11}$$

$$m_{ij} = \frac{m_i m_j}{m_i + m_j} \tag{3.12}$$

Of all these collisions, only those with energy above activation energy E^* will lead to chemical reactions. For a Boltzmann distribution, the fraction of molecules of species having energy in excess of E_i^* is given by:

$$\frac{n_i^*}{n_i} = \exp(-E_i^* / R^* T) \tag{3.13}$$

In addition to the energy factor, a *generic* or *steric factor p* also has to be considered in calculating the probability of a collision leading to chemical reaction. The steric factor p takes into account the geometry of collision. This factor is generally less than 1 but there are exceptions. The probability P is given by:

$$P = p \frac{n_i^*}{n_i} \frac{n_j^*}{n_j} = p \exp\left(-\frac{E_i^* + E_j^*}{R^* T} \right) = p \exp\left(-\frac{E_a}{R^* T} \right) \tag{3.14}$$

where E_a is the activation energy.

Thus, $-\frac{d[X_i]}{dt}$ = the number of collisions of molecules of species i and j per unit volume per unit time. X is the probability that a collision leads to a reaction:

$$X\,(\text{kmol of } i/\text{number of molecules of } i) = Z_{ij} P N_{av}^{-1} \tag{3.15}$$

So, using Equations (3.9) to (3.12) in Equation (3.15):

$$-\frac{d[X_i]}{dt} = \Pi D_{ij}^2 \left(\frac{8k_B T}{m_{ij}} \right)^{\frac{1}{2}} n_i n_j p \exp\left(-\frac{E_a}{R^* T} \right) N_{av}^{-1} \tag{3.16}$$

Furthermore, Equation (3.16) can be elaborated as:

$$-\frac{d[X_i]}{dt} = \Pi D_{ij}^2 \left(\frac{8k_B T}{m_{ij}} \right)^{\frac{1}{2}} [X_i] N_{av} [X_j] N_{av} p \exp\left(-\frac{E_a}{R^* T} \right) N_{av}^{-1} \tag{3.17}$$

Finally:

$$-\frac{d[X_i]}{dt} = A T^{\frac{1}{2}} \exp\left(-\frac{E_a}{R^* T} \right) [X_i][X_j] \tag{3.18}$$

$$A = \Pi D_{ij}^2 \left(\frac{8k_B}{m_{ij}} \right)^{\frac{1}{2}} N_{av} p$$

where the activation energy E_a represents the energy that colliding molecules must possess for chemical reaction to occur.

During a chemical reaction, the reactants have to pass through a highly energised, activated complex state before being converted to products. The additional energy needed for the reactants to reach the activated state is the activation energy of the forward reaction. After passing through the activated state, the products are formed and the energy level decreases to the level of that of the products. The difference between the energy levels of the activated complex state and the products is the activated energy of backward reaction. Since the difference between the activation energies of the forward and backward reactions is equal to the difference between the potential (stored) energy of the reactants and the products, one can write:

$$E_f - E_b = \Delta h_c < 0 \text{ for exothermic reactions} \tag{3.19}$$

$$E_f - E_b = \Delta h_c > 0 \text{ for endothermic reactions} \tag{3.20}$$

The backward reaction for an endothermic process is a slow process because of the higher activation energy required.

3.5 UNIMOLECULAR REACTIONS

A unimolecular reaction describes a chemical process in which the reactant undergoes an *isomerisation* or *decomposition*. A generic representation of unimolecular reactions is given by:

$$A \rightarrow B \text{ (isomerisation)} \tag{3.21}$$

$$A \rightarrow B + C \text{ (decomposition)} \tag{3.22}$$

Unimolecular reactions require the reactant to acquire sufficient energy before the reaction can occur. This generally occurs through collision with another molecule. Thus, a more accurate representation of a unimolecular reaction is:

$$A + M \rightarrow B + M \tag{3.23}$$

M represents an arbitrary species known as *third body* that does not undergo any chemical change in this reaction. Thus, really unimolecular reactions are effectively second order in nature. Unimolecular reactions are first order at high pressure; at lower pressures, they are second order (depends on the concentration of third body) molecules as well. Unimolecular reactions can be very important in combustion. For example, during ignition of methane, the reaction:

$$\mathrm{CH_4} + M \Leftrightarrow \mathrm{CH_3} + \mathrm{H} + M \tag{3.24}$$

often starts a reaction chain.

3.6 TERMOLECULAR REACTION

Termolecular reactions involve three reactant molecules and are often the reverse of decomposition type unimolecular reactions at low pressures. The general representation of termolecular reaction is:

$$A + B + M \Leftrightarrow C + M \tag{3.25}$$

Recombination reactions, which are examples of termolecular reactions, are:

$$\mathrm{H} + \mathrm{H} + \mathrm{M} \Leftrightarrow \mathrm{H_2} + \mathrm{M} \tag{3.26}$$

and

$$\mathrm{H} + \mathrm{OH} + \mathrm{M} \Leftrightarrow \mathrm{H_2O} + \mathrm{M} \tag{3.27}$$

In reactions between radicals, the third body absorbs a large portion of the energy liberated in the reaction. Without this, the newly formed molecule would dissociate back. Termolecular reactions are generally slow compared to bimolecular or unimolecular reactions.

3.7 CHAIN REACTIONS

A sequence of elementary reactions involving production and destruction of radicals is known as a reaction chain. A reaction chain generally starts with formation of a radical from a reaction between stable species. The radical thus formed can give rise to more radicals. Formation of stable species from two or more radicals terminates the chain. Elementary reactions can be classified according to their rate in the reaction chain as follows.

Chain initiation reactions are those involving formation of a free radical from stable species. An example is:

$$H_2O_2 + H \rightarrow OH + OH + H \tag{3.28}$$

Chain propagation reactions are those in which one radical reacts with a stable species forming another radical and a stable species. An example is:

$$H_2 + OH \rightarrow H_2O + H \tag{3.29}$$

Chain branching reactions are those involving reaction between a radical and a stable species leading to formation of two radicals, for example:

$$O + H_2 \rightarrow H + OH \tag{3.30}$$

Chain terminating reactions involve destruction of radicals leading to formation of stable species, for example:

$$H + OH + M \rightarrow H_2O + M \tag{3.31}$$

Chain propagation and chain branching reactions are mostly bimolecular in nature, while chain termination reactions are often termolecular.

3.8 CHEMICAL TIME SCALES

Let us consider a ***unimolecular reaction***:

$$A \rightarrow B \tag{3.32}$$

The rate of consumption of A is given by:

$$\frac{d[X_A]}{dt} = -k_{uni}[A] \tag{3.33}$$

Integrating, we have:

$$[X_A](t) = [X_A]_o \exp(-k_{uni}t) \tag{3.34}$$

A chemical reaction time can be defined as the time in which the concentration of A decreases to a value equal to times its initial value. Thus:

$$\frac{[X_A](\tau_{chem})}{[X_A]_o} = \frac{1}{e} \tag{3.35}$$

Thus, one can write:

$$\exp(-k_{uni}\tau_{chem}) = \frac{1}{e} \tag{3.36}$$

Taking log of both sides, we have:

$$-k_{uni}\tau_{chem} = -1 \tag{3.37}$$

that is:

$$\tau_{chem} = \frac{1}{k_{uni}} \tag{3.38}$$

For a ***bimolecular reaction***, we can follow the reaction as follows:

$$A + B \rightarrow C + D \tag{3.39}$$

In this reaction, the rate of consumption of A is given by:

$$\frac{d[X_A]}{dt} = -k_{bimolecular}[X_A][X_B] \tag{3.40}$$

Similarly:

$$\frac{d[X_B]}{dt} = -k_{bimolecular}[X_A][X_B] \tag{3.41}$$

For every mole of A destroyed, 1 mole of B is destroyed. Thus:

$$[X_A]_O - [X_A] = [X_B]_O - [X_B] \tag{3.42}$$

that is:

$$[X_B] = [X_B]_O - [X_A]_O + [X_A] \tag{3.43}$$

Hence:

$$\frac{d[X_A]}{dt} = -k_{bimolecular}[X_A]\{[X_B]_O - [X_A]_O + [X_A]\} \tag{3.44}$$

$$\frac{d[X_A]}{[X_A]\{[X_B]_O - [X_A]_O + [X_A]\}} = -k_{bimolecular}dt \tag{3.45}$$

Using partial fractions,

$$\frac{d[X_A]}{\{[X_B]_O - [X_A]_O\}}\left\{\frac{1}{[X_A]} - \frac{1}{\{[X_B]_O - [X_A]_O + [X_A]\}}\right\} = -k_{bimolecular}dt \tag{3.46}$$

Integrating this, we have:

$$\ln[X_A] - \ln\{[X_B]_O - [X_A]_O + [X_A]\} = -k_{bimolecular}\{[X_B]_O - [X_A]_O\}t + \ln c \tag{3.47}$$

Let us consider the following:

$$c = \ln[X_A] - \ln[X_B] \tag{3.48}$$

Hence:

$$\ln\frac{[X_A]}{[X_A]_O} - \ln\frac{[X_A] + [X_B]_O - [X_A]_O}{[X_B]_O} = -k_{bimolecular}\{[X_B]_O - [X_A]_O\}t \tag{3.49}$$

$$\ln\frac{[X_A]}{[X_A]_O} - \ln\left\{\frac{[X_A]}{[X_B]_O} + 1 - \frac{[X_A]_O}{[X_B]_O}\right\} = -k_{bimolecular}\{[X_B]_O - [X_A]_O\}t \tag{3.50}$$

At $t = \tau_{chem}$

$$\ln\frac{1}{e}-\ln\left\{\frac{1}{e}\frac{[X_A]_O}{[X_B]_O}+1-\frac{[X_A]_O}{[X_B]_O}\right\}=-k_{bimolecular}\left\{[X_B]_O-[X_A]_O\right\}\tau_{chem} \quad (3.51)$$

$$-\ln e-\ln\left\{e+(1-e)\frac{[X_A]_O}{[X_B]_O}\right\}+\ln e=-k_{bimolecular}\left\{[X_B]_O-[X_A]_O\right\}\tau_{chem} \quad (3.52)$$

$$\tau_{chem}=\frac{\ln\left[e+(1-e)\frac{[X_A]_O}{[X_B]_O}\right]}{\left\{[X_B]_O-[X_A]_O\right\}k_{bimolecular}} \quad (3.53)$$

If one of the reactants has much less concentration than the other, as in a lean mixture:

$$\frac{[X_A]_O}{[X_B]_O}\to 0$$

Hence:

$$\tau_{chem}\approx\frac{1}{[X_B]_O k_{bimolecular}} \quad (3.54)$$

Now let us consider a ***termolecular reaction***:

$$A+B+M\to C+M \quad (3.55)$$

In this reaction, we have:

$$\frac{d[X_A]}{dt}=-k_{ter}[X_A][X_B][X_M] \quad (3.56)$$

Since third body concentration does not change with time, one can write:

$$\frac{d[X_A]}{dt}=-k_{ter}[X_M][X_A][X_B]=-k^*[X_A][X_B] \quad (3.57)$$

Hence, for $[X_B]>>[X_A]$

$$\tau_{chem}=\frac{1}{[X_B]_O k^*}=\frac{1}{k_{ter}[X_M][X_B]_O} \quad (3.58)$$

3.9 RELATION BETWEEN KINETIC RATE COEFFICIENTS AND EQUILIBRIUM CONSTANTS

Let us consider a reversible reaction:

$$aA + bB \Leftrightarrow cC + dD \tag{3.59}$$

The rate of formation of A is given by:

$$\frac{d[X_A]}{dt} = -k_f[X_A]^a[X_B]^b + k_b[X_C]^c[X_D]^d \tag{3.60}$$

The variables k_f and k_b are forward and reverse rate coefficients, respectively.

At equilibrium, the compositions of the species do not change with time. Hence, at equilibrium:

$$\frac{d[X_A]}{dt} = 0 \tag{3.61}$$

That is:

$$-k_f[X_A]^a[X_B]^b + k_b[X_C]^c[X_D]^d = 0 \tag{3.62}$$

This leads to:

$$\frac{k_f}{k_b} = \frac{[X_C]^c[X_D]^d}{[X_A]^a[X_B]^b} = k_c \tag{3.63}$$

3.10 MULTISTEP MECHANISM

Let us consider the following steps:

$$HCN + O \Leftrightarrow NCO + H \tag{R1}$$

$$NCO + H \Leftrightarrow NH + CO \tag{R2}$$

$$NH + H \Leftrightarrow N + H_2 \tag{R3}$$

$$N + OH \Leftrightarrow NO + H \tag{R4}$$

Knowing the rates of each elementary reaction, we can express the net production or destruction rate of any species involved in a series of elementary steps.

For example, considering the reactions R1–R4, the net rate of production of a hydrogen atom will be:

$$\frac{d[X_H]}{dt} = k_{1_f}[X_{HCN}][X_O] - k_{1_b}[X_{NCO}][X_H] - k_{2_f}[X_{NCO}][X_H] + k_{2_b}[X_{NH}][X_{CO}] \\ - k_{3_f}[X_{NH}][X_H] + k_{3_b}[X_N][X_{H_2}] + k_{4_f}[X_N][X_{OH}] - k_{4_b}[X_{NO}][X_H] \quad (3.64)$$

Let us consider a reaction chain as follows:

$$H + O_2 \Leftrightarrow O + OH \quad (R5)$$

$$O + H_2 \Leftrightarrow H + OH \quad (R6)$$

$$H_2 + OH \Leftrightarrow H_2O + H \quad (R7)$$

In reality, a reaction mechanism may consist of thousands of such elementary reactions involving hundreds of species. So it is convenient to develop a compact notation for a mechanism consisting of N species and M reactions. This compact representation is given by:

$$\sum_{j=1}^{N} \nu'_{ji} X_j \Leftrightarrow \sum_{j=1}^{N} \nu''_{ji} X_j \;\; \forall i = 1, 2, \ldots, M \quad (3.65)$$

The net rate of production of species j is given by:

$$\omega_j = \sum_{i=1}^{M} \nu_{ji} q_i \;\; \forall j = 1, 2, \ldots, N \quad (3.66)$$

ν'_{ji}, ν''_{ji} are the stoichiometric coefficients on the reactant and product side, respectively.

The net stoichiometric coefficient, ν_{ji} is expressed by:

$$\nu_{ji} = \nu''_{ji} - \nu'_{ji} \quad (3.67)$$

q_i is defined as the rate of progress variable for ith reaction, expressed as:

$$q_i = k_{f_i} \prod_{j=1}^{N} [X_j]^{\nu'_{ji}} - k_{b_i} \prod_{j=1}^{N} [X_j]^{\nu''_{ji}} \quad (3.68)$$

Considering the reactions (R5–R7), we have the following species and reactions:

j	Species	I	Reaction
1	H	1	R1
2	O_2	2	R2
3	O	3	R3
4	OH		
5	H_2		
6	H_2O		

The stoichiometric coefficient on the reaction side is given by:

$$\nu'_{ji} = \begin{bmatrix} 1 & 1 & 0 & 0 & 0 & 0 \\ 0 & 0 & 1 & 0 & 1 & 0 \\ 0 & 0 & 0 & 1 & 1 & 0 \end{bmatrix}$$

The stoichiometric coefficient on the product side is expressed by:

$$\nu''_{ji} = \begin{bmatrix} 0 & 0 & 1 & 1 & 0 & 0 \\ 1 & 0 & 0 & 1 & 0 & 0 \\ 1 & 0 & 0 & 0 & 0 & 1 \end{bmatrix}$$

Thus:

$$q_1 = k_{f_1}[H]^1[O_2]^1[O]^0[OH]^0[H_2]^0[H_2O]^0 - k_{b_1}[H]^0[O_2]^0[O]^1[OH]^1[H_2]^0[H_2O]^0$$

$$= k_{f_1}[H][O_2] - k_{b_1}[O][OH] \tag{3.69}$$

Similarly:

$$q_2 = k_{f_2}[H]^0[O_2]^0[O]^1[OH]^0[H_2]^1[H_2O]^0 - k_{b_2}[H]^1[O_2]^0[O]^0[OH]^1[H_2]^0[H_2O]^0$$

$$= k_{f_2}[O][H_2] - k_{b_2}[H][OH] \tag{3.70}$$

$$q_3 = k_{f_3}[H]^0[O_2]^0[O]^0[OH]^1[H_2]^1[H_2O]^0 - k_{b_3}[H]^1[O_2]^0[O]^0[OH]^0[H_2]^0[H_2O]^1$$

$$= k_{f_3}[OH][H_2] - k_{b_3}[H][H_2O] \tag{3.71}$$

$$\frac{d[H]}{dt} = \omega_H = \sum_{i=1}^{3} \nu_{Hi} q_i$$

$$= \nu_{H1} q_1 + \nu_{H2} q_2 + \nu_{H3} q_3$$

$$= (\nu''_{H1} - \nu'_{H1}) q_1 + (\nu''_{H2} - \nu'_{H2}) q_2 + (\nu''_{H3} - \nu'_{H3}) q_3$$

$$= (0-1)\left\{k_{f1}[X_H][X_{O_2}] - k_{b1}[X_O][X_{OH}]\right\} + (1-0)\left\{k_{f2}[X_O][X_{H_2}] - k_{b2}[X_H][X_{OH}]\right\}$$
$$+ (1-0)\left\{k_{f3}[X_{H_2}][X_{OH}] - k_{b3}[X_H][X_{H_2O}]\right\}$$

$$= -\left\{k_{f1}[X_H][X_{O_2}] - k_{b1}[X_O][X_{OH}]\right\} + \left\{k_{f2}[X_O][X_{H_2}] - k_{b2}[X_H][X_{OH}]\right\}$$
$$+ \left\{k_{f3}[X_{H_2}][X_{OH}] - k_{b3}[X_H][X_{H_2O}]\right\}$$

$$= -k_{f1}[X_H][X_{O_2}] + k_{b1}[X_O][X_{OH}] + k_{f2}[X_O][X_{H_2}] - k_{b2}[X_H][X_{OH}]$$
$$+ k_{f3}[X_{H_2}][X_{OH}] - k_{b3}[X_H][X_{H_2O}] \quad (3.72)$$

Chemical kinetics calculations ordinarily involve solutions of as many differential equations as there are species. This implies solution of a large number of differential equations, which can be expensive computationally. Moreover, different species have widely different rates of formation, which makes the system stiff. This further complicates the problem. Often computational efforts can be reduced by replacing some of the differential equations by algebraic equations. It further helps if the differential equations replaced represent fast changing species. Two common approaches are ***steady state approximation*** and ***partial equilibrium approximation.***

3.11 STEADY STATE APPROXIMATION

In chain reactions, many highly reactive intermediate species or radicals are formed. After an initial transient, during which phase the concentration of the radicals rapidly builds up, these species are destroyed very fast, as quickly as they are formed. Thus, although the rates of formation and consumption of these species are high, the species concentration and hence its rate of change is low. Thus, for such a species, we can express the following:

$$\omega_j = \frac{d[X_j]}{dt} = \omega_j^+ - \omega_j^- \quad (3.73)$$

The steady state approximation implies that:

$$\left|\frac{d\left[X_j\right]}{dt}\right| << (\omega_j^+, \omega_j^-) \tag{3.74}$$

This in turn implies:

$$\omega_j^+ \approx \omega_j^- \tag{3.75}$$

It is important to remember, however, that steady state approximation does imply $\frac{d[X_j]}{dt} = 0$ exactly. It only satisfies the following:

$$\omega_j^+ \approx \omega_j^- >> \left|\frac{d[X_j]}{dt}\right| \tag{3.76}$$

Let us illustrate the application of steady state approximation with one example.

$$\mathrm{O} + \mathrm{N_2} \xrightarrow{k_1} \mathrm{NO} + \mathrm{N} \tag{3.77}$$

$$\mathrm{N} + \mathrm{O_2} \xrightarrow{k_2} \mathrm{NO} + \mathrm{O} \tag{3.78}$$

$$\frac{d[X_N]}{dt} = k_1[X_\mathrm{O}][X_{\mathrm{N_2}}] - k_2[X_\mathrm{N}][X_{\mathrm{O_2}}] \tag{3.79}$$

The first reaction, in Equation (3.77), is slow; the second, in Equation (3.78) is very fast. The *N* atom is consumed as soon as it is formed. Thus, the *N* concentration attains a steady state:

$$\frac{d[X_N]}{dt} = k_1[X_\mathrm{O}][X_{\mathrm{N_2}}] - k_2[X_\mathrm{N}][X_{\mathrm{O_2}}] = 0$$

$$\Rightarrow k_1[X_\mathrm{O}][X_{N_2}] = k_2[X_\mathrm{N}][X_{\mathrm{O_2}}]$$

$$[X_N]_{ss} = \frac{k_1[X_\mathrm{O}][X_{\mathrm{N_2}}]}{k_2[X_{\mathrm{O_2}}]} \tag{3.80}$$

3.12 PARTIAL EQUILIBRIUM APPROXIMATION

In a reaction chain, some reactions are much faster or slower than others. For example, recombination reactions are much slower than the chain branching or chain propagating reactions. Thus, in the time scale of the slow reaction, the fast reactions have time to reach a state of equilibrium. Mathematically, the implication is that both forward and backward reaction rates are comparable and much larger than the net reaction rate:

$$q_i << k_{f_i} \prod_{j=1}^{N} [X_j]^{\nu'_{ji}} \approx k_{b_i} \prod_{j=1}^{N} [X_j]^{\nu''_{ji}}$$

This does not imply, however, that q_i is small in comparison with ω_j and that ω_j can be calculated by setting $q_i = 0$.

Let us illustrate with an example. Recall (R5–R7) again:

$$H + O_2 \Leftrightarrow O + OH$$

$$O + H_2 \Leftrightarrow H + OH$$

$$H_2 + OH \Leftrightarrow H_2O + H$$

These reactions are assumed to be much faster than other accompanying reactions and hence equilibrium can be assumed. Thus:

$$k_{f_1}[X_H][X_{O_2}] = k_{b_1}[X_{OH}][X_O]$$

$$\frac{[X_{OH}][X_O]}{[X_H][X_{O_2}]} = \frac{k_{f_1}}{k_{b_1}} = k_{c_1} \tag{3.81}$$

Similarly:

$$\frac{[X_{OH}][X_H]}{[X_O][X_{H_2}]} = \frac{k_{f_2}}{k_{b_2}} = k_{c_2} \tag{3.82}$$

and

$$\frac{[X_H][X_{H_2O}]}{[X_{OH}][X_{H_2}]} = \frac{k_{f_3}}{k_{b_3}} = k_{c_3} \tag{3.83}$$

From the above equations, $[X_H]$, $[X_O]$ and $[X_{OH}]$ can be obtained in terms of $[X_{H_2}]$, $[X_{O_2}]$ and $[X_{H_2O}]$. It is important to note that steady state approximation applies to a species, while partial equilibrium applies to a reaction.

EXERCISES

3.1 Classify the following reactions as global or elementary and provide your reasoning. In the case of elementary reactions, mention whether they are unimolecular, bimolecular or termolecular and also specify their role in the reaction chain (chain initiating, chain terminating, etc.).

$$2H_2 + O_2 \rightarrow 2H_2O$$

$$H_2 + M \rightarrow H + H + M$$

$$H + O_2 \rightarrow O + OH$$

$$H_2 + OH \rightarrow H_2O + H$$

$$H + H + M \rightarrow H_2 + M$$

$$H + O + M \rightarrow OH + M$$

$$O_3 \rightarrow O_2 + O$$

3.2 Consider the reaction chain involved in H_2O-catalyzed kinetics of CO oxidation:

$CO + O_2 \Leftrightarrow CO_2 + O$ (forward and reverse reaction rates: $k_{1,f}$ and $k_{1,r}$)

$O + H_2O \Leftrightarrow OH + OH$ (forward and reverse reaction rates: $k_{2,f}$ and $k_{2,r}$)

$CO + OH \Leftrightarrow CO_2 + H$ (forward and reverse reaction rates: $k_{3,f}$ and $k_{3,r}$)

$H + O_2 \Leftrightarrow OH + O$ (forward and reverse reaction rates: $k_{4,f}$ and $k_{4,r}$)

Calculate the neat rate of production of O.

3.3 Consider the following fast forward and reverse reactions involving radicals in combustion of hydrogen:

$H + O_2 \Leftrightarrow OH + O$ (forward and reverse reaction rates: $k_{1,f}$ and $k_{1,r}$)

$O + H_2 \Leftrightarrow OH + H$ (forward and reverse reaction rates: $k_{2,f}$ and $k_{2,r}$)

$OH + H_2 \Leftrightarrow H_2O + H$ (forward and reverse reaction rates: $k_{3,f}$ and $k_{3,r}$)

Use partial equilibrium approximation to derive algebraic expressions for molar concentrations of O, H and OH in terms of O_2, H_2 and H_2O.

3.4 Consider the prediction of NO in the combustion products in an expanding piston cylinder. Find an expression for the rate of change of concentration of NO, and take into consideration the temporal variations in volume $V=V(t)$.

3.5 *A four-step reaction mechanism for oxidation of methane is as follows:

$$CH_4+2H+H_2O \rightleftharpoons CO+4H_2$$

$$CO+H_2O \rightleftharpoons CO_2+H_2$$

$$H+H+M \rightleftharpoons H_2+M$$

$$O_2+3H_2 \rightleftharpoons 2H+H_2O$$

Derive a three-step reaction mechanism and the corresponding reaction rates by assuming that H is in steady state. Can we derive a two-step mechanism by assuming that the second reaction in the four-reaction mechanism is in partial equilibrium?

3.6 Consider the NO formation in an octane-air mixture in a combustion engine. The combustion process may be modelled as a polytropic compression ($n = 1.3$) followed by an isobaric adiabatic composition.

a. Determine the equilibrium composition of combustion products by considering a lean air-fuel mixture ($\phi = 0.9$) as a function of the compression ratio in the range 6–14 for an initial mixture temperature and pressure of 298 K and 1 bar. Consider dissociation of CO_2 as the only dissociation reaction.

b. Refer to the global mechanism in Exercise 3.3 and determine the instantaneous NO concentration as a function of time for compression ratios of 6, 10 and 14.

Chapter 4

Simple Reactor Models

Combustion calculations, in general, involve a combination of thermodynamics, chemical kinetics and transport phenomena. Insight can often be obtained, however, through use of simple models combining basic conservation principles with chemical kinetics without invoking the complexities due to heat or mass diffusion. These models often neglect spatial variations, leading to lumped models. Such models can be described by ordinary differential equations for unsteady models and algebraic (though non-linear) equations for steady models. Some common reactor models are fixed mass constant pressure and constant volume reactors, well-stirred reactors and plug flow reactors. Apart from providing insight into fundamental aspects of combustion, practical combustion systems can be modelled using combinations of these models.

4.1 CONSTANT PRESSURE REACTORS

Let us consider reactants contained in a piston-cylinder kind of arrangement such that the pressure inside remains constant in time. The reactants are uniformly mixed and react uniformly within the reactor. Thus, there is no temperature or concentration gradient and the state of the system is described by a single set of variables. The dynamics can be described by a set of first order ordinary differential equations. Starting from the First Law of Thermodynamics, one can write:

$$\dot{Q} - \dot{W} = m\frac{du}{dt} \tag{4.1}$$

Here, $\dot{Q}$ and $\dot{W}$ are rates of heat and work transfer during the process, and m and u are the mass and specific internal energy of the system.

Using the definition of enthalpy, one obtains at constant pressure:

$$\frac{du}{dt} = \frac{dh}{dt} - P\frac{dv}{dt} \tag{4.2}$$

Assuming $\dot{W} = Pm\frac{dv}{dt}$, the energy equation becomes:

$$\dot{Q} = m\frac{dh}{dt}$$

For a mixture of ideal gases, mixture enthalpy h can be defined as:

$$h = \frac{H}{m} = \frac{\sum_i n_i h_i^*}{m} \tag{4.3}$$

Hence, for a fixed mass reactor:

$$\frac{dh}{dt} = \frac{1}{m}\left[\sum_i h_i^* \frac{dn_i}{dt} + \sum_i n_i \frac{dh_i^*}{dt}\right] \tag{4.4}$$

For ideal gases as $h_i^* = h_i^*(T)$ is a function of temperature only:

$$\frac{dh_i^*}{dt} = \frac{dh_i^*}{dT}\frac{dT}{dt} = C_{p_i}^* \frac{dT}{dt} \tag{4.5}$$

The number of moles of species i, n_i is related to its molar concentration $[X_i]$ as follows:

$$n_i = V[X_i] \tag{4.6}$$

Similarly, defining ω_i as the rate of production of species i per unit volume, we have:

$$\frac{dn_i}{dt} = V\omega_i \tag{4.7}$$

Note that $[X_i]$ changes with time due to changes in both n_i and V. Thus:

$$\frac{d[X_i]}{dt} = \frac{d}{dt}\left(\frac{n_i}{V}\right) = \frac{1}{V}\frac{dn_i}{dt} - \frac{n_i}{V^2}\frac{dV}{dt} \tag{4.8}$$

Substituting the results of Equations (4.6) and (4.7) in Equation (4.8), we obtain:

$$\frac{d[X_i]}{dt} = \omega_i - \frac{[X_i]}{V}\frac{dV}{dt} \tag{4.9}$$

Again, from the First Law of Thermodynamics:

$$\frac{\dot{Q}}{\dot{m}} = \frac{dh}{dt} = \frac{1}{m}\left[\sum_{i=1}^{N} h_i^* \frac{dn_i}{dt} + \sum_{i=1}^{N} n_i C_{p_i}^* \frac{dT}{dt}\right]$$

(4.10)

Equation (4.10) can be simplified to temporal variation of temperature T as:

$$\frac{dT}{dt} = \frac{\dfrac{\dot{Q}}{V} - \displaystyle\sum_{i=1}^{N} h_i^* \omega_i}{\displaystyle\sum_{i=1}^{N} C_{p_i}^* [X_i]}$$

(4.11)

From the ideal gas equation of state, one can write at constant pressure, using Equations (4.6) and (4.7):

$$\frac{1}{V}\frac{dV}{dt} = \frac{1}{T}\frac{dT}{dt} + \frac{1}{\displaystyle\sum_{i=1}^{N} n_i}\sum_{i=1}^{N}\frac{dn_i}{dt} = \frac{1}{T}\frac{dT}{dt} + \frac{1}{\displaystyle\sum_{i=1}^{N}[X_i]}\sum_{i=1}^{N}\omega_i$$

(4.12)

Substituting Equation (4.12) in Equation (4.9), one obtains:

$$\frac{d[X_i]}{dt} = \omega_i - [X_i]\left[\frac{1}{T}\frac{dT}{dt} + \frac{\displaystyle\sum_{i=1}^{N}\omega_i}{\displaystyle\sum_{i=1}^{N}[X_i]}\right]$$

(4.13)

The equations require specification of initial conditions, as follows:

At $\qquad\qquad\qquad t = 0 : T = T_0; [X_i] = [X_i]_0$

In the above equations, volume V and molar specific enthalpy h_i^* are obtained as:

$$V = \frac{m}{\displaystyle\sum_{i=1}^{N}[X_i]MW_i}$$

$$h_i^* = \Delta h_{f_i}^* + \int_{T_{ref}}^{T} C_{p_i}^*(T)dT$$

Thus, the model can be summarised as:

$$\frac{dT}{dt}=\frac{\dfrac{\dot{Q}}{V}-\sum_{i=1}^{N}h_i^{*}\omega_i}{\sum_{i=1}^{N}C_{p_i}^{*}[X_i]}$$

4.2 CONSTANT VOLUME REACTOR

Proceeding along similar lines as in the previous case, one can write for a constant volume case where $\dot{W}=0$:

$$\dot{Q}=m\frac{du}{dt} \tag{4.14}$$

For an ideal gas mixture, we have:

$$u=\frac{\sum_{i=1}^{N}n_iu_i^{*}}{m}=\frac{\sum_{i=1}^{N}n_i\left(h_i^{*}-R^{*}T\right)}{m} \tag{4.15}$$

Hence, using Equations (4.14) and (4.15), one can write:

$$\begin{aligned}\frac{\dot{Q}}{m}=\frac{du}{dt}&=\frac{1}{m}\left[\sum_{i=1}^{N}\left(h_i^{*}-R^{*}T\right)\frac{dn_i}{dt}+\sum_{i=1}^{N}n_i\left(\frac{dh_i^{*}}{dt}-R^{*}\frac{dT}{dt}\right)\right]\\&=\frac{1}{m}\left[V\sum_{i=1}^{N}\left(h_i^{*}-R^{*}T\right)\omega_i+V\sum_{i=1}^{N}[X_i]\left(C_{p_i}^{*}-R^{*}\right)\frac{dT}{dt}\right]\end{aligned} \tag{4.16}$$

The above relation can be rearranged as:

$$\frac{dT}{dt}=\frac{\dfrac{\dot{Q}}{V}-\sum_{i=1}^{N}\left(h_i^{*}-R^{*}T\right)\omega_i}{\sum_{i=1}^{N}[X_i]\left(C_{p_i}^{*}-R^{*}\right)} \tag{4.17}$$

Also, from Equation (4.6), we obtain:

$$\frac{d[X_i]}{dt} = \frac{1}{V}\frac{dn_i}{dt} = \omega_i \tag{4.18}$$

For a constant volume reactor, pressure in the reactor increases with time. From the ideal gas equation of state, we can write:

$$\frac{dP}{dt} = R^* T \sum_{i=1}^{N} \omega_i + R^* \sum_{i=1}^{N} [X_i] \frac{dT}{dt} \tag{4.19}$$

The system of equations needs the following initial conditions:

At $$t = 0 : T = T_0; [X_i] = [X_i]_0 ; P = P_0 \tag{4.20}$$

4.3 WELL-STIRRED REACTOR

Well-stirred (or perfectly stirred) reactors are a class of idealised reactors in which all the chemical species are perfectly mixed such that there is no spatial variation inside the reactor and the reactor can be assumed to be in a uniform state. Unlike the two previous reactor models, this model is not a fixed mass model. Well-stirred models are used for studying fundamental combustion characteristics like flame blowout.

Applying the law of conservation of mass to the reactor, we have:

$$\frac{dm_{cv}}{dt} = \dot{m}_{in} - \dot{m}_{out} \tag{4.21}$$

The mass within the control volume is given by $m_{cv} = \rho V$, where ρ is the density of the mixture within the control volume of volume V. Thus, one can write from Equation (4.21):

$$\frac{d\rho}{dt} = \frac{\dot{m}_{in}}{V} - \frac{\dot{m}_{out}}{V} \tag{4.22}$$

Applying the law of conservation of mass to a species i, we have:

$$\frac{dm_{i,cv}}{dt} = \dot{m}_{i,in} - \dot{m}_{i,out} + \dot{m}_i''' V \tag{4.23}$$

The first three terms denote the net rate of accumulation and rates of inflow and outflow of species *i*. The last term in Equation (4.23) denotes the rate of production of species *i*. The rate of production of species *i* is given by:

$$\dot{m}_i''' = \omega_i MW_i \tag{4.24}$$

In terms of mass fractions, Equation (4.23) becomes:

$$\frac{d}{dt}(\rho y_i V) = \dot{m}_{in} y_{i,in} - \dot{m}_{out} y_{i,out} + \omega_i MW_i V \tag{4.25}$$

For a well-stirred reactor, $y_{i,out} = y_i$. Hence, one can write:

$$\frac{d}{dt}(\rho y_i) = \frac{\dot{m}_{in}}{V} y_{i,in} - \frac{\dot{m}_{out}}{V} y_i + +\omega_i MW_i \tag{4.26}$$

Expanding the left-hand side of Equation (4.26) and using Equation (4.22), the species conservation equation becomes:

$$\rho \frac{dy_i}{dt} = \frac{\dot{m}_{in}}{V}(y_{i,in} - y_i) + \omega_i MW_i \tag{4.27}$$

The mass fraction is related to species molar concentration as follows:

$$y_i = \frac{MW_i [X_i]}{\sum_{j=1}^{N} MW_j [X_j]} \tag{4.28}$$

Energy balance within the reactor gives:

$$\frac{d}{dt}(m_{cv} u_{cv}) = \sum_{j=1}^{N} \dot{m}_{j,in} h_{j,in} - \sum_{j=1}^{N} \dot{m}_{j,out} h_{j,out} + \dot{Q} \tag{4.29}$$

Noting that $h_{j,out} = h_j$ and using the relations $m_{cv} = \rho V$ in terms of mass fractions, one can write:

$$\frac{d}{dt}(\rho V u_{cv}) = \sum_{j=1}^{N} \dot{m}_{in} y_{j,in} h_{j,in} - \sum_{j=1}^{N} \dot{m}_{out} y_j h_j + \dot{Q} \tag{4.30}$$

The internal energy can be written as:

$$u_{cv} = \sum_{j=1}^{N} y_j \left(h_j - R_j T\right) = \sum_{j=1}^{N} y_j h_j - \bar{R}T \tag{4.31}$$

Differentiating u_{cv} with respect to time and using ideal gas properties, we get:

$$\frac{du_{cv}}{dt} = \sum_{j=1}^{N} \left(h_j - R_j T\right)\frac{dy_j}{dt} + \sum_{j=1}^{N} y_j \left(C_{p_j} - R_j\right)\frac{dT}{dt} \tag{4.32}$$

Substituting Equations (4.22), (4.31) and (4.32) in Equation (4.29), we have:

$$\begin{aligned}&\rho\left[\sum_{j=1}^{N}\left(h_j - R_j T\right)\frac{dy_j}{dt} + \sum_{j=1}^{N} y_j\left(C_{p_j} - R_j\right)\frac{dT}{dt}\right] \\ &+\left(\frac{\dot{m}_{in}}{V} - \frac{\dot{m}_{out}}{V}\right)\left(\sum_{j-1}^{N} y_j h_j - \bar{R}\right) = \sum_{j=1}^{N}\frac{\dot{m}_{in}}{V} y_{j,in} h_{j,in} - \sum_{j=1}^{N}\frac{\dot{m}_{out}}{V} y_j h_j + \frac{\dot{Q}}{V}\end{aligned} \tag{4.33}$$

Rearranging Equation (4.33), we obtain:

$$\frac{dT}{dt} = \frac{1}{\bar{C}_v}\left[\begin{aligned}&\left\{-\sum_{j=1}^{N}\left(h_j - R_j T\right)\frac{dy_j}{dt}\right\} + \frac{\dot{m}_{in}}{\rho V}\sum_{j=1}^{N}\left(y_{j,in} h_{j,in} - y_j h_j\right) \\ &-\left(\frac{\dot{m}_{in}}{\rho V} - \frac{\dot{m}_{out}}{\rho V}\right)\bar{R} + \frac{\dot{Q}}{\rho V}\end{aligned}\right] \tag{4.34}$$

This model of unsteady well-stirred reactors was used by Lieuwen and Zinn [1] to study thermoacoustic instability for gas turbine combustors. Sometimes, the mass accumulation effect inside the reactor is not considered, which implies $\frac{d\rho}{dt} = \frac{\dot{m}_{in}}{V} - \frac{\dot{m}_{out}}{V} = 0$. The resulting simplified model was used by many researchers (e.g., Yi and Gutmark [2]) for developing models to control instability and lean blowout in gas turbine combustors.

Steady well-stirred reactors are often used, which can be obtained by dropping all transient terms.

4.4 PLUG FLOW REACTOR

The plug flow reactor (PFR) is another idealised reactor. However, in this reactor, spatial variation in the flow direction is considered, though conditions are assumed to be steady. The major assumptions are as follows:

1. Steady state steady flow
2. Perfect mixing leading to uniform properties in transverse direction
3. Negligible diffusion in axial direction
4. Ideal frictionless flow and ideal gas equation of state

From the conservation of mass, one can write:

$$\frac{d}{dx}(\rho v_x A)=0 \tag{4.35}$$

From the conservation of momentum, one can write:

$$\rho v_x \frac{dv_x}{dx}+\frac{dP}{dx}=0 \tag{4.36}$$

From the conservation of energy, one can write:

$$\rho v_x \frac{d}{dx}\left(h+\frac{v_x^2}{2}\right)+q''\wp=0 \tag{4.37}$$

From species conservation, one can write:

$$\rho v_x \frac{dy_i}{dx}-\omega_i MW_i=0 \tag{4.38}$$

From Equation (4.35), one can write:

$$\frac{1}{\rho}\frac{d\rho}{dx}+\frac{1}{v_x}\frac{dv_x}{dx}+\frac{1}{A}\frac{dA}{dx}=0 \tag{4.39}$$

Since $\dot{m}=\rho v_x$, from Equation (4.37), one obtains:

$$\frac{dh}{dx}+v_x\frac{dv_x}{dx}+\frac{q''\wp}{\dot{m}}=0 \tag{4.40}$$

For ideal gas, one can write:

$$\frac{dh}{dx}=\sum_{j=1}^{N} y_j \frac{dh_j}{dT}\frac{dT}{dx}+\sum_{j=1}^{N} h_j \frac{dy_j}{dx}=\sum_{j=1}^{N} y_j C_{p_j}\frac{dT}{dx}$$
$$+\sum_{j=1}^{N} h_j \frac{dy_j}{dx}=\bar{C}_p\frac{dT}{dx}+\sum_{j=1}^{N} h_j \frac{dy_j}{dx} \tag{4.41}$$

Also, the ideal gas equation of state gives $P=\rho R_{mix}T=\frac{\rho R_{mix}T}{MW_{mix}}$. Hence, one can write:

$$\frac{1}{P}\frac{dP}{dx}=\frac{1}{\rho}\frac{d\rho}{dx}+\frac{1}{T}\frac{dT}{dx}-\frac{1}{MW_{mix}}\frac{dMW_{mix}}{dx} \tag{4.42}$$

Since $\frac{1}{MW_{mix}}=\sum_{j=1}^{N}\frac{y_j}{MW_j}$, one can write:

$$\frac{1}{MW_{mix}}\frac{dMW_{mix}}{dx}=-MW_{mix}\sum_{j=1}^{N}\frac{1}{MW_j}\frac{dy_j}{dx} \tag{4.43}$$

From Equation (4.39), one can write:

$$v_x\frac{dv_x}{dx}=-\frac{v_x^2}{A}\frac{dA}{dx}-\frac{v_x^2}{\rho}\frac{d\rho}{dx} \tag{4.44}$$

From Equation (4.38), one can write:

$$\frac{dy_i}{dx}=\frac{\omega_i MW_i}{\rho v_{xi}} \tag{4.45}$$

Substituting Equations (4.41) through (4.45) in Equation (4.40), one obtains:

$$\frac{dT}{dx}=\frac{v_x^2}{A\bar{C}_p}\frac{dA}{dx}+\frac{v_x^2}{\rho\bar{C}_p}\frac{d\rho}{dx}-\sum_{j=1}^{N} h_j\frac{\omega_j MW_j}{\rho v_x\bar{C}_p}-\frac{q''\wp}{\dot{m}\bar{C}_p} \tag{4.46}$$

Substituting Equations (4.36), (4.43), (4.45) and (4.46) in Equation (4.42), one obtains:

$$\frac{d\rho}{dx}=\frac{\rho}{P}\left(-\rho v_x\frac{dv_x}{dx}\right)-\frac{\rho}{T}\left(\frac{v_x^2}{A\bar{C}_p}\frac{dA}{dx}+\frac{v_x^2}{\rho\bar{C}_p}\frac{d\rho}{dx}-\sum_{j=1}^{N}h_j\frac{\omega_j MW_j}{\rho v_x\bar{C}_p}-\frac{q''\wp}{\dot{m}\bar{C}_p}\right)$$
$$-\rho MW_{mix}\sum_{j=1}^{N}\frac{1}{MW_j}\frac{\omega_j MW_j}{\rho v_x} \tag{4.47}$$

Substituting Equation (4.44) in Equation (4.47), one obtains:

$$P\frac{d\rho}{dx}=\left(\frac{\rho^2 v_x^2}{A}-\frac{\rho^2 R^*}{MW_{mix}}\frac{v_x^2}{A\bar{C}_p}\right)\frac{dA}{dx}+\left(\rho-\frac{P}{\bar{C}_pT}\right)v_x^2\frac{d\rho}{dx}+\frac{q''\wp}{v_x\bar{C}_p}\frac{P}{T}$$
$$+\sum_{j=1}^{N}\omega_j MW_j\left[\frac{h_jP}{v_x\bar{C}_pT}-\frac{MW_{mix}P}{MW_jv_x}\right]$$

On rearranging, one obtains:

$$\frac{d\rho}{dx}=\frac{\left(\frac{\rho^2 v_x^2}{A}-\frac{\rho^2 R^*}{MW_{mix}}\frac{v_x^2}{A\bar{C}_p}\right)\frac{dA}{dx}+\sum_{j=1}^{N}\omega_j MW_j\frac{\rho R^*}{v_x\bar{C}_pMW_{mix}}\left[h_j-\frac{MW_{mix}}{MW_j}\bar{C}_pT\right]+\frac{q''\wp}{v_x\bar{C}_p}\frac{\rho R^*}{MW_{mix}}}{P\left(1+\frac{v_x^2}{\bar{C}_pT}\right)-\rho v_x^2} \tag{4.48}$$

For both the perfectly stirred reactor and the plug flow reactor, one can define a flowtime or residence time τ_R. For perfectly stirred reactors, τ_R is defined as $\tau_R=\rho V/\dot{m}$. For plug flow reactors, residence time is obtained from $\frac{d\tau_R}{dx}=\frac{1}{v_x}$.

EXERCISES

4.1 Water spray is added to reactants to reduce the flame temperature and NO_x production. Develop a model for a well-stirred reactor in which water spray is added to the reacting mixture. For simplicity, assume that the water droplets undergo complete evaporation in the reactor.

4.2 Develop a model for a well-stirred reactor considering convective and radiative heat loss from the combustor given by $Q_c = \mu_c(T - T_\infty)$ and $Q_r = \mu_r \rho V(y_{f,i} - y_f)(T^4 - T_\infty^4)$, respectively.

4.3 Repeat Exercise 4.1 for a plug flow reactor. For simplicity, assume that the droplet velocity is equal to that of the gas at all locations.

REFERENCES

1. Lieuwen, T., Neumeier, Y. and Zinn, B. T., The role of unmixedness and chemical kinetics in driving combustion instabilities in lean premixed combustors, *Combustion Science and Technology*, **135**, 193–211, 1998.
2. Yi, T., Gutmark, E. J. and Walker, B. K., Stability and control of lean blowout in chemical kinetics-controlled combustion systems, *Combustion Science and Technology*, **181**, 226–244, 2009.

Chapter 5

Conservation Equations

5.1 REYNOLDS TRANSPORT THEOREM

We shall now look at the transport of any generic extensive property N through a control volume. The intensive property corresponding to N is φ (it is the value of N per unit mass) (Figure 5.1).

So, the rate of change of the property value for the system is given as:

$$\frac{N_{t+\Delta t} - N_t}{\Delta t} = \frac{(N_{II})_{t+\Delta t} + (N_{III})_{t+\Delta t} - (N_{II})_t - (N_I)_t}{\Delta t} \tag{5.1}$$

Now,

$$(N_{II})_{t+\Delta t} = \left(\iiint_{II} \varphi \rho dv \right)_{t+\Delta t} \tag{5.2}$$

Similarly, other quantities can be written and substituting those in Equation (5.1), we get:

$$\frac{N_{t+\Delta t} - N_t}{\Delta t} = \frac{\left(\iiint_{II} \varphi \rho dv + \iiint_{III} \varphi \rho dv \right)_{t+\Delta t} - \left(\iiint_{II} \varphi \rho dv + \iiint_{I} \varphi \rho dv \right)_t}{\Delta t}$$

$$= \left[\left\{ \left(\iiint_{II} \varphi \rho dv \right)_{t+\Delta t} - \left(\iiint_{II} \varphi \rho dv \right)_t \right\} + \left\{ \left(\iiint_{III} \varphi \rho dv \right)_{t+\Delta t} - \left(\iiint_{I} \varphi \rho dv \right)_t \right\} \right] / \Delta t \tag{5.3}$$

Taking limit $\Delta t \to 0$, the RHS of the above equation can be written as $\frac{\partial}{\partial t} \iiint_{CV} \varphi \rho dv + \textit{Net efflux of N from CV}$

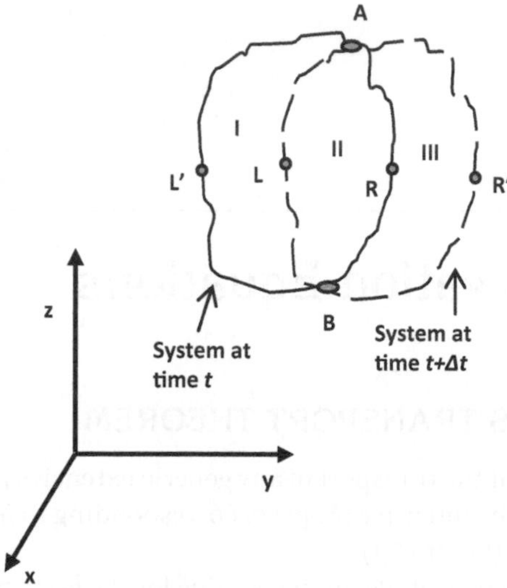

Figure 5.1 shows a system at two time instants. AL'BRA is the condition of the system at time t and ALBR'A is the condition of the same system after Δt time. Region II is the intersection of the two areas. One can conclude that the medium in region I is flowing into region II, while region III is coming out from region II, during the time duration Δt.

V is the velocity vector; n is unit normal on the small surface area dA. The system boundary is moved by $V\Delta t$ amount in time Δt:

$$\text{Volume flow} = V\Delta t\, dA\, cos\alpha = V \cdot dA\,\Delta t = n \cdot V\,\Delta t\, dA$$

$$\text{Property flow} = V\Delta t\, dA\, cos\alpha = \rho\left(n \cdot V\,\Delta t\, dA\right)\varphi$$

$$\text{Exact rate of flow at time } t \text{ at } dA = \lim_{\Delta t \to 0} \frac{\rho\left(n \cdot V\,\Delta t\, dA\right)\varphi}{\Delta t}$$

$$\text{Total outflow at time } t = \iint_{ARB} n \cdot V \cdot \rho\varphi dA$$

$$\text{Total inflow at time } t = -\iint_{ALB} n \cdot V \cdot \rho\varphi dA$$

$$\text{Net efflux from } CV \text{ at time } t = \text{outflow} - \text{inflow} = \oiint_{CS} n \cdot \left(\rho\varphi V\right) dA$$

$$\frac{DN}{Dt} = I = \frac{\partial}{\partial t}\iiint_{CV} \rho\varphi dv + \oiint_{CS} n \cdot \left(\rho\varphi V\right) dA \tag{5.4}$$

We can write, $\frac{DN}{Dt}=I$, where I is the rate of change of φ due to external and internal agencies. Equation (5.4) is known as Reynolds transport equation.

5.2 CONSERVATION OF MASS

To derive the mass conservation equation from Reynolds, we should take the following:

$$N = mass,\ \varphi = 1,\ I = 0$$

Substituting these in Equation (5.4), we get:

$$\frac{\partial}{\partial t}\iiint_{CV} \rho dv + \oiint_{CS} n\cdot(\rho V)dA = 0 \tag{5.5}$$

Using the Gauss divergence theorem, the above equation can be written as:

$$\frac{\partial}{\partial t}\iiint_{CV} \rho dv + \iiint_{CV} \nabla\cdot(\rho V)dv = 0 \tag{5.6}$$

$$\iiint_{CV}\left[\frac{\partial \rho}{\partial t}+\nabla\cdot(\rho V)\right]dv = 0 \tag{5.7}$$

The integration is true for any volume in space. So, to satisfy the equation, $\left[\frac{\partial \rho}{\partial t}+\nabla\cdot(\rho V)\right]$ must be equal to zero. We can take the differential balance at each point:

$$\frac{\partial \rho}{\partial t}+\nabla\cdot(\rho V) = 0 \tag{5.8}$$

Or we can write:

$$\left[\frac{\partial \rho}{\partial t}+V\cdot\nabla\rho\right]+\rho(\nabla\cdot V) = 0 \tag{5.9}$$

The term within the parenthesis is the substantial derivative. We can say:

$$\frac{D\rho}{Dt}+\rho(\nabla\cdot V) = 0 \tag{5.10}$$

TABLE 5.1 CONTINUITY EQUATION IN DIFFERENT COORDINATE SYSTEMS

Rectangular coordinates (x, y, z):

$$\frac{\partial \rho}{\partial t}+\frac{\partial}{\partial x}(\rho v_x)+\frac{\partial}{\partial y}(\rho v_y)+\frac{\partial}{\partial z}(\rho v_z)=0$$

Cylindrical coordinates (r, θ, z):

$$\frac{\partial \rho}{\partial t}+\frac{1}{r}\frac{\partial}{\partial r}(\rho r v_r)+\frac{1}{r}\frac{\partial}{\partial \theta}(\rho v_\theta)+\frac{\partial}{\partial z}(\rho v_z)=0$$

Spherical coordinates $(\mathrm{r}, \theta, \phi)$:

$$\frac{\partial \rho}{\partial t}+\frac{1}{r^2}\frac{\partial}{\partial r}\left(\rho r^2 v_r\right)+\frac{1}{r\,sin\theta}\frac{\partial}{\partial \theta}(\rho v_\theta\, sin\theta)+\frac{1}{r\,sin\theta}\frac{\partial}{\partial \phi}(\rho v_\phi)$$

If the flow is incompressible flow, density will not vary with space and time. So, $\frac{D\rho}{Dt}=0$. The continuity equation for incompressible flow can be written as:

$$\nabla \cdot V = 0 \tag{5.11}$$

Table 5.1 shows the continuity equation in different coordinate system.

5.3 CONSERVATION OF SPECIES

In combustion science, we deal with several species which have different properties. We need to equate the continuity equation for each and every species. Let us consider a species i for this analysis.

$$N = mass\,of\ ith\,species = Total\,mass \times y_i \tag{5.12}$$

$$\varphi = y_i \tag{5.13}$$

$$I = \iiint_{CV} \dot{m}_i''' \, dv - \oiint_{CS} n \cdot \dot{m}_i ds \tag{5.14}$$

The first term in Equation (5.14) accounts for the volumetric generation of ith species. $\dot{m}_i'''$ denotes generation of ith species per unit volume per unit time.

The second term is the contribution of species mass flux across the control surface. $\dot{m}_i$ is the species mass flux in the direction of surface normal n. The negative sign is signifying incoming mass, as the surface normal is in outward direction.

Equation (5.4) can be written, after substitution, as follows:

$$\frac{\partial}{\partial t}\iiint_{CV} \rho y_i dv + \oiint_{CS} n\cdot(\rho y_i V)dA = \iiint_{CV} \dot{m}_i''' \, dv - \oiint_{CS} n\cdot\dot{m}_i ds \quad (5.15)$$

Using the divergence theorem, one can write:

$$\frac{\partial}{\partial t}\iiint_{CV} \rho y_i dv + \iiint_{CV} \nabla\cdot(\rho y_i V)\, dv = \iiint_{CV} \dot{m}_i''' \, dv - \iiint_{CV} \nabla\cdot\dot{m}_i dv \quad (5.16)$$

Mass flux of ith species can be substituted, using Fick's law, as $\dot{m}_i = -\rho\mathfrak{D}(\nabla y_i)$, where $\mathfrak{D}$ is the diffusion coefficient, and Equation (5.16) can be written as follows:

$$\frac{\partial}{\partial t}\iiint_{CV} \rho y_i dv + \iiint_{CV} \nabla\cdot(\rho y_i V)\, dv = \iiint_{CV} \dot{m}_i''' \, dv - \iiint_{CV} \nabla\cdot[-\rho\cdot\dot{\mathfrak{D}}(\nabla y_i)]\, dv \quad (5.17)$$

The integration is true for any volume in space, and we can write the same in differential form as:

$$\frac{\partial(\rho y_i)}{\partial t} + \nabla\cdot(\rho y_i V) = \nabla\cdot\left[\rho\dot{\mathfrak{D}}(\nabla y_i)\right] + \dot{m}_i''' \quad (5.18)$$

Equation (5.18) is known as the species conservation equation. The LHS of the equation can be simplified as:

$$\frac{\partial(\rho y_i)}{\partial t} + \nabla\cdot(\rho y_i V) = \rho\frac{\partial y_i}{\partial t} + y_i\frac{\partial \rho}{\partial t} + y_i\nabla\cdot(\rho V) + \rho V\cdot\nabla y_i = \rho\frac{\partial y_i}{\partial t} + \rho V\cdot\nabla y_i \quad (5.19)$$

Consider the highlighted part equal to zero from Equation (5.8). Thus, Equation (5.18) can be written as:

$$\rho\frac{\partial y_i}{\partial t} + \rho V\cdot\nabla y_i = \nabla\cdot\left[\rho\mathfrak{D}(\nabla y_i)\right] + \dot{m}_i''' \quad (5.20)$$

Table 5.2 shows different forms of the species conservation equation.

TABLE 5.2 CONSERVATION OF ITH SPECIES FOR CONSTANT ρ AND $\mathfrak{D}$

Rectangular coordinates (x, y, z):
$$\frac{\partial y_i}{\partial t}+\left(v_x\frac{\partial y_i}{\partial x}+v_y\frac{\partial y_i}{\partial y}+v_z\frac{\partial y_i}{\partial z}\right)=\mathfrak{D}\left(\frac{\partial^2 y_{iA}}{\partial x^2}+\frac{\partial^2 y_i}{\partial y^2}+\frac{\partial^2 y_i}{\partial z^2}\right)+\dot{m}_i'''$$
Cylindrical coordinates (r,θ,Z):
$$\frac{\partial y_i}{\partial t}+\left(v_r\frac{\partial y_i}{\partial r}+v_\theta\frac{1}{r}\frac{\partial y_i}{\partial \theta}+v_z\frac{\partial y_i}{\partial z}\right)=\mathfrak{D}\left(\frac{1}{r}\frac{\partial}{\partial r}\left(r\frac{\partial y_i}{\partial r}\right)+\frac{1}{r^2}\frac{\partial^2 y_i}{\partial \theta^2}+\frac{\partial^2 y_i}{\partial z^2}\right)+\dot{m}_i'''$$
Spherical coordinates (r,θ,ϕ):
$$\frac{\partial y_i}{\partial t}+\left(v_r\frac{\partial y_i}{\partial r}+v_\theta\frac{1}{r}\frac{\partial y_i}{\partial \theta}+v_\phi\frac{1}{r\sin\theta}\frac{\partial y_i}{\partial \phi}\right)$$ $$=\mathfrak{D}\left(\frac{1}{r^2}\frac{\partial}{\partial r}\left(r^2\frac{\partial y_i}{\partial r}\right)+\frac{1}{r^2\sin\theta}\frac{\partial}{\partial\theta}\left(\sin\theta\frac{\partial y_i}{\partial\theta}\right)+\frac{1}{r^2\sin^2\theta}\frac{\partial^2 y_i}{\partial\phi^2}\right)+\dot{m}_i'''$$

5.4 CONSERVATION OF MOMENTUM

For derivation of the momentum conservation equation, we need to substitute the following in the Reynolds transport theorem:

$$N = mV = Momentum \tag{5.21}$$

$$\varphi = V \tag{5.22}$$

$$I = Body\ Force + Surface\ Force = \iiint_{CV} \rho b dv + \iint_{CS} n\cdot\overline{T} ds \tag{5.23}$$

Substituting Equation (5.22) and (5.23) in Equation (5.4), we have:

$$\frac{\partial}{\partial t}\iiint_{CV}\rho V dv + \iiint \nabla\cdot(\rho VV)\, dv = \iiint_{CV}\rho b dv + \iiint_{CV}\nabla\cdot\overline{T} dv \tag{5.24}$$

We can write this in differential form:

$$\frac{\partial(\rho V)}{\partial t}+\nabla\cdot(\rho VV)-\rho b-\nabla\cdot\overline{T}=0 \tag{5.25}$$

$$\frac{\partial(\rho V)}{\partial t}+\nabla\cdot(\rho VV)=\rho b+\nabla\cdot\left[-p\overline{I}+\overline{\tau}\right] \tag{5.26}$$

where p is the pressure, $\overline{I}$ is the identity tensor and $\overline{\tau}$ is the stress tensor. Equation (5.26) can be written as:

$$\frac{\partial(\rho V)}{\partial t}+\nabla\cdot(\rho VV)=-\nabla p+\rho b+\nabla\cdot\overline{\tau} \tag{5.27}$$

With the help of the continuity equation from Equation (5.8), we get:

$$\rho\frac{\partial V}{\partial t}+\rho V\,\nabla\cdot V=-\nabla p+\rho b+\nabla\cdot\overline{\tau} \tag{5.28}$$

This equation can also be written as follows:

$$\rho\frac{\partial}{\partial t}(V_i)+\rho V_j\frac{\partial}{\partial x_j}(V_i)=-\frac{\partial}{\partial x_i}p+\rho b_i+\frac{\partial}{\partial x_j}\tau_{ij} \tag{5.29}$$

Now, we need to know the form for the stress tensor:

$$\overline{\tau}=\lambda[\nabla.V]\overline{I}+\mu\left[(\nabla V)+(\nabla V)^T\right] \tag{5.30}$$

Detailed discussion regarding this may be found in [1–5]. For convenience, we shall consider Equation (5.30) in rectangular Cartesian system:

$$\tau_{xx}=2\mu\frac{\partial u}{\partial x}-\frac{2}{3}\mu\left(\frac{\partial u}{\partial x}+\frac{\partial v}{\partial y}+\frac{\partial w}{\partial z}\right) \tag{5.31}$$

$$\tau_{yy}=2\mu\frac{\partial v}{\partial y}-\frac{2}{3}\mu\left(\frac{\partial u}{\partial x}+\frac{\partial v}{\partial y}+\frac{\partial w}{\partial z}\right) \tag{5.32}$$

$$\tau_{xx}=2\mu\frac{\partial w}{\partial z}-\frac{2}{3}\mu\left(\frac{\partial u}{\partial x}+\frac{\partial v}{\partial y}+\frac{\partial w}{\partial z}\right) \tag{5.33}$$

$$\tau_{xy}=\tau_{yx}=\mu\left(\frac{\partial u}{\partial y}+\frac{\partial v}{\partial x}\right) \tag{5.34}$$

$$\tau_{xz}=\tau_{zx}=\mu\left(\frac{\partial u}{\partial z}+\frac{\partial w}{\partial x}\right) \tag{5.35}$$

$$\tau_{zy}=\tau_{yz}=\mu\left(\frac{\partial w}{\partial y}+\frac{\partial v}{\partial z}\right) \tag{5.36}$$

The momentum equation, also called the Navier Stokes equation is shown in different forms and coordinate systems in Tables 5.3 – 5.5. Stress tensor in different coordinate systems is shown in Tables 5.6 – 5.8.

TABLE 5.3 MOMENTUM CONSERVATION EQUATION IN RECTANGULAR CARTESIAN COORDINATE (x, y, z)

In terms of τ:

x-component:

$$\rho\left(\frac{\partial v_x}{\partial t}+v_x\frac{\partial v_x}{\partial x}+v_y\frac{\partial v_x}{\partial y}+v_z\frac{\partial v_x}{\partial z}\right)=-\frac{\partial p}{\partial x}-\left(\frac{\partial \tau_{xx}}{\partial x}+\frac{\partial \tau_{yx}}{\partial y}+\frac{\partial \tau_{zx}}{\partial z}\right)+\rho g_x$$

y-component:

$$\rho\left(\frac{\partial v_y}{\partial t}+v_x\frac{\partial v_y}{\partial x}+v_y\frac{\partial v_y}{\partial y}+v_z\frac{\partial v_y}{\partial z}\right)=-\frac{\partial p}{\partial y}-\left(\frac{\partial \tau_{xy}}{\partial x}+\frac{\partial \tau_{yy}}{\partial y}+\frac{\partial \tau_{zy}}{\partial z}\right)+\rho g_y$$

z-component:

$$\rho\left(\frac{\partial v_z}{\partial t}+v_x\frac{\partial v_z}{\partial x}+v_y\frac{\partial v_z}{\partial y}+v_z\frac{\partial v_z}{\partial z}\right)=-\frac{\partial p}{\partial z}-\left(\frac{\partial \tau_{xz}}{\partial x}+\frac{\partial \tau_{yz}}{\partial y}+\frac{\partial \tau_{zz}}{\partial z}\right)+\rho g_z$$

In terms of velocity gradients for a Newtonian fluid with constants ρ and μ:

x-component:

$$\rho\left(\frac{\partial v_x}{\partial t}+v_x\frac{\partial v_x}{\partial x}+v_y\frac{\partial v_x}{\partial y}+v_z\frac{\partial v_x}{\partial z}\right)=-\frac{\partial p}{\partial x}+\mu\left(\frac{\partial^2 v_x}{\partial x^2}+\frac{\partial^2 v_x}{\partial y^2}+\frac{\partial^2 v_x}{\partial z^2}\right)+\rho g_x$$

y-component:

$$\rho\left(\frac{\partial v_y}{\partial t}+v_x\frac{\partial v_y}{\partial x}+v_y\frac{\partial v_y}{\partial y}+v_z\frac{\partial v_y}{\partial z}\right)=-\frac{\partial p}{\partial y}+\mu\left(\frac{\partial^2 v_y}{\partial x^2}+\frac{\partial^2 v_y}{\partial y^2}+\frac{\partial^2 v_y}{\partial z^2}\right)+\rho g_y$$

z-component:

$$\rho\left(\frac{\partial v_z}{\partial t}+v_x\frac{\partial v_z}{\partial x}+v_y\frac{\partial v_z}{\partial y}+v_z\frac{\partial v_z}{\partial z}\right)=-\frac{\partial p}{\partial z}+\mu\left(\frac{\partial^2 v_z}{\partial x^2}+\frac{\partial^2 v_z}{\partial y^2}+\frac{\partial^2 v_z}{\partial z^2}\right)+\rho g_z$$

TABLE 5.4 MOMENTUM CONSERVATION EQUATION IN CYLINDRICAL COORDINATE (r,θ,z)

In terms of τ:

r-component:

$$\rho\left(\frac{\partial v_r}{\partial t}+v_r\frac{\partial v_r}{\partial r}+\frac{v_\theta}{r}\frac{\partial v_r}{\partial \theta}-\frac{v_\theta^2}{r}+v_z\frac{\partial v_r}{\partial z}\right)=-\frac{\partial p}{\partial r}-\left(\frac{1}{r}\frac{\partial}{\partial r}(r\tau_{rr})+\frac{1}{r}\frac{\partial \tau_{r\theta}}{\partial \theta}-\frac{\tau_{\theta\theta}}{r}+\frac{\partial \tau_{rz}}{\partial z}\right)+\rho g_r$$

θ-component:

$$\rho\left(\frac{\partial v_\theta}{\partial t}+v_r\frac{\partial v_\theta}{\partial r}+\frac{v_\theta}{r}\frac{\partial v_\theta}{\partial \theta}+\frac{v_r v_\theta}{r}+v_z\frac{\partial v_\theta}{\partial z}\right)=-\frac{1}{r}\frac{\partial p}{\partial \theta}-\left(\frac{1}{r^2}\frac{\partial}{\partial r}\left(r^2\tau_{r\theta}\right)+\frac{1}{r}\frac{\partial \tau_{\theta\theta}}{\partial \theta}+\frac{\partial \tau_{\theta z}}{\partial z}\right)+\rho g_\theta$$

z-component:

$$\rho\left(\frac{\partial v_z}{\partial t}+v_r\frac{\partial v_z}{\partial r}+\frac{v_\theta}{r}\frac{\partial v_z}{\partial \theta}+v_z\frac{\partial v_z}{\partial z}\right)=-\frac{\partial p}{\partial z}-\left(\frac{1}{r}\frac{\partial}{\partial r}(r\tau_{rz})+\frac{1}{r}\frac{\partial \tau_{\theta z}}{\partial \theta}+\frac{\partial \tau_{zz}}{\partial z}\right)+\rho g_z$$

In terms of velocity gradients for a Newtonian fluid with constant ρ and μ:

r-component:

$$\rho\left(\frac{\partial v_r}{\partial t}+v_r\frac{\partial v_r}{\partial r}+\frac{v_\theta}{r}\frac{\partial v_r}{\partial \theta}-\frac{v_\theta^2}{r}+v_z\frac{\partial v_r}{\partial z}\right)$$

$$=-\frac{\partial p}{\partial r}+\mu\left[\frac{\partial}{\partial r}\left(\frac{1}{r}\frac{\partial}{\partial r}(r v_r)\right)+\frac{1}{r^2}\frac{\partial^2 v_r}{\partial \theta^2}-\frac{2}{r^2}\frac{\partial v_\theta}{\partial \theta}+\frac{\partial^2 v_r}{\partial z^2}\right]+\rho g_r$$

θ-component:

$$\rho\left(\frac{\partial v_\theta}{\partial t}+v_r\frac{\partial v_\theta}{\partial r}+\frac{v_\theta}{r}\frac{\partial v_\theta}{\partial \theta}+\frac{v_r v_\theta}{r}+v_z\frac{\partial v_\theta}{\partial z}\right)$$

$$=-\frac{1}{r}\frac{\partial p}{\partial \theta}+\mu\left[\frac{\partial}{\partial r}\left(\frac{1}{r}\frac{\partial}{\partial r}(r v_\theta)\right)+\frac{1}{r^2}\frac{\partial^2 v_\theta}{\partial \theta^2}+\frac{2}{r^2}\frac{\partial v_r}{\partial \theta}+\frac{\partial^2 v_\theta}{\partial z^2}\right]+\rho g_\theta$$

z-component:

$$\rho\left(\frac{\partial v_z}{\partial t}+v_r\frac{\partial v_z}{\partial r}+\frac{v_\theta}{r}\frac{\partial v_z}{\partial \theta}+v_z\frac{\partial v_z}{\partial z}\right)=-\frac{\partial p}{\partial z}+\mu\left[\frac{1}{r}\frac{\partial}{\partial r}\left(r\frac{\partial v_z}{\partial r}\right)+\frac{1}{r^2}\frac{\partial^2 v_z}{\partial \theta^2}+\frac{\partial^2 v_z}{\partial z^2}\right]+\rho g_z$$

TABLE 5.5 MOMENTUM CONSERVATION EQUATION IN SPHERICAL COORDINATE (r,θ,ϕ)

In terms of τ:

r-component:

$$\rho\left(\frac{\partial v_r}{\partial t}+v_r\frac{\partial v_r}{\partial r}+\frac{v_\theta}{r}\frac{\partial v_r}{\partial \theta}+\frac{v_\phi}{r\sin\theta}\frac{\partial v_r}{\partial \phi}-\frac{{v_\theta}^2+{v_\phi}^2}{r}\right)$$

$$=-\frac{\partial p}{\partial r}-\left(\frac{1}{r^2}\frac{\partial}{\partial r}\left(r^2\tau_{rr}\right)+\frac{1}{r\sin\theta}\frac{\partial}{\partial \theta}\left(\tau_{r\theta}\sin\theta\right)+\frac{1}{r\sin\theta}\frac{\partial \tau_{r\phi}}{\partial \phi}-\frac{\tau_{\theta\theta}+\tau_{\phi\phi}}{r}\right)+\rho g_r$$

θ-component:

$$\rho\left(\frac{\partial v_\theta}{\partial t}+v_r\frac{\partial v_\theta}{\partial r}+\frac{v_\theta}{r}\frac{\partial v_\theta}{\partial \theta}+\frac{v_\phi}{r\sin\theta}\frac{\partial v_\theta}{\partial \phi}+\frac{v_r v_\theta}{r}-\frac{{v_\phi}^2\cot\theta}{r}\right)$$

$$=-\frac{1}{r}\frac{\partial p}{\partial \theta}-\left(\frac{1}{r^2}\frac{\partial}{\partial r}\left(r^2\tau_{r\theta}\right)+\frac{1}{r\sin\theta}\frac{\partial}{\partial \theta}\left(\tau_{\theta\theta}\sin\theta\right)+\frac{1}{r\sin\theta}\frac{\partial \tau_{\theta\phi}}{\partial \phi}+\frac{\tau_{r\theta}}{r}-\frac{\cot\theta}{r}\tau_{\phi\phi}\right)+\rho g_\theta$$

ϕ-component:

$$\rho\left(\frac{\partial v_\phi}{\partial t}+v_r\frac{\partial v_\phi}{\partial r}+\frac{v_\theta}{r}\frac{\partial v_\phi}{\partial \theta}+\frac{v_\phi}{r\sin\theta}\frac{\partial v_\phi}{\partial \phi}+\frac{v_\phi v_r}{r}+\frac{v_\theta v_\phi\cot\theta}{r}\right)$$

$$=-\frac{1}{r\sin\theta}\frac{\partial p}{\partial \phi}-\left(\frac{1}{r^2}\frac{\partial}{\partial r}\left(r^2\tau_{r\phi}\right)+\frac{1}{r}\frac{\partial}{\partial \theta}\left(\tau_{\theta\phi}\right)+\frac{1}{r\sin\theta}\frac{\partial \tau_{\phi\phi}}{\partial \phi}+\frac{\tau_{r\phi}}{r}+\frac{2\cot\theta}{r}\tau_{\theta\phi}\right)+\rho g_\phi$$

In terms of velocity gradients for a Newtonian fluid with constant ρ and μ:

r-component:

$$\rho\left(\frac{\partial v_r}{\partial t}+v_r\frac{\partial v_r}{\partial r}+\frac{v_\theta}{r}\frac{\partial v_r}{\partial \theta}+\frac{v_\phi}{r\sin\theta}\frac{\partial v_r}{\partial \phi}-\frac{{v_\theta}^2+{v_\phi}^2}{r}\right)$$

$$=-\frac{\partial p}{\partial r}+\mu\left(\frac{1}{r^2}\frac{\partial^2}{\partial r^2}\left(r^2 v_r\right)+\frac{1}{r^2\sin\theta}\frac{\partial}{\partial \theta}\left(\sin\theta\frac{\partial v_r}{\partial \theta}\right)+\frac{1}{r^2 sin^2\theta}\frac{\partial^2 v_r}{\partial \phi^2}\right)+\rho g_r$$

(*Continued*)

TABLE 5.5 (*Continued*) MOMENTUM CONSERVATION EQUATION IN SPHERICAL COORDINATE (r,θ,ϕ)

θ-component:

$$\rho\left(\frac{\partial v_\theta}{\partial t}+v_r\frac{\partial v_\theta}{\partial r}+\frac{v_\theta}{r}\frac{\partial v_\theta}{\partial \theta}+\frac{v_\phi}{r\sin\theta}\frac{\partial v_\theta}{\partial \phi}+\frac{v_r v_\theta}{r}-\frac{v_\phi{}^2\cot\theta}{r}\right)$$

$$=-\frac{1}{r}\frac{\partial p}{\partial \theta}+\mu\left(\frac{1}{r^2}\frac{\partial}{\partial r}\left(r^2\frac{\partial v_\theta}{\partial r}\right)+\frac{1}{r^2}\frac{\partial}{\partial \theta}\left(\frac{1}{\sin\theta}\frac{\partial}{\partial \theta}(v_\theta\sin\theta)\right)+\frac{1}{r^2 sin^2\theta}\frac{\partial^2 v_\theta}{\partial \phi^2}\right.$$

$$\left.+\frac{2}{r^2}\frac{\partial v_r}{\partial \theta}-\frac{2}{r^2}\frac{\cos\theta}{sin^2\theta}\frac{\partial v_\phi}{\partial \phi}\right)+\rho g_\theta$$

ϕ-component:

$$\rho\left(\frac{\partial v_\phi}{\partial t}+v_r\frac{\partial v_\phi}{\partial r}+\frac{v_\theta}{r}\frac{\partial v_\phi}{\partial \theta}+\frac{v_\phi}{r\sin\theta}\frac{\partial v_\phi}{\partial \phi}+\frac{v_\phi v_r}{r}+\frac{v_\theta v_\phi\cot\theta}{r}\right)$$

$$=-\frac{1}{rsin\theta}\frac{\partial p}{\partial \phi}+\mu\left(\frac{1}{r^2}\frac{\partial}{\partial r}\left(r^2\frac{\partial v_\phi}{\partial r}\right)+\frac{1}{r^2}\frac{\partial}{\partial \theta}\left(\frac{1}{\sin\theta}\frac{\partial}{\partial \theta}(v_\phi\sin\theta)\right)+\frac{1}{r^2 sin^2\theta}\frac{\partial^2 v_\phi}{\partial \phi^2}\right.$$

$$\left.+\frac{2}{r^2 sin^2\theta}\frac{\partial v_r}{\partial \phi}+\frac{2}{r^2}\frac{\cos\theta}{sin^2\theta}\frac{\partial v_\theta}{\partial \phi}\right)+\rho g_\phi$$

TABLE 5.6 STRESS TENSOR COMPONENTS OF NEWTONIAN FLUIDS IN RECTANGULAR CARTESIAN COORDINATE (x,y,z)

$$\tau_{xx}=-\mu\left[2\frac{\partial v_x}{\partial x}-\frac{2}{3}(\nabla.v)\right]$$

$$\tau_{yy}=-\mu\left[2\frac{\partial v_y}{\partial y}-\frac{2}{3}(\nabla.v)\right]$$

$$\tau_{zz}=-\mu\left[2\frac{\partial v_z}{\partial z}-\frac{2}{3}(\nabla.v)\right]$$

$$\tau_{xy}=\tau_{yx}=-\mu\left[\frac{\partial v_x}{\partial y}+\frac{\partial v_y}{\partial x}\right]$$

$$\tau_{yz}=\tau_{zy}=-\mu\left[\frac{\partial v_y}{\partial z}+\frac{\partial v_z}{\partial y}\right]$$

$$\tau_{zx}=\tau_{xz}=-\mu\left[\frac{\partial v_z}{\partial x}+\frac{\partial v_x}{\partial z}\right]$$

$$(\nabla.v)=\frac{\partial v_x}{\partial x}+\frac{\partial v_y}{\partial y}+\frac{\partial v_z}{\partial z}$$

TABLE 5.7 STRESS TENSOR COMPONENTS OF NEWTONIAN FLUIDS IN CYLINDRICAL COORDINATE (r,θ,z)

$$\tau_{rr}=-\mu\left[2\frac{\partial v_r}{\partial r}-\frac{2}{3}(\nabla.v)\right]$$

$$\tau_{\theta\theta}=-\mu\left[2\left(\frac{1}{r}\frac{\partial v_\theta}{\partial\theta}+\frac{v_r}{r}\right)-\frac{2}{3}(\nabla.v)\right]$$

$$\tau_{zz}=-\mu\left[2\frac{\partial v_z}{\partial z}-\frac{2}{3}(\nabla.v)\right]$$

$$\tau_{r\theta}=\tau_{\theta r}=-\mu\left[r\frac{\partial}{\partial r}\left(\frac{v_\theta}{r}\right)+\frac{1}{r}\frac{\partial v_r}{\partial\theta}\right]$$

$$\tau_{\theta z}=\tau_{z\theta}=-\mu\left[\frac{\partial v_\theta}{\partial z}+\frac{1}{r}\frac{\partial v_z}{\partial\theta}\right]$$

$$\tau_{zr}=\tau_{rz}=-\mu\left[\frac{\partial v_z}{\partial r}+\frac{\partial v_r}{\partial z}\right]$$

$$(\nabla.v)=\frac{1}{r}\frac{\partial}{\partial r}(rv_r)+\frac{1}{r}\frac{\partial v_\theta}{\partial\theta}+\frac{\partial v_z}{\partial z}$$

TABLE 5.8 STRESS TENSOR COMPONENTS OF NEWTONIAN FLUIDS IN SPHERICAL COORDINATE (r,θ,ϕ)

$$\tau_{rr}=-\mu\left[2\frac{\partial v_r}{\partial r}-\frac{2}{3}(\nabla.v)\right]$$

$$\tau_{\theta\theta}=-\mu\left[2\left(\frac{1}{r}\frac{\partial v_\theta}{\partial\theta}+\frac{v_r}{r}\right)-\frac{2}{3}(\nabla.v)\right]$$

$$\tau_{\phi\phi}=-\mu\left[2\left(\frac{1}{r\sin\theta}\frac{\partial v_\phi}{\partial\phi}+\frac{v_r}{r}+\frac{v_\theta\cot\theta}{r}\right)-\frac{2}{3}(\nabla.v)\right].$$

$$\tau_{r\theta}=\tau_{\theta r}=-\mu\left[r\frac{\partial}{\partial r}\left(\frac{v_\theta}{r}\right)+\frac{1}{r}\frac{\partial v_r}{\partial\theta}\right]$$

$$\tau_{\theta\phi}=\tau_{\phi\theta}=-\mu\left[\frac{\sin\theta}{r}\frac{\partial}{\partial\theta}\left(\frac{v_\phi}{\sin\theta}\right)+\frac{1}{rsin\theta}\frac{\partial v_\theta}{\partial\phi}\right]$$

$$\tau_{\phi r}=\tau_{r\phi}=-\mu\left[\frac{1}{r\,sin\theta}\frac{\partial v_r}{\partial\phi}+r\frac{\partial}{\partial r}\left(\frac{v_\phi}{r}\right)\right].$$

$$(\nabla.v)=\frac{1}{r^2}\frac{\partial}{\partial r}\left(r^2v_r\right)+\frac{1}{r\sin\theta}\frac{\partial}{\partial\theta}(v_\theta\sin\theta)+\frac{1}{r\sin\theta}\frac{\partial v_\phi}{\partial\phi}$$

5.5 CONSERVATION OF ENERGY

We shall first consider the conservation of internal and kinetic energy. We can take the following:

$$N = Internal\ Energy\,(IE) + Kinetic\ Energy\,(KE) \tag{5.37}$$

$$\varphi = u + \frac{V \cdot V}{2} \tag{5.38}$$

$$I = \iiint_{CV} \dot{q}''' dv + \iiint_{CV} \rho b \cdot V dv + \iint_{CS} n \cdot \bar{T} \cdot V ds + \iint_{CS} -(n \cdot q) ds \tag{5.39}$$

The first term in the RHS of Equation (5.39) denotes energy generation in the control volume. $\dot{q}'''$ is the energy generation per unit volume per unit time. Rate of working of body force is shown in the second term, which can also be called the potential energy. The third term denotes rate of working by the surface forces. The last term accounts for the rate of contact heat transfer across the surface (conduction energy transfer).

If we substitute Equations (5.38) and (5.39) in the Reynolds transport equation in Equation (5.4) and use divergence theorem, we get:

$$\begin{aligned} &\frac{\partial}{\partial t} \iiint_{CV} \rho \left(u + \frac{V^2}{2} \right) dv + \iiint_{CV} \nabla \cdot \left[\rho V \left(u + \frac{V^2}{2} \right) \right] dv \\ &= \iiint_{CV} \dot{q}''' dv + \iiint_{CV} \rho b \cdot V\, dv + \iiint_{CV} \nabla \cdot \left(\bar{T} \cdot V \right) dv - \iiint_{CV} \nabla \cdot q\, dv \end{aligned} \tag{5.40}$$

The above integral relation is true for any volume, so we can take the differential form. Substituting $\bar{T} = -p\bar{I} + \bar{\tau}$ in the differential form, we get:

$$\begin{aligned} &\frac{\partial}{\partial t} \left[\rho \left(u + \frac{V^2}{2} \right) \right] + \nabla \cdot \left[\rho V \left(u + \frac{V^2}{2} \right) \right] \\ &= \dot{q}''' - \nabla \cdot q + \rho b \cdot V + \nabla \cdot (\bar{\tau} V) + \nabla \cdot (-pV) \end{aligned} \tag{5.41}$$

Equation (5.41) is the conservation equation for total energy (internal and mechanical). To get the mechanical energy conservation equation, we can take a dot product of momentum equation [Equation (5.28)] with velocity. So, mechanical energy balance equation can be written as:

$$V. \left[\rho \frac{DV}{Dt} \right] = V. \left[-\nabla p + \rho b + \nabla \cdot \bar{\tau} \right] \tag{5.42}$$

Or

$$\frac{\partial}{\partial t}\left(\frac{1}{2}\rho V^2\right)+\nabla\cdot\left[\rho V\left(\frac{1}{2}V^2\right)\right]=-(\nabla p)\cdot V+\rho b\cdot V+(\nabla\cdot\bar{\tau})\cdot V \tag{5.43}$$

For symmetric $\bar{\tau}$, we can write:

$$\bar{\tau}:\nabla V=\nabla\cdot(\bar{\tau}\cdot V)-V\cdot(\nabla\cdot\bar{\tau}) \tag{5.44}$$

So Equation (5.43) can be written as:

$$\frac{\partial}{\partial t}\left(\frac{1}{2}\rho V^2\right)+\nabla\cdot\left[\rho V\left(\frac{1}{2}V^2\right)\right]=-(\nabla p)\cdot V+\rho b\cdot V+\nabla\cdot(\bar{\tau}\cdot V)-\bar{\tau}:\nabla V \tag{5.45}$$

Subtracting mechanical energy Equation (5.45) from total energy balance Equation (5.41), we get:

$$\frac{\partial}{\partial t}(\rho\,\mathrm{u})+\nabla\cdot(\rho V\,\mathrm{u})=-p(\nabla\cdot V)+\bar{\tau}:\nabla V+\dot{q}'''-\nabla\cdot q \tag{5.46}$$

The above equation is the conservation equation for internal energy. The first term in the LHS is rate of gain of internal energy per unit volume. The second term is the rate of advective loss of internal energy per unit volume. The first two terms in the RHS denote rate of inter-conversion of mechanical and internal energy. The first term is the reversible rate of inter-conversion; the second term is the irreversible rate. The third term is the rate of internal energy generation, and the last one is the rate of internal energy gain through conduction. Rearranging this equation and considering continuity equation, we get:

$$\rho\frac{D\mathrm{u}}{Dt}=-p(\nabla\cdot V)+\bar{\tau}:\nabla V+\dot{q}'''-\nabla\cdot q \tag{5.47}$$

$$\begin{aligned}\rho\frac{D\mathrm{u}}{Dt}+\rho\frac{\partial}{\partial t}(pv)+\rho V\cdot\nabla(pv)&=-p(\nabla\cdot V)+\bar{\tau}:\nabla V+\dot{q}'''-\nabla\cdot q\\&\quad+\rho\frac{\partial}{\partial t}(pv)+\rho V\cdot\nabla(pv)\end{aligned} \tag{5.48}$$

$$\rho\frac{Dh}{Dt}=-p(\nabla\cdot V)+\bar{\tau}:\nabla V+\dot{q}'''-\nabla\cdot q+\rho\frac{\partial}{\partial t}(pv)+\rho V\cdot\nabla(pv) \tag{5.49}$$

where h is the specific enthalpy, $h=\mathrm{u}+pv$.

Now,

$$-p(\nabla\cdot V)+\rho\frac{\partial}{\partial t}(pv)+\rho V\cdot\nabla(pv)$$

$$=-p(\nabla\cdot V)+\rho p\frac{\partial v}{\partial t}+\rho v\frac{\partial p}{\partial t}+\rho vV\cdot\nabla p+\rho pV\cdot\nabla v$$

$$=-p(\nabla\cdot V)+\rho p\frac{Dv}{Dt}+\frac{Dp}{Dt} \tag{5.50}$$

Again,

$$\rho\frac{Dv}{Dt}=\rho\frac{D(1/\rho)}{Dt}=-\frac{\rho}{\rho^2}\frac{D\rho}{Dt}=-\frac{1}{\rho}\left[\frac{\partial\rho}{\partial t}+V\cdot\nabla\rho\right] \tag{5.51}$$

From the continuity equation, we have:

$$\frac{\partial\rho}{\partial t}+\nabla\cdot\rho V=0 \tag{5.52}$$

$$\frac{\partial\rho}{\partial t}+V\cdot\nabla\rho+\rho\nabla\cdot V=0 \tag{5.53}$$

$$\frac{\partial\rho}{\partial t}+V\cdot\nabla\rho=-\rho\nabla\cdot V \tag{5.54}$$

So,

$$\rho\frac{Dv}{Dt}=\frac{1}{\rho}\rho\nabla\cdot V=\nabla\cdot V \tag{5.55}$$

From Equation (5.50), we get:

$$-p(\nabla\cdot \mathrm{V})+\rho\frac{\partial}{\partial t}(pv)+\rho V\cdot\nabla(pv)$$

$$=-p(\nabla\cdot \mathrm{V})+\rho p\frac{Dv}{Dt}+\frac{Dp}{Dt}$$

$$=-p(\nabla\cdot \mathrm{V})+p(\nabla\cdot \mathrm{V})+\frac{Dp}{Dt}$$

$$=\frac{Dp}{Dt} \tag{5.56}$$

So, from Equation (5.49), we get:

$$\rho\frac{Dh}{Dt}=\frac{Dp}{Dt}+\overline{\tau}:\nabla V+\dot{q}'''-\nabla\cdot q \tag{5.57}$$

Again,

$$h = h(T, p)$$

$$Dh = \left(\frac{\partial h}{\partial t}\right)_p DT + \left(\frac{\partial h}{\partial p}\right)_T Dp = C_P DT + \left(\frac{\partial h}{\partial p}\right)_T Dp \tag{5.58}$$

From the second *Tds* relation, *dh* = *Tds* + *vdp*.

$$\therefore \left(\frac{\partial h}{\partial p}\right)_T = T\left(\frac{\partial s}{\partial p}\right)_T + v = v(1 - T\beta) \tag{5.59}$$

where

$$\beta = \frac{1}{v}\left(\frac{\partial v}{\partial T}\right)_p$$

So,

$$Dh = C_P DT + v(1 - T\beta)Dp \tag{5.60}$$

or

$$\rho\frac{Dh}{Dt} = \rho C_P \frac{DT}{Dt} + v(1 - T\beta)\frac{Dp}{Dt} \tag{5.61}$$

Substituting Equation (5.61) in Equation (5.57), we get:

$$\rho C_P \frac{DT}{Dt} = T\beta\frac{Dp}{Dt} + \overline{\tau} : \nabla V + \dot{q}''' - \nabla \cdot q \tag{5.62}$$

Equations (5.57) and (5.62) are two forms of enthalpy balance equations. The second term in the RHS is the viscous dissipation term and is normally written as $\mu\Phi_V$. The last term can be taken as the conduction term. Thus, the equation can be written as follows:

$$\rho C_P \frac{DT}{Dt} = T\beta\frac{Dp}{Dt} + \mu\Phi_V + \dot{q}''' + \nabla \cdot (k\nabla T) \tag{5.63}$$

Viscous dissipation in different coordinate systems is shown in Table 5.9.

TABLE 5.9 FORMS OF $\mu\Phi_v$ IN DIFFERENT COORDINATES

Rectangular coordinates (x, y, z):

$$\Phi_v = 2\left[\left(\frac{\partial v_x}{\partial x}\right)^2 + \left(\frac{\partial v_y}{\partial y}\right)^2 + \left(\frac{\partial v_z}{\partial z}\right)^2\right] + \left[\frac{\partial v_y}{\partial x} + \frac{\partial v_x}{\partial y}\right]^2$$

$$+\left[\frac{\partial v_z}{\partial y} + \frac{\partial v_y}{\partial z}\right]^2 + \left[\frac{\partial v_x}{\partial z} + \frac{\partial v_z}{\partial x}\right]^2 - \frac{2}{3}\left[\frac{\partial v_x}{\partial x} + \frac{\partial v_y}{\partial y} + \frac{\partial v_z}{\partial z}\right]^2.$$

Cylindrical coordinates (r, θ, Z):

$$\Phi_v = 2\left[\left(\frac{\partial v_r}{\partial r}\right)^2 + \left(\frac{1}{r}\frac{\partial v_\theta}{\partial \theta} + \frac{v_r}{r}\right)^2 + \left(\frac{\partial v_z}{\partial z}\right)^2\right]$$

$$+\left[r\frac{\partial}{\partial r}\left(\frac{v_\theta}{r}\right) + \frac{1}{r}\frac{\partial v_r}{\partial \theta}\right]^2 + \left[\frac{1}{r}\frac{\partial v_z}{\partial \theta} + \frac{\partial v_\theta}{\partial z}\right]^2$$

$$+\left[\frac{\partial v_r}{\partial z} + \frac{\partial v_z}{\partial r}\right]^2 - \frac{2}{3}\left[\frac{1}{r}\frac{\partial}{\partial r}(r v_r) + \frac{1}{r}\frac{\partial v_\theta}{\partial \theta} + \frac{\partial v_z}{\partial z}\right]^2$$

Spherical coordinates (r, θ, ϕ):

$$\Phi_v = 2\left[\left(\frac{\partial v_r}{\partial r}\right)^2 + \left(\frac{1}{r}\frac{\partial v_\theta}{\partial \theta} + \frac{v_r}{r}\right)^2 + \left(\frac{1}{r\sin\theta}\frac{\partial v_\phi}{\partial \phi} + \frac{v_r}{r} + \frac{v_\theta \cot\theta}{r}\right)^2\right]$$

$$+\left[r\frac{\partial}{\partial r}\left(\frac{v_\theta}{r}\right) + \frac{1}{r}\frac{\partial v_r}{\partial \theta}\right]^2 + \left[\frac{\sin\theta}{r}\frac{\partial}{\partial \theta}\left(\frac{v_\phi}{\sin\theta}\right) + \frac{1}{r\sin\theta}\frac{\partial v_\theta}{\partial \phi}\right]^2$$

$$+\left[\frac{1}{r\sin\theta}\frac{\partial v_r}{\partial \phi} + r\frac{\partial}{\partial r}\left(\frac{v_\phi}{r}\right)\right]^2 - \frac{2}{3}\left[\frac{1}{r^2}\frac{\partial}{\partial r}\left(r^2 v_r\right) + \frac{1}{r\sin\theta}\frac{\partial}{\partial \theta}(v_\theta \sin\theta) + \frac{1}{r\sin\theta}\frac{\partial v_\phi}{\partial \phi}\right]^2$$

5.6 ENTROPY BALANCE EQUATION

Entropy is also a conserved quantity. Here, we shall try to formulate the entropy balance equation for a multi-species system.

The change in enthalpy of the system can be augmented by the enthalpy transported across the boundary with different species. So, we can write:

$$\rho C_P \frac{DT}{Dt} = T\beta \frac{Dp}{Dt} + \bar{\tau} : \nabla V + \dot{q}''' - \nabla \cdot q$$
$$dh = Tds + vdp + \sum_i \left. \frac{\partial h}{\partial y_i} \right|_{s,p,y_{j \neq i}} dy_i \tag{5.64}$$

By definition of chemical potential, we can write:

$$\left. \frac{Dh}{Dy_i} \right|_{s,p,y_{j \neq i}} = \left. \frac{Dg}{Dy_i} \right|_{T,p,y_{j \neq i}} = \mu_i \tag{5.65}$$

Substituting Equation (5.64) in Equation (5.65), we get:

$$dh = Tds + vdp + \sum_i \mu_i dy_i \tag{5.66}$$

Taking the substantial derivative on the basis of Equation (5.66), we can write:

$$\frac{Dh}{Dt} = T\frac{Ds}{Dt} + v\frac{Dp}{Dt} + \sum_i \mu_i \frac{Dy_i}{Dt} \tag{5.67}$$

Rearranging, we see:

$$T\frac{Ds}{Dt} = \frac{Dh}{Dt} - v\frac{Dp}{Dt} - \sum_i \mu_i \frac{Dy_i}{Dt} \tag{5.68}$$

So, we can write:

$$T\frac{Ds}{Dt} = \frac{1}{T}\left[\rho \frac{Dh}{Dt} - \frac{Dp}{Dt}\right] - \frac{1}{T}\sum_i \mu_i \frac{Dy_i}{Dt} \tag{5.69}$$

From the energy equation, we can write:

$$\rho \frac{Dh}{Dt} - \frac{Dp}{Dt} = \bar{\tau} : \nabla \mathrm{v} - \nabla .\mathrm{q} - \sum_i f_i . j_i \tag{5.70}$$

Here, f_i and j_i are body force and mass flux due to ith species, and q is the heat flux vector.

Similarly, from species balance equation, we get:

$$\rho\frac{Dy_i}{Dt} = -\nabla.\mathrm{j_i} + \omega_i \tag{5.71}$$

Here, ω_i is the species generation term.

Substituting, Equations (5.70) and (5.71) in Equation (5.69), we get:

$$\rho\frac{Ds}{Dt} = \frac{1}{T}\left[\bar{\tau}:\nabla\mathrm{v} - \nabla.\mathrm{q} - \sum_i f_i.j_i\right] - \sum_i \frac{\mu_i}{T}\left[-\nabla.\mathrm{j_i} + \omega_i\right] \tag{5.72}$$

After a suitable rearrangement, we can write:

$$\rho\frac{Ds}{Dt} = -\nabla.\frac{\mathrm{q}}{\mathrm{T}} + \sum_i \nabla.\left(\frac{\mu_i \mathrm{j_i}}{T}\right) - \frac{\mathrm{q}.\nabla\mathrm{T}}{\mathrm{T}^2} - \frac{1}{\mathrm{T}}\sum_i \mathrm{j_i}.\nabla\mu_i$$
$$+ \sum_i \mu_i \mathrm{j_i}.\frac{\nabla\mathrm{T}}{T^2} + \frac{\bar{\tau}:\nabla\mathrm{v}}{T} + \frac{\sum_i f_i.j_i}{T} - \sum_i \frac{\mu_i\omega_i}{T} \tag{5.73}$$

Now, we can write:

$$q = q^c + \sum_i j_i h_i = q^c + \sum_i j_i(\mu_i + Ts_i) \tag{5.74}$$

where the heat flux is expressed as a sum of conductive flux (q^c) and enthalpy flux due to species transport. After some rearrangements in Equation (5.74), Equation (5.73) takes the following form:

$$\rho\frac{Ds}{Dt} = -\nabla.\left(\frac{q^c}{\mathrm{T}}\right) - \sum_i \nabla.(\mathrm{j_i s_i}) - \left(-\frac{q^c.\nabla\mathrm{T}}{\mathrm{T}^2}\right)$$
$$-\frac{1}{\mathrm{T}}\sum_i \mathrm{j_i}.(\mathrm{s_i}\nabla\mathrm{T} + \nabla\mu_i) + \frac{\bar{\tau}:\nabla\mathrm{v}}{T} + \frac{\sum_i f_i j_i}{T} - \sum_i \frac{\mu_i\omega_i}{T} \tag{5.75}$$

The first two terms in the right side of the above equation are the entropy transport terms. The first one is due to heat conduction; the second term is due to entropy transport through different species flow across the boundary. Other terms in the right side are generation terms. The third term is entropy generation due to heat transfer. The fourth term is due to generation related to species mass transfer. Viscous dissipation related generation is the fifth one. The sixth term will exist if there is a difference in body force in different species. The last term is due to chemical reaction.

REFERENCES

1. Bird, R. B., Stewart, W. E. and Lightfoot, E. N., *Transport Phenomena*, New York: John Wiley & Sons, 2007.
2. Schlichting, H. and Gersten, K., *Boundary Layer Theory*, New York: Springer, 2016.
3. Srinath, L., *Advanced Mechanics of Solids*, New York: Tata McGraw-Hill, 2008.
4. Shames, I. H., *Mechanics of Fluids*, New York: McGraw-Hill, 2003.
5. Burmeister, L. C., *Convective Heat Transfer*, New York: John Wiley & Sons, 1993.

Chapter 6

Laminar Premixed Flames

A flame is a localised region of intense chemical reaction. This region of intense chemical reaction is characterised by emission of heat and generally light as well. If the fuel and the oxidiser are initially separated and subsequently transported towards each other, after ignition, the flame always locates itself at a position where the fuel and the oxidiser are present in stoichiometric proportions. Such flames are called non-premixed flames. On the other hand, if the reactants are mixed a priori, after ignition, the reaction front propagates through the unburnt mixture with a definite speed. Such flames are known as premixed flames. Common examples of premixed flames are flame in a Bunsen burner with air holes open and spark ignition engines (carburettor engines).

6.1 PHYSICAL DESCRIPTION

As mentioned before, a premixed flame propagates through the unburned mixture at a definite speed. Thus, it can be considered as a front separating burned and unburned mixture. At this level of resolution, the flame is a surface of infinitesimal thickness across which there is a jump of mixture properties from unburned gas temperature T_u to burned gas temperature T_b and fuel concentration in the unburned region y_u to the fuel concentration in the burned region, $y_b = 0$, This level of resolution is known as the hydrodynamic or flame sheet level.

The next, more detailed level of resolution is the transport-dominated level. At this level, the flame is expanded to a finite thickness over which the temperature of the reactants increases from T_u to T_b and the fuel concentration drops from y_u to 0. This is due to transport of heat and fuel species from and to the reaction zone by convection and diffusion. However, the chemical reaction is limited to a region close to the maximum temperature, and also the reaction is completed within a very thin region due to the high reaction rate. At this level of resolution, the chemical reaction is confined to an infinitesimal region: the reaction sheet. The flame is thus made of a finite preheat zone and an infinitesimal reaction sheet. This level of resolution is the transport or reaction sheet level.

The third and the most detailed level of resolution considers a small but finite reaction zone (δ_R) towards the end of the preheat zone (δ_D). In the reaction zone, δ_R ($<<\delta_D$), a highly peaked reaction profile occurs. The rapid rate of property change within the reaction zone implies that diffusion has a much greater influence than the convection. This level of resolution is referred to as the chemical level.

6.2 RANKINE–HUGONIOT RELATIONS

As discussed in the previous section, at the hydrodynamic scale, a premixed flame may be considered as a wave propagating through a reacting mixture and separating the unburned region from the burned region. Let us consider a one-dimensional planar flame propagating as a wave through a quiescent reacting mixture with a velocity V_u. At the hydrodynamic level of resolution, the length scales are much larger than the preheat zone where the diffusion fluxes are important. The entire flame, including the preheat zone itself, appears as an infinitesimally thin surface. Hence, on a flame-fitted coordinate system, the diffusive fluxes vanish at the boundaries of a typical domain encapsulating the flame. Integrating the transport equations across the flame on a hydrodynamic length scale, the following conservation relations are obtained.

Mass:

$$\rho_u V_u = \rho_b V_b = f \tag{6.1}$$

Momentum:

$$\rho_u V_u^2 + p_u = \rho_b V_b^2 + p_b \tag{6.2}$$

Energy:

$$h_u + \frac{V_u^2}{2} = h_b + \frac{V_b^2}{2} \tag{6.3}$$

6.2.1 Rayleigh Lines

Rearranging Equation (6.2) and using Equation (6.1), one can write:

$$p_b - p_u = \rho_b V_b^2 - \rho_u V_u^2 = f^2\left(\frac{1}{\rho_u} - \frac{1}{\rho_b}\right) = -f^2\left(v_b - v_u\right) \tag{6.4}$$

After some algebraic manipulations, one obtains, in a dimensionless form:

$$\hat{p}-1=-\gamma M_u^2(\hat{v}-1) \tag{6.5}$$

Here, $\hat{p}=p_b/p_u; \hat{v}=v_b/v_u$

Here, the upstream Mach number is defined as:

$$M_u^2=\frac{V_u^2}{c_u^2}=\frac{V_u^2}{\frac{\gamma p_u}{\rho_u}}=\frac{\rho_u^2 V_u^2 v_u}{\gamma p_u}=\frac{f^2 v_u}{\gamma p_u} \tag{6.6}$$

Equation (6.5) can be recast as:

$$M_u^2=-\frac{\hat{p}-1}{\gamma(\hat{v}-1)} \tag{6.7}$$

It readily follows that:

$$M_b^2=-\frac{(\hat{p}-1)}{\gamma(\hat{v}-1)}\frac{\hat{v}}{\hat{p}} \tag{6.8}$$

Equation (6.5) represents a straight line in $\hat{p}-\hat{v}$ plane passing through the point (1,1) with a slope $-\gamma M_u^2$. These straight lines are known as **Rayleigh lines**. Since $M_u^2>0$, the straight line always has a negative slope. As all feasible solutions must satisfy Rayleigh lines, solutions with a simultaneous decrease or increase in pressure on crossing the wave (flame) are not possible.

6.2.2 Hugoniot Lines

From Equation (6.3), one can write:

$$h_b-h_u=\frac{1}{2}(V_u^2-V_b^2)=\frac{1}{2}\left(\frac{\rho_u^2 V_u^2}{\rho_u^2}-\frac{\rho_b^2 V_b^2}{\rho_b^2}\right)=\frac{1}{2}f^2(v_b-v_u)(v_b+v_u) \tag{6.9}$$

Combining Equations (6.4) and (6.9), one obtains:

$$h_b-h_u=\frac{1}{2}(p_b-p_u)(v_b+v_u) \tag{6.10}$$

For heat release q_c per unit mass, one can write:

$$h_b-h_u=-q_c+C_p(T_b-T_u)=-q_c+\frac{C_p}{R}\left(\frac{p_b}{\rho_b}-\frac{p_u}{\rho_u}\right) \tag{6.11}$$

Combining Equations (6.10) and (6.11), one can write:

$$\frac{1}{2}(p_b - p_u)(v_b + v_u) = -q_c + \frac{\gamma}{\gamma - 1}(p_b v_b - p_u v_u)$$

After some algebraic operations, the above equation can be expressed as follows:

$$\left(\hat{p} + \frac{\gamma - 1}{\gamma + 1}\right)\left(\hat{v} - \frac{\gamma - 1}{\gamma + 1}\right) = \frac{4\gamma}{(\gamma + 1)^2} + 2\hat{q}_c\left(\frac{\gamma - 1}{\gamma + 1}\right) \tag{6.12}$$

In the above equation $\hat{q}_c = \frac{q_c}{p_u v_u}$.

Equation (6.12) represents a rectangular hyperbola in the $\hat{p} - \hat{v}$ plane with asymptotes of $\hat{p} \to -\frac{\gamma-1}{\gamma+1}$ and $\hat{v} \to \frac{\gamma-1}{\gamma+1}$. These curves are known as **Hugoniot curves**. In the absence of heat release ($\hat{q}_c = 0$), the curve passes through the point (1,1).

6.2.3 Detonation and Deflagration Waves

Since both Rayleigh lines and Hugoniot curves are obtained from conservation equations, any feasible solution must satisfy both. Thus, the feasible solutions reduce to the roots of the set of simultaneous equations Equations (6.5) and (6.12). The results are known as Rankine–Hugoniot relations.

Algebraic manipulations leading to the solutions of Equations (6.5) and (6.12) can lead to significant physical insights. Considerable algebraic manipulations lead to the following roots of Equations (6.5) and (6.12):

$$\hat{v} - 1 = \frac{1 - M_u^2}{(\gamma + 1)M_u^2}\left[1 \pm \sqrt{1 - \frac{2(\gamma^2 - 1)}{\gamma}\frac{M_u^2}{(1 - M_u^2)^2}\hat{q}_c}\right] \tag{6.13}$$

$$\hat{p} - 1 = -\frac{\gamma(1 - M_u^2)}{(\gamma + 1)}\left[1 \pm \sqrt{1 - \frac{2(\gamma^2 - 1)}{\gamma}\frac{M_u^2}{(1 - M_u^2)^2}\hat{q}_c}\right] \tag{6.14}$$

From these equations, we see that, for each set of parameters, there can be two solutions. Since $\frac{2(\gamma^2-1)}{\gamma}\frac{M_u^2}{(1-M_u^2)^2}\hat{q} > 0$ for all conditions, the term under the root sign is always less than 1. Hence the term inside the box bracket is always positive. Hence, depending on whether $M_u^2 > 1$ or $M_u^2 < 1$, one can obtain $\hat{v} < 1$ or $\hat{v} > 1$ and $\hat{p} > 1$ or $\hat{p} < 1$. The supersonic solution ($M_u > 1$) is known as ***detonation wave***; the subsonic solution ($M_u < 1$) is known as ***deflagration wave***. It can be readily seen that, in general, each Rayleigh line intersects a given Hugoniot curve at two points, both either in the subsonic or in the supersonic branch. The solution closer to the point (1,1) is referred to as a weak solution; the one farther away from the (1,1) point is called the strong solution. It can be shown from entropy and other considerations that weak detonations and strong deflagrations do not exist. For deflagrations, only weak solutions exist for which $\hat{p}$ is only slightly less than 1. Thus, the pressure jump across a deflagration wave is small. This allows one to neglect pressure changes across flames in subsonic cases. On the other hand, in case of detonations, only strong solutions exist, so there is a significant increase in pressure across the flame.

6.2.4 Chapman–Jouguet Waves

An inspection of the Rankine–Hugoniot relations reveals that there exists a combination of parameters (γ, q_c, M_u) for which the Rayleigh line is tangential to the Hugoniot curve. This is the minimum Rayleigh line for which a solution exists. Beyond this line, no solution exists. This solution is known as ***Chapman–Jouguet (CJ) point***. There exists two such points, one each on the supersonic and the subsonic branch, respectively. The CJ point on the supersonic branch is known as the upper CJ point, while that on the subsonic branch is called the lower CJ point. It is obvious that, at the CJ point:

$$\left(\frac{dp}{dv}\right)_{Raleigh} = \left(\frac{dp}{dv}\right)_{Hugoniot} \tag{6.15}$$

It can be shown that Equation (6.15) is equivalent to $-\frac{(\hat{p}-1)}{\gamma_u(\hat{v}-1)}\frac{\hat{v}}{\hat{p}} = M_b^{\,2} = 1$. Hence downstream conditions at CJ points are sonic. From Equation (6.15), it can be shown that $\left(\frac{dp}{dv}\right)_{Rayleigh} < \left(\frac{dp}{dv}\right)_{Hugoniot}$ implies $-\frac{(\hat{p}-1)}{\gamma_u(\hat{v}-1)}\frac{\hat{v}}{\hat{p}} < 1$; that is, subsonic flow downstream of the wave. The slope of the Hugoniot curve is generally greater than that of the Rayleigh line for strong detonations and weak deflagrations, so $M_b^{\,2} < 1$ for these cases, while the opposite is true for weak detonations and strong deflagrations.

6.3 FLAME PROPAGATION AND FLAME SPEED

A premixed flame is a self-sustaining localised combustion zone that propagates at subsonic speeds[4]. Thus, the speed at which the combustion zone propagates is crucial. The velocity of the flame relative to fixed coordinates, $\mathbf{V}_{flame}$ is known as the flame propagation speed. However, the mixture through which the flame propagates can also have some velocity. For example, the fluid inside the combustion chamber of a spark ignition engine is in a highly turbulent motion. The normal component of the velocity of the fluid relative to the flame is important for calculating the mass flux across the moving flame. If $\mathbf{V}_{flow}$ is the local flow velocity at the flame location, the velocity of the flame relative to the flow is defined as the flame displacement speed, S_u. Thus, the flame displacement speed is defined as:

$$S_u = \left(\mathbf{V}_{flame} - \mathbf{V}_{flow}\right) \cdot \hat{\mathbf{n}} \tag{6.16}$$

where $\hat{\mathbf{n}}$ is the unit normal to the flame surface. This flame displacement speed is referred to as flame speed or burning velocity. Although the two terms are often used synonymously, the former generally is used to denote flame motion, and the latter is used as a flame property.

To illustrate the concept of flame speed, we consider two extreme cases. First, let us consider a quiescent mixture in a tube. If the mixture is ignited at one end, the flame propagates through the quiescent mixture to the other end. Here the flow velocity is zero. Hence the flame displacement speed is obtained as:

$$S_u = \mathbf{V}_{flame} \cdot \hat{\mathbf{n}} \tag{6.17}$$

Now, let us consider the other extreme of a flame stabilised on a burner. The flame is stationary and hence its propagation speed is given by:

$$S_u = -\mathbf{V}_{flow} \cdot \hat{\mathbf{n}} \tag{6.18}$$

In fact, the second method is a standard way of calculating flame speed.

The flame in a Bunsen burner consists of an inner cone of premixed flame and often an outer cone of non-premixed flame. If V_u denotes the fluid velocity just upstream of the flame and the flow is considered to remain parallel to the burner axis till it reaches the flame, the flame speed may be expressed as:

$$S_u = V_u \mathrm{Sin}\alpha \tag{6.19}$$

Here α is the local angle between the velocity vector and the flame surface. The assumption of the flow remaining parallel to the burner axis till it reaches the flames implies neglecting thermal expansion of the fluid upstream of the flame causing a flow divergence. In general, in Equation (6.19), both S_u and V_u can vary with the radial coordinate, r. Thus, the angle α would also vary with the radial position, that is:

$$\alpha(r) = \mathrm{Sin}^{-1}\left[\frac{S_u(r)}{V_u(r)}\right] \tag{6.20}$$

Again, the slope of the flame surface at any radial position is given by:

$$\frac{dz}{dr} = \tan\alpha(r) = \tan\left\{\text{Sin}^{-1}\left[\frac{S_u(r)}{V_u(r)}\right]\right\} \tag{6.21}$$

Once the values of V_u and S_u at any radial location are known, the local slope of the flame surface is readily obtained. The flame profile can be obtained by integrating Equation (6.21) and using the condition that the flame is attached to the burner rim; that is, $z(R)=0$, where R is the radius of the Bunsen burner. Thus, the flame profile is obtained as:

$$z(r) = \int_R^r \frac{dz}{dr}dr = -\int_r^R \tan\left\{\text{Sin}^{-1}\left[\frac{S_u(r)}{V_u(r)}\right]\right\}dr \tag{6.22}$$

However, if S_u remains constant across the burner and V_u is uniform, the slope of the flame becomes constant and hence the flame is conical.

6.4 DETERMINATION OF FLAME SPEED

6.4.1 Determination of Flame Speed by Bunsen Flame Method

The analysis of the Bunsen burner described in the previous section forms the basis for a simple method of determination of flame speed. In the simplest version of this method, V_u is taken as the average mixture velocity at the exit of the burner tube. Since the flow is assumed to remain parallel to the burner axis, and flow divergence due to thermal expansion upstream of the flame is neglected, the area of the stream tube of the reactants remains constant. Hence the average velocity measured at the tube exit prevails up to the upstream boundary of the flame. The average velocity V_u is obtained as:

$$V_u = \frac{Q}{A} \tag{6.23}$$

Here, Q is the volumetric flow rate of the reacting mixture through the burner tube, measured with the help of a rotameter or mass flow meter or mass flow controller, and A is the cross-sectional area of the burner tube, calculated from the knowledge of the burner diameter. From Equation (6.19), one can determine the flame speed. The angle α is determined through image processing. In the simplest form of image processing, a straight line is fitted to the edge of the conical flame, and the angle between this fitted line and the burner axis is determined. However, advancement of measurement techniques allow more accurate determination of the flame speed. The flow velocity can be determined at the flame location using laser-based diagnostic tools like particle image velocimetry (PIV). Similarly, the local curvature and slope of the flame can be determined using suitable customised or general-purpose image processing software. This image processing can be done using software obtained commercially or through open source or can also be developed in-house. This simple method of determining flame speed, however, has some limitations. The flame speed is affected by factors like divergence of flow field, curvature of the flame front and heat loss, all of which are present in the Bunsen flame. To eliminate these effects, one needs to determine the flame speed for a planar flame, propagating through a stationary mixture without heat loss. The flame speed under these conditions is often referred to as unstretched adiabatic flame speed (S_u^0). This is difficult to determine directly from this method.

Example 6.1

A premixed methane-air flame is stabilised on a burner of diameter 8 mm. The methane flow rate is 0.72 lpm and the equivalence ratio of the mixture is 0.9. If the flame height is measured to be 3 cm, find the flame speed of the mixture.

Solution

Volume flow rate of methane = 0.72 lpm

Stoichiometric air-fuel ratio of methane by volume = $2 \times 4.76 = 9.52$

Hence, actual air-fuel ratio by volume = $9.52/0.9 = 10.58$

Hence total volume flow rate of the reactant mixture = $0.72 \times (10.58 + 1) =$ 8.33 lpm = $1.39 \times 10^{-4} m^3/s$

Velocity of mixture, $V_u = 1.39 \times 10^{-4}/\pi(0.004)^2 = 2.76$ m/s

Angle $\alpha = \tan^{-1}(4/30) = 7.59°$

Hence, flame speed $S_u = V_u \text{Sin}\alpha = 0.364$ m/s

6.4.2 Determination of Flame Speed by Flat Flame Burner Method

If V_u is adjusted so that it becomes equal to the flame speed, that is, $S_u = V_u$, the flame becomes flat and parallel to the burner. A burner which produces such flames is known as a flat flame burner. Flat flame burners require uniform flow velocity, which is achieved with a porous burner. The flat flame eliminates the problems of flame curvature and flow divergence. However, the preheat zone of the flame starts at the porous burner surface. Hence there is significant heat loss to the burner, which affects the flame speed. Botha and Spalding [1] varied the mixture flow rate and determined the cooling rate needed to produce a flat flame. Since $S_u = V_u$, the velocity of the unburned mixture gave the flame speed under the given conditions of heat loss. The experiments were conducted to determine flame speeds at different values of heat loss and the results extrapolated to zero heat loss to obtain the adiabatic flame speed for flat flame, S_u^0.

6.4.3 Determination of Flame Speed by Spherically Propagating Flame Method

A conceptually simple method of flame speed determination involves propagation of a spherical flame through a quiescent mixture. For a quiescent mixture, the flame speed is equal to the flame propagation speed. If we assume that spherical symmetry exists, the flame propagation speed is given by $\frac{dr_f}{dt}$, where r_f is the instantaneous flame radius. If optical access is provided, the direct method of determining $\frac{dr_f}{dt}$ is by processing of high-speed video images. But optical access can be extremely difficult in many cases, especially at high pressures. An alternative technique involves relating $\frac{dr_f}{dt}$ to time traces of pressure. Since pressure is spatially uniform inside the combustion chamber, even a single pressure sensor can be sufficient.

If we assume that ignition takes place at the centre of a spherical vessel of radius R filled with a homogeneous reacting mixture, overall mass conservation gives:

$$\frac{4}{3}\pi\left[\left(R^3 - r_f^{\ 3}\right)\rho_u + r_f^{\ 3}\rho_b\right] = \frac{4}{3}\pi R^3 \rho_{u0} \tag{6.24}$$

If we assume isentropic compression of the gases due to the expanding flame, the instantaneous densities of the unburned and burned gases, ρ_u and ρ_b, respectively, are related to the initial densities of unburned and burned gases, ρ_{u_0} and ρ_{b_0}, respectively, as:

$$\frac{\rho_u}{\rho_{u0}} = \frac{\rho_b}{\rho_{b0}} = \left(\frac{p}{p_0}\right)^{1/\gamma} \tag{6.25}$$

Substituting Equation (6.25) in Equation (6.24), one obtains:

$$\left[\left(R^3 - r_f^{\ 3}\right)\rho_{u0} + r_f^{\ 3}\rho_{b0}\right]\left(\frac{p}{p_0}\right)^{1/\gamma} = R^3 \rho_{u0} \tag{6.26}$$

After some algebraic manipulations, the above result can be expressed in a dimensionless form as:

$$\left[1+(\bar{\rho}_b-1)\bar{r}_f^{\,3}\right]\bar{p}^{1/\gamma}=1 \tag{6.27}$$

Here, the dimensionless (overbar) quantities are defined as $\bar{r}_f=\frac{r_f}{R};\bar{p}=\frac{p}{p_0};\bar{\rho}_b=\frac{\rho_{b_0}}{\rho_{u_0}}$.

Differentiating Equation (6.27) with respect to time, the rate of growth of the dimensionless flame radius is obtained as follows:

$$\frac{d\bar{r}_f}{dt}=\left[3\gamma\left(\bar{\rho}_b-1\right)^{1/3}\bar{p}^{(1+1/\gamma)}\left(1-\bar{p}^{-1/\gamma}\right)^{2/3}\right]\frac{d\bar{p}}{dt} \tag{6.28}$$

Recognising that the rate of change of r_f measures the rate of growth of the burned gas volume, the flame speed measured at the burned end of the flame is given by $s_b=\frac{dr_f}{dt}=R\frac{d\bar{r}_f}{dt}$.

Using conservation of mass, one can obtain the flame speed, measured at the upstream boundary of the flame, as:

$$s_u=\frac{\rho_b}{\rho_u}s_u=\bar{\rho}_b s_b$$

The limitation of this method is the influence of flame curvature on flame speed. However, since flame speeds are obtained at different instants of time, and therefore at different flame radii, the flame speeds obtained at different instants of time can be extrapolated to zero curvature $\left(r_f\rightarrow\infty\right)$ to obtain unstretched planar flame speed, $s_u^{\,0}$.

6.5 SIMPLIFIED ANALYSIS

The earliest analysis of laminar premixed flames was done by Mallard and Le Chatelier [2] in 1883. Spalding [3] developed a simplified analysis, which elucidates the essential physics. The present analysis closely follows that analysis. The major assumptions of the analysis [4] are as follows:

1. One-dimensional, constant area steady flow.
2. Viscous dissipation, radiative heat transfer and changes in kinetic and potential energy are neglected.
3. Since the pressure change across the flame is small, it is neglected and constant pressure is assumed across the flame.

4. Thermal and species diffusion follow the Fourier law of heat conduction and Fick's law of mass diffusion, respectively. Binary diffusion and unity Lewis number are assumed.
5. Constant and identical value of specific heat is assumed for all the species.
6. The mixture is assumed to be stoichiometric or lean so that all the fuel is consumed. The reaction is described by single step kinetics.

The statements of conservation of mass, momentum and energy can be expressed as follows.

Mass:

$$\frac{d}{dx}(\rho V_x)=0 \tag{6.29}$$

Integrating Equation (6.29), one obtains $\rho V_x = const = f$

Species:

$$\frac{df_i}{dx}=\omega_i \tag{6.30}$$

Using Fick's law of diffusion, this can be expressed as:

$$\frac{d}{dx}\left(fy_i-\rho D\frac{dy_i}{dx}\right)=\omega_i \tag{6.31}$$

If it is assumed that 1 kg fuel reacts with ν kg oxidiser to produce (ν + 1) kg products, one can write:

$$\omega_F=\frac{\omega_{Ox}}{\nu}=-\frac{\omega_{\mathrm{Pr}}}{\nu+1} \tag{6.32}$$

Energy:

Using Equation (6.32), the source term in energy equation can be written as:

$$\begin{aligned} fC_p\frac{dT}{dx}-\frac{d}{dx}\left(\rho DC_p\frac{dT}{dx}\right) &= -\sum_i \omega_i\Delta h_{f_i} \\ -\sum_i \omega_i\Delta h_{f_i} &= -\omega_F\Delta h_{f_F}-\omega_{Ox}\Delta h_{f_{Ox}}-\omega_{\mathrm{Pr}}\Delta h_{f_{\mathrm{Pr}}} \\ &= -\omega_F\left[\Delta h_{f_F}+\nu\Delta h_{f_{Ox}}-(\nu+1)\right]\Delta h_{f_{\mathrm{Pr}}}=-\omega_F\left(-\Delta h_c\right) \end{aligned} \tag{6.33}$$

Using Equation (6.33) and $Le=\frac{k}{\rho DC_p}=1$, the energy equation can be written as:

$$f\frac{dT}{dx}-\frac{1}{C_p}\frac{d}{dx}\left(k\frac{dT}{dx}\right)=\frac{\omega_F\Delta h_c}{C_p} \tag{6.34}$$

Far upstream and downstream of the flame, the diffusion fluxes vanish. Also, the temperature approaches the unburned and burned gas temperatures (as seen in the hydrodynamic scale), respectively. Thus, the boundary conditions are:

$$x \to -\infty : T \to T_u; \frac{dT}{dx} \to 0$$

$$x \to \infty : T \to T_b; \frac{dT}{dx} \to 0 \qquad (6.35)$$

Integrating Equation (6.34) from $x \to -\infty$ to $x \to \infty$ and using Equation (6.35), one obtains:

$$f(T_b - T_u) = \frac{\Delta h_c}{C_p} \int_{-\infty}^{\infty} \omega_F dx$$

However, the reaction term is non-zero only within the flame. Taking the origin at the upstream boundary of the flame, the reaction term is non-zero only in the region described by $0 \le x \le \delta_D$. Thus, the above equation can be written as:

$$f(T_b - T_u) = \frac{\Delta h_c}{C_p} \int_{0}^{\delta_D} \omega_F dx$$

The temperature inside the flame monotonically increases with x. Assuming a linear temperature variation within the flame, that is, $\frac{dT}{dx} = \frac{T_b - T_u}{\delta_D}$, a change of variables from x to T can be effected such that $dT = \frac{T_b - T_u}{\delta_D} dx$. Thus, the above equation can be written as:

$$f(T_b - T_u) = \frac{\Delta h_c}{C_p} \frac{\delta_D}{(T_b - T_u)} \int_{T_u}^{T_b} \omega_F dT$$

Defining the average reaction rate $\bar{\omega}_F$ as $\bar{\omega}_F = \frac{1}{(T_b - T_u)} \int_{T_u}^{T_b} \omega_F dT$, one can write:

$$f(T_b - T_u) = \frac{\Delta h_c}{C_p} \delta_D \bar{\omega}_F \qquad (6.36)$$

Note that the temperature profile is obtained in terms of f and δ_D, which are themselves unknowns. In addition to Equation (6.36), one needs another condition to evaluate these two unknowns.

Since a chemical reaction is limited to a very thin region, $\delta_R(<<\delta_D)$ close to the downstream boundary of the flame ($x=\delta_D$), it is readily observed that $\int_{-\infty}^{\delta_D/2}\omega_F dx=0$. Integrating Equation (6.34) from $x\rightarrow-\infty$ to $x=\frac{\delta_D}{2}$, one can use Equation (6.35) to write the following:

$$f\left[T\Big|_{x=\frac{\delta_D}{2}}-T_u\right]-k\frac{dT}{dx}\Big|_{x=\frac{\delta_D}{2}}=0 \tag{6.37}$$

Assuming linear temperature profile within the flame:

$$T\Big|_{x=\frac{\delta_D}{2}}=\frac{T_u+T_b}{2}$$
$$\frac{dT}{dx}\Big|_{x=\frac{\delta_D}{2}}=\frac{T_b-T_u}{\delta_D} \tag{6.38}$$

Substituting Equation (6.38) into Equation (6.37), one can write:

$$f\frac{\delta_D}{2}-\frac{k}{C_p}=0 \tag{6.39}$$

Solving Equations (6.35) and (6.39), one obtains:

$$f=\left[\frac{2k(\Delta h_c)}{C_p^{\,2}(T_b-T_u)}\bar{\omega}_F\right]^{1/2} \text{ and } \delta_D=\frac{2k}{fC_p}$$

Now, for stoichiometric or lean mixture, all the fuel is consumed. Equating heat release from combustion to sensible enthalpy change, $-\Delta h_c=(\nu+1)C_p(T_b-T_u)$. Substituting this in the expression for f, the following relation is obtained:

$$f=\left[\frac{2k}{C_p}\{-(\nu+1)\}\bar{\omega}_F\right]^{1/2}$$

Since flame speed is given by the velocity of the unburned fluid relative to the flame, the flame speed is equal to the volume flux of the unburned mixture across the flame. Thus, the expression for flame speed can be obtained as:

$$S_u=\frac{f}{\rho_u}=\left[\frac{2k}{\rho_u C_p}\{-(\nu+1)\}\frac{\bar{\omega}_F}{\rho_u}\right]^{1/2}=\left[-2\alpha(\nu+1)\frac{\bar{\omega}_F}{\rho_u}\right]^{1/2} \tag{6.40}$$

The flame thickness can likewise be expressed as:

$$\delta_D = \frac{2k}{fC_p} = \left[-\frac{2\rho_u \alpha}{(\nu+1)\bar{\omega}_F} \right]^{1/2} \tag{6.41}$$

From Equations (6.40) and (6.41), one can write:

$$\delta_D = \frac{2\alpha}{S_u} \tag{6.42}$$

Equations (6.40) and (6.41) show that, as the reaction rate increases, the flame speed, as expected, increases but the flame thickness decreases. It is also noted that the density of unburned gas and thermal diffusivity are two properties that play a major role in the determination of flame speed and flame thickness.

6.6 FACTORS AFFECTING FLAME SPEED

Flame speed depends on both physical and chemical properties of the reacting mixture and operating conditions like temperature and pressure.

6.6.1 Dependence on Temperature

The dependence of flame speed, and consequently flame thickness, on temperature comes through dependence of reaction rate and transport properties like thermal diffusivity on temperature. Dependence of thermal diffusivity on temperature and pressure can be simplified as [4]:

$$\alpha \propto T_u \bar{T}^{0.75} P^{-1} \text{ where } \bar{T} = (T_u + T_b)/2 \tag{6.43}$$

Expressing reaction rate in terms of burned gas temperature, one can write:

$$\frac{\bar{\omega}_F}{\rho_u} \propto T_u T_b^{-n} P^{n-1} \exp\left(-E_a / R^* T_b\right) \tag{6.44}$$

Substituting the above temperature dependence in the expressions for flame speed and flame thickness (Equations 6.40 and 6.41), the temperature and pressure dependence can be expressed as follows:

$$S_u \propto \bar{T}^{0.375} T_u T_b^{-n/2} \exp\left(-E_a / R^* T_b\right) P^{(n-2)/2} \tag{6.45}$$

$$\delta_D \propto \bar{T}^{0.375} T_b^{n/2} \exp\left(E_a / R^* T_b\right) P^{-n/2} \tag{6.46}$$

The above relations show that both flame speed and flame thickness are strongly dependent on temperature. In general, flame speed increases with increase in temperature.

6.6.2 Dependence on Pressure

Equation (6.45) shows that for a global reaction rate of 2, flame speed does not change with pressure. However, experimental results generally show a negative dependence.

6.6.3 Dependence on Fuel Type

Experimental studies show roughly similar trends in flame speeds for C_3–C_6 hydrocarbons [4], while C_2H_2 and C_2H_4 have higher velocities but the flame speed of CH_4 is lower. Hydrogen has a much higher flame speed due to high thermal and mass diffusivities of hydrogen and rapid reaction kinetics for hydrogen.

6.7 FLAME QUENCHING AND IGNITION

In the analyses carried out in the previous sections, heat loss from the flame was neglected. However, heat loss from flames is an important factor in determining the structure and dynamics of the flame and plays a major role in quenching and ignition of flames.

6.7.1 Quenching of Ducted Flame Due to Heat Loss

In case of flames inside ducts or tubes, heat loss to the duct wall is very often the deciding factor for the stability of the flame. The basic condition for quenching of the flame is that the heat loss from the flame exceeds the heat generated inside the flame. This would lead to a lowering of the temperature of the flame. Therefore, a critical condition exists when the heat loss and heat release balance each other. This would be the limiting condition for survival of the flame. Although a lower flame temperature implies reduction in both heat release due to chemical reaction and heat loss from the flame, the former is a much stronger function of temperature. Hence the decrease in rate of heat release significantly exceeds the decrease in the rate of heat loss, causing a further reduction in temperature till the temperature becomes too low for the reaction to sustain.

The analysis presented below is for a flame between two infinite parallel plates with a finite spacing D between them. Considering a flame of thickness

δ_D, the heat release from the flame per unit span of the plates is given by $Q \times D \times \delta_D \times 1$ where $Q = \bar{\omega}_F \Delta h_c$ is the heat released per unit volume of the flame. Similarly, the heat lost from the two walls is given by $-k(\delta_D \times 1)\frac{dT}{dx} \times 2$. The factor 2 accounts for heat loss from both walls. At the limiting condition of flame quenching, one can write:

$$(\bar{\omega}_F \Delta h_c)(D\delta_D) = -2k\delta_D \left.\frac{dT}{dx}\right|_{wall} \tag{6.47}$$

To calculate $\left.\frac{dT}{dx}\right|_{wall}$, one can assume that the gas temperature changes from wall temperature T_w at the wall to the burned gas temperature T_b at the centreline, where the effects of cooling from the wall are minimum. In the conduction limit this would imply a linear variation of temperature, so that $\left.\frac{dT}{dx}\right|_{wall} = \frac{T_b - T_w}{D/2}$. However, due to fluid motion, $\frac{dT}{dx}$ would be much higher at the wall and $T \to T_b$ within a much shorter distance. This effect is incorporated by assuming the temperature approaches the burned gas temperature within a distance D/b from the wall, where b is an arbitrary constant (>>2). Thus, the temperature gradient would be expressed as: $\left.\frac{dT}{dx}\right|_{wall} = \frac{T_b - T_w}{D/b}$

Substituting this value of temperature gradient at the wall in Equation (6.43), one obtains:

$$D^2 = \frac{2kb(T_b - T_w)}{\bar{\omega}_F \Delta h_c}$$

Since $-\Delta h_c = (\nu + 1)C_p(T_b - T_u)$, one can write:

$$D^2 = -\frac{2kb(T_b - T_w)}{\bar{\omega}_F (\nu + 1)C_p(T_b - T_u)}$$

Assuming $T_u = T_w$, and using the definition of S_u, the expression for D is obtained as:

$$D = \frac{2\alpha\sqrt{b}}{S_u} = \sqrt{b}\delta \tag{6.48}$$

This value of spacing is the minimum gap needed between the two walls for the flame to survive and is known as **quenching distance.** It can be seen here that the quenching distance has to be several times the flame thickness as $b >> 2$.

6.7.2 Minimum Energy for Ignition

A homogeneous fuel-air mixture is often ignited by a heat source. Although electric spark is the most common external energy source for ignition, some experiments have also used focused laser beams to supply the energy. For homogeneous reacting mixtures, the energy source generally ignites a small localised volume of the mixture. Under favourable conditions, the flame kernel thus formed expands outward, consuming the reacting mixture. For flame generated by ignition to be sustained, the heat generated within the flame volume must exceed the heat lost from the flame surface to the surrounding unburned mixture. Since heat loss is a surface phenomenon and the heat is generated from the entire flame volume, a high volume-to-surface ratio helps to sustain the flame. On the other hand, energy needed for ignition increases with the volume ignited. Hence, there exists a minimum volume that has to be ignited for sustenance of the flame and a corresponding minimum energy that has to be supplied.

Assuming a homogeneous reacting mixture and a point source of ignition, the ignited volume would be a sphere of radius R. The critical radius of the ignited volume satisfies the condition that the heat generated by the reaction is equal to the heat loss. Since quiescent mixture is considered, heat loss from the flame kernel is by conduction only. Thus, one can write:

$$\frac{4}{3}\pi R_{crit}^{\ 3}(-\bar{\omega}_F)(-\Delta h_c) = 4\pi R_{crit}^{\ 2} k \left|\frac{dT}{dR}\right|_{R_{crit}} \tag{6.49}$$

Substituting the value of the temperature gradient $\left|\frac{dT}{dR}\right|_{R_{crit}} = \frac{T_b - T_u}{R_{crit}}$ the expression for critical radius becomes:

$$R_{crit}^{\ 2} = 3k\frac{(T_b - T_u)}{(-\omega_F)(-\Delta h_c)}$$

Since $-\Delta h_c = (\nu + 1)C_p(T_b - T_u)$, the critical radius can be finally expressed as using Equation (6.40):

$$R_{crit}^{\ 2} = \frac{3k}{C_p(-\bar{\omega}_F)(\nu + 1)} = \frac{6\alpha^2}{S_u^2} \tag{6.50}$$

Alternatively, using Equation (6.42), one can express the critical radius in terms of flame thickness as:

$$R_{crit} = \frac{\sqrt{6}}{2}\delta_D \tag{6.51}$$

Comparing Equations (6.48) and (6.51), one can see that the critical radius for ignition is of the order of the flame thickness, while the quenching distance is several times larger than the flame thickness.

The energy needed to produce a kernel of burned gas of critical radius is obtained as:

$$E_{ignition} = \frac{4}{3}\pi R_{crit}^{3} \rho_b C_p \left(T_b - T_u\right) \tag{6.52}$$

6.8 FLAME PROPAGATION IN MICROSCALE COMBUSTORS

In the previous section, we saw that there exists a critical dimension of the duct below which a premixed flame gets extinguished. However, there is a strong motivation for development of combustors of extremely small sizes, which can act as power sources for microscale devices. The size limitations in such devices often demand construction of combustors, whose characteristic dimension is less than the critical dimension for quenching. One of the ways in which flame is sustained in a combustor, whose characteristic dimension is smaller than the quenching distance, is by means of heat recirculation. **Heat recirculation** refers to transfer of some energy from the burned gas region to the unburned mixture. This transfer of energy preheats the reacting mixture, leading to stronger reaction, which prevents extinction of the flame in spite of high heat loss due to high volume-to-surface ratio. Under certain circumstances, the heat recirculation to the upstream mixture can lead to preheating of the reactant mixture to such an extent that the temperature of the burned gases exceeds the flame temperature of planar adiabatic premixed flames in spite of heat loss from the flames. These flames are called **super-adiabatic flames**.

Leach et al. [5] developed a simple model for heat recirculation in combustor walls by extending the classical analysis of Mallard and Le Chatelier [2] to include the resistances to different heat transfer paths available for heat recirculation. Their analysis showed that. for no heat transfer or large burner dimensions, the results of Mallard and Le Chatelier [2] were retrieved. However, for small burner dimensions and large heat transfer from the flame, the flame speed was always less than that of planar adiabatic flames. The flame speed is determined by the ratio of the thermal conductivities of the gas and the material of the structure and relative areas for axial heat conduction.

6.9 FLAME STABILITY, LIFT-OFF, BLOWOUT AND FLASHBACK

Premixed flames suffer from serious hazards like possibilities of different kinds of instabilities like flashback, lift-off and blowout. Flashback refers to a condition when the flame propagates towards the unburned mixture. In the case of burner flames, this condition implies entry of the flame into the burner tube carrying reacting mixtures and propagation upstream along the tube. This condition not only disturbs the stability of the flame stabilised on the burner but is also a potential safety hazard as the flame can move upstream and ignite a large volume of reactants in the mixing chamber where the fuel and air are mixed. This can lead to explosion. Since flashback occurs when the flame propagates into the unburned mixture, the flame propagation velocity exceeds the local flow velocity. Thus, mixtures with high flame speed are more prone to flashback.

Lift-off, on the other hand, refers to a condition where the flame detaches itself from the burner and moves downstream away from the burner. In lift-off, the flames are stabilised at some distance above the burner and are known as lifted flame. Lifted flames can be undesirable in practical applications for several reasons [4]. First, with the flame stabilised away from the burner, some fuel may escape unburned. Second, as it is difficult to stabilise the flame at a specific location, poor heat transfer may result from the flame. These flames can also be noisy and susceptible to external perturbations. Flame lift-off occurs when the flame propagates in the direction of the fluid motion. Hence, in these situations, local flow velocity exceeds the flame speed. In case of jet flames, as one moves away from the burner rim, the jet expands and its velocity decreases. The lifted flame positions itself at a height where the flame speed balances the local flow velocity. As the flow velocity at the burner exit increases, the flame has to position itself at a greater distance from the burner. Beyond a certain distance, the flame can no longer be stabilised and it blows off completely. This phenomenon is known as **blowout**. Blowout is an undesirable phenomenon as it can lead to shutdown or fatal accidents. Since flame speed decreases as one moves away from stoichiometric mixture towards lean or rich side, lean mixtures used for low emission characteristics are often susceptible to blowout. Such blowout is called **lean blowout**.

Since flame speed plays a major role in flame stabilisation, there are strong fuel effects on both flashback and blowout. With the current focus on shifting from use of fossil fuels to biofuels and other renewable fuels like hydrogen-based fuels, there is a strong need for designing fuel-flexible burners, that is, burners which can use a wide variety of fuels. For design of fuel-flexible burners, issues like flashback and blowout need to be addressed carefully [6].

6.10 FLAME STRETCH

Flame speed is affected by several factors like flow non-uniformity, flame curvature and flame/flow unsteadiness [7]. It is possible to combine these effects in terms of a single parameter called **flame stretch**. Classically, flame stretch is defined on a flame surface, which implies consideration of the flame at the hydrodynamic scale. At this level the flame is considered an infinitesimal surface separating the unburned and burned gases.

The evolution of the flame surface can be represented by $\mathbf{r}(\xi,\eta,t)$, where ξ and η are two orthonormal curvilinear coordinates on the surface. The elemental arcs, *ds* in ξ and η directions, can be represented by their projections on the tangential plane at $\mathbf{r}(\xi,\eta,t)$. Hence, one can write:

$$ds_\xi \hat{\mathbf{e}}_\xi = \frac{\partial \mathbf{r}}{\partial \xi} d\xi$$

$$ds_\eta \hat{\mathbf{e}}_\eta = \frac{\partial \mathbf{r}}{\partial \eta} d\eta$$

Hence, the area of the infinitesimal surface, approximated as the area of the parallelogram on the tangential plane, is:

$$A(t) = \left[ds_\xi ds_\eta \hat{\mathbf{e}}_\xi \times \hat{\mathbf{e}}_\eta \right] \cdot \hat{\mathbf{n}} = \left(\frac{\partial \mathbf{r}}{\partial \xi} \times \frac{\partial \mathbf{r}}{\partial \eta} \right) \cdot \hat{\mathbf{n}} d\xi d\eta \tag{6.53}$$

At time $t + \Delta t$, the flame surface can be represented as:

$$\mathbf{r}(\xi,\eta,t+\Delta t) = \mathbf{r}(\xi,\eta,t) + \mathbf{V}_{flame}(\xi,\eta,t)\Delta t$$

The sides of the new parallelogram are:

$$ds'_\xi \hat{\mathbf{e}}_\xi = \frac{\partial \mathbf{r}}{\partial \xi} d\xi + \frac{\partial \mathbf{V}_{flame}}{\partial \xi} d\xi \Delta t$$

$$ds'_\eta \hat{\mathbf{e}}_\eta = \frac{\partial \mathbf{r}}{\partial \eta} d\eta + \frac{\partial \mathbf{V}_{flame}}{\partial \eta} d\eta \Delta t$$

Hence the new area is given by:

$$A(t+\Delta t) = \left[ds'_\xi \, ds'_\eta \, \hat{\mathbf{e}}_\xi \times \hat{\mathbf{e}}_\eta \right] \cdot \hat{\mathbf{n}} = \left[\left(\frac{\partial \mathbf{r}}{\partial \xi} + \frac{\partial \mathbf{V}_{flame}}{\partial \xi} \Delta t \right) \times \left(\frac{\partial \mathbf{r}}{\partial \eta} + \frac{\partial \mathbf{V}_{flame}}{\partial \eta} \Delta t \right) \right] \cdot \hat{\mathbf{n}} d\xi d\eta \tag{6.54}$$

The flame stretch is defined as the fractional rate of change of an area element on the flame surface, that is:

$$\kappa=\frac{1}{A}\frac{dA}{dt}=\frac{1}{A}\lim_{\Delta t\to 0}\frac{A(t+\Delta t)-A(t)}{\Delta t}=\left[ds'_{\xi}ds'_{\eta}\,\hat{\mathbf{e}}_{\xi}\times\hat{\mathbf{e}}_{\eta}\right]\cdot\hat{\mathbf{n}}$$

$$=\frac{\left(\dfrac{\partial\mathbf{V}_{flame}}{\partial\xi}\times\dfrac{\partial\mathbf{r}}{\partial\eta}+\dfrac{\partial\mathbf{r}}{\partial\xi}\times\dfrac{\partial\mathbf{V}_{flame}}{\partial\eta}\right)\cdot\hat{\mathbf{n}}}{\left(\dfrac{\partial\mathbf{r}}{\partial\xi}\times\dfrac{\partial\mathbf{r}}{\partial\eta}\right)\cdot\hat{\mathbf{n}}} \tag{6.55}$$

Resolving $\mathbf{V}_{flame}$ into tangential and normal components as $\mathbf{V}_{flame}=\mathbf{V}_{flame,t}+\left(\mathbf{V}_{flame}\cdot\hat{\mathbf{n}}\right)\hat{\mathbf{n}}$, one can write the following using the vector identity $\left(\hat{\mathbf{n}}\times\frac{\partial\mathbf{r}}{\partial\xi}\right)\cdot\hat{\mathbf{n}}=0$:

$$\left(\frac{\partial\mathbf{V}_{flame}}{\partial\xi}\times\frac{\partial\mathbf{r}}{\partial\eta}\right)\cdot\hat{\mathbf{n}}=\left[\frac{\partial\mathbf{V}_{flame,t}}{\partial\xi}\times\frac{\partial\mathbf{r}}{\partial\eta}+\frac{\partial}{\partial\xi}\left\{\left(\mathbf{V}_{flame}\cdot\hat{\mathbf{n}}\right)\hat{\mathbf{n}}\right\}\times\frac{\partial\mathbf{r}}{\partial\eta}\right]\cdot\hat{\mathbf{n}}$$

$$=\left[\frac{\partial\mathbf{V}_{flame,t}}{\partial\xi}\times\frac{\partial\mathbf{r}}{\partial\eta}+\left(\mathbf{V}_{flame}\cdot\hat{\mathbf{n}}\right)\frac{\partial\hat{\mathbf{n}}}{\partial\xi}\times\frac{\partial\mathbf{r}}{\partial\eta}\right]\cdot\hat{\mathbf{n}}$$

Similarly:

$$\left(\frac{\partial\mathbf{r}}{\partial\xi}\times\frac{\partial\mathbf{V}_{flame}}{\partial\eta}\right)\cdot\hat{\mathbf{n}}=\left[\frac{\partial\mathbf{r}}{\partial\xi}\times\frac{\partial\mathbf{V}_{flame,t}}{\partial\eta}+\frac{\partial\mathbf{r}}{\partial\xi}\times\frac{\partial}{\partial\eta}\left\{\left(\mathbf{V}_{flame}\cdot\hat{\mathbf{n}}\right)\hat{\mathbf{n}}\right\}\right]\cdot\hat{\mathbf{n}}$$

$$=\left[\frac{\partial\mathbf{r}}{\partial\xi}\times\frac{\partial\mathbf{V}_{flame,t}}{\partial\eta}+\left(\mathbf{V}_{flame}\cdot\hat{\mathbf{n}}\right)\frac{\partial\mathbf{r}}{\partial\xi}\times\frac{\partial\hat{\mathbf{n}}}{\partial\eta}\right]\cdot\hat{\mathbf{n}}$$

The numerator of Equation (6.55) becomes:

$$\left[\left(\frac{\partial\mathbf{V}_{flame,t}}{\partial\xi}\times\frac{\partial\mathbf{r}}{\partial\eta}+\frac{\partial\mathbf{r}}{\partial\xi}\times\frac{\partial\mathbf{V}_{flame,t}}{\partial\eta}\right)+\left(\mathbf{V}_{flame}\cdot\hat{\mathbf{n}}\right)\left(\frac{\partial\hat{\mathbf{n}}}{\partial\xi}\times\frac{\partial\mathbf{r}}{\partial\eta}+\frac{\partial\mathbf{r}}{\partial\xi}\times\frac{\partial\hat{\mathbf{n}}}{\partial\eta}\right)\right]\cdot\hat{\mathbf{n}}$$

For the orthogonal coordinate system:

$$\frac{\partial\mathbf{r}}{\partial\xi}=\hat{\mathbf{e}}_{\xi};\quad\frac{\partial\mathbf{r}}{\partial\eta}=\hat{\mathbf{e}}_{\eta};\quad\hat{\mathbf{e}}_{\xi}\times\hat{\mathbf{e}}_{\eta}=\hat{\mathbf{n}}$$

Using the cyclic triple product rule of vectors $(\mathbf{b}\times\mathbf{c})\cdot\mathbf{a}=(\mathbf{c}\times\mathbf{a})\cdot\mathbf{b}=(\mathbf{a}\times\mathbf{b})\cdot\mathbf{c}$, one can write:

$$\left(\frac{\partial \mathbf{V}_{flame,t}}{\partial \xi}\times\frac{\partial \mathbf{r}}{\partial \eta}\right)\cdot\hat{\mathbf{n}}=\left(\frac{\partial \mathbf{V}_{flame,t}}{\partial \xi}\times\hat{\mathbf{e}}_\eta\right)\cdot\hat{\mathbf{n}}=\frac{\partial \mathbf{V}_{flame,t}}{\partial \xi}\cdot\left(\hat{\mathbf{e}}_\eta\times\hat{\mathbf{n}}\right)$$

$$=\frac{\partial \mathbf{V}_{flame,t}}{\partial \xi}\cdot\left(\hat{\mathbf{e}}_\eta\times\hat{\mathbf{n}}\right)=\frac{\partial \mathbf{V}_{flame,t}}{\partial \xi}\cdot\hat{\mathbf{e}}_\xi \tag{6.56}$$

With similar manipulations for other terms, the numerator in the expression for the flame stretch becomes:

$$\left(\frac{\partial \mathbf{V}_{flame,t}}{\partial \xi}\cdot\hat{\mathbf{e}}_\xi+\frac{\partial \mathbf{V}_{flame,t}}{\partial \eta}\cdot\hat{\mathbf{e}}_\eta\right)+\left(\mathbf{V}_{flame}\cdot\hat{\mathbf{n}}\right)\left(\frac{\partial \hat{\mathbf{n}}}{\partial \xi}\cdot\hat{\mathbf{e}}_\xi+\frac{\partial \hat{\mathbf{n}}}{\partial \eta}\cdot\hat{\mathbf{e}}_\eta\right)$$

The denominator in the expression for flame stretch becomes:

$$\left(\frac{\partial \mathbf{r}}{\partial \xi}\times\frac{\partial \mathbf{r}}{\partial \eta}\right)\cdot\hat{\mathbf{n}}=\left(\hat{\mathbf{e}}_\xi\times\hat{\mathbf{e}}_\eta\right)\cdot\hat{\mathbf{n}}=1$$

Note that the operator $\frac{\partial}{\partial \xi}\cdot\hat{\mathbf{e}}_\xi+\frac{\partial}{\partial \eta}\cdot\hat{\mathbf{e}}_\eta$ represents the gradient operator in the plane of the flame and can denoted as the tangential gradient operator over the flame surface ∇_t. Thus, the expression for flame stretch becomes:

$$\kappa=\left(\frac{\partial \mathbf{V}_{flame,t}}{\partial \xi}\cdot\hat{\mathbf{e}}_\xi+\frac{\partial \mathbf{V}_{flame,t}}{\partial \eta}\cdot\hat{\mathbf{e}}_\eta\right)+\left(\mathbf{V}_{flame}\cdot\hat{\mathbf{n}}\right)\left(\frac{\partial \hat{\mathbf{n}}}{\partial \xi}\cdot\hat{\mathbf{e}}_\xi+\frac{\partial \hat{\mathbf{n}}}{\partial \eta}\cdot\hat{\mathbf{e}}_\eta\right)$$

$$=\left(\nabla_t\cdot\mathbf{V}_{flame,t}\right)+\left(\mathbf{V}_{flame}\cdot\hat{\mathbf{n}}\right)\left(\nabla_t\cdot\hat{\mathbf{n}}\right) \tag{6.57}$$

It is reasonable to assume that the tangential component of flame velocity is equal to the tangential component of the local flow velocity; that is, $\mathbf{V}_{flame,t}=\mathbf{V}_{flow,t}$. Denoting the component of the gradient operator normal to the flame surface as ∇_n and noting that $\nabla_n\cdot\mathbf{V}_{flame,t}=0$ and also that $\nabla_n\cdot\hat{\mathbf{n}}=0$, one can express flame stretch as:

$$\kappa=\left(\nabla\cdot\mathbf{V}_{flow,t}\right)+\left(\mathbf{V}_{flame}\cdot\hat{\mathbf{n}}\right)\left(\nabla\cdot\hat{\mathbf{n}}\right)$$

The flow velocity can be expressed as $\mathbf{V}_{flow}=\mathbf{V}_{flow,t}+\left(\mathbf{V}_{flow}\cdot\hat{\mathbf{n}}\right)\hat{\mathbf{n}}$. Substituting this in the expression for flame stretch, one can calculate flame stretch as follows:

$$\kappa=\left(\nabla\cdot\mathbf{V}_{flow}\right)-\nabla\cdot\left[\left(\mathbf{V}_{flow}\cdot\hat{\mathbf{n}}\right)\hat{\mathbf{n}}\right]+\left(\mathbf{V}_{flame}\cdot\hat{\mathbf{n}}\right)\left(\nabla\cdot\hat{\mathbf{n}}\right) \tag{6.58}$$

The above expression can be further modified using the following vector and tensor identities:

$$\mathbf{a}(\mathbf{b}\cdot\mathbf{c})=\mathbf{ab}\cdot\mathbf{c}$$

$$\underline{\underline{\mathbf{A}}}:\nabla\mathbf{v}=\nabla\cdot(\underline{\underline{\mathbf{A}}}\cdot\mathbf{v})-\mathbf{v}\cdot(\nabla\cdot\underline{\underline{\mathbf{A}}})$$

$$\nabla\cdot(\mathbf{uv})=\mathbf{u}\cdot\nabla\mathbf{v}+\mathbf{v}\cdot\nabla\mathbf{u}$$

Using the above identities, one can write:

$$\nabla\cdot\left[(\mathbf{V}_{flow}\cdot\hat{\mathbf{n}})\hat{\mathbf{n}}\right]=\nabla\cdot(\hat{\mathbf{n}}\hat{\mathbf{n}}\cdot\mathbf{V}_{flow})=\hat{\mathbf{n}}\hat{\mathbf{n}}:\nabla\mathbf{V}_{flow}+\mathbf{V}_{flow}\cdot\left[\nabla\cdot(\hat{\mathbf{n}}\hat{\mathbf{n}})\right]$$

$$=\hat{\mathbf{n}}\hat{\mathbf{n}}:\nabla\mathbf{V}_{flow}+(\mathbf{V}_{flow}\cdot\hat{\mathbf{n}})(\nabla\cdot\hat{\mathbf{n}})+\mathbf{V}_{flow}\cdot(\hat{\mathbf{n}}\cdot\nabla\hat{\mathbf{n}})$$

Noting that $\hat{\mathbf{n}}\cdot\nabla\hat{\mathbf{n}}=0$, the expression for stretch becomes:

$$\kappa=(\nabla\cdot\mathbf{V}_{flow})-\hat{\mathbf{n}}\hat{\mathbf{n}}:\nabla\mathbf{V}_{flow}+\left[(\mathbf{V}_{flame}-\mathbf{V}_{flow})\cdot\hat{\mathbf{n}}\right](\nabla\cdot\hat{\mathbf{n}})$$

Using the expression for flame displacement speed, the final expression for stretch becomes:

$$\kappa=(\nabla\cdot\mathbf{V}_{flow})-\hat{\mathbf{n}}\hat{\mathbf{n}}:\nabla\mathbf{V}_{flow}+S_u(\nabla\cdot\hat{\mathbf{n}}) \tag{6.59}$$

The first two terms on the RHS represent the effect of tangential strain rate (i.e., rate of variation of flow velocity parallel to the flame surface (often referred to as hydrodynamic strain rate); the last term represents the effect of curvature.

6.10.1 Identification of Flame Surface

The analysis of flame stretch discussed above requires identification of flame surface. In the hydrodynamic limit, the flame is identified as a surface and its evolution is described by the G equation [8,9]. The geometry of the flame surface is described in terms of a scalar function as:

$$G(\mathbf{r},t)=0$$

Any iso-surface of the scalar G would suffice as the definition of the flame surface, but the choice of $G(\mathbf{r},t)=0$ is convenient as the scalar function changes sign in moving from the unburned to the burned gas zone. On the flame surface one can write:

$$\frac{dG}{dt}=\frac{\partial G}{\partial t}+\mathbf{V}_{flame}\cdot\nabla G=0 \tag{6.60}$$

Here $\mathbf{V}_{flame} = \frac{dr}{dt}$ is the flame propagation velocity. Using the definition of flame speed, this can be expressed as:

$$\frac{dG}{dt} = \frac{\partial G}{\partial t} + \mathbf{V}_{flow}\Big|_{G=0^-} \cdot \nabla G = S_u \hat{\mathbf{n}} \cdot \nabla G \tag{6.61}$$

Using the definition of unit normal $\hat{\mathbf{n}} = -\nabla G / |\nabla G|$, the above equation takes the form:

$$\frac{\partial G}{\partial t} + \mathbf{V}_{flow}\Big|_{G=0^-} \cdot \nabla G = S_u \frac{\nabla G}{|\nabla G|} \cdot \nabla G = S_u |\nabla G| \tag{6.62}$$

The above equation shows the coupling between the flame front and the flow through the term $\mathbf{V}_{flow}$.

However, to include the flame stretch in a more complete analysis, where the diffusion and reaction phenomena are also included, stretch is computed on an iso-surface of a transported scalar like fuel mass fraction. Denoting the transported scalar as ϕ, the equation for the iso-surface becomes:

$$\frac{d\phi}{dt} = \frac{\partial \phi}{\partial t} + \mathbf{V}_{flame} \cdot \nabla \phi = \frac{\partial \phi}{\partial t} + \mathbf{V}_{flow} \cdot \nabla \phi - S_u |\nabla \phi| = 0 \tag{6.63}$$

Thus, the flame speed can also be expressed as:

$$S_u = \frac{1}{|\nabla \phi|}\left[\frac{\partial \phi}{\partial t} + \mathbf{V}_{flow} \cdot \nabla \phi\right] \tag{6.64}$$

Using the general convective-diffusive transport equation, $\frac{\partial \phi}{\partial t} + \mathbf{V}_{flow} \cdot \nabla \phi = \nabla \cdot \left[\Gamma_\phi \nabla \phi\right] + \omega_\phi$, where Γ_ϕ and ω_ϕ denote the diffusion coefficient and the source term, respectively, an alternate expression for flame speed can be given as:

$$S_u = \frac{1}{|\nabla \phi|}\left[\nabla \cdot \left(\Gamma_\phi \nabla \phi\right) + \omega_\phi\right] \tag{6.65}$$

The general practice is to choose the surface location in an arbitrary but consistent manner within the flame zone as revealed in the transport or reaction scale. This introduces some arbitrariness, however, in the definition of flame speed. The definition of flame speed becomes precise and unique only in the hydrodynamic limit when the whole flame shrinks to an infinitesimal surface [10].

The importance of flame stretch primarily stems from the dependence of flame speed on flame stretch. For weakly stretched flames, linear theory [11,12] suggests the following relationship:

$$S_u = S_u^0 + \kappa L$$

The coefficient L is known as Markstein length. This parameter depends on thermal expansion and the Lewis number. In particular, at a critical Lewis number near unity, it changes sign. As is obvious from the above equation, flame speed of a positively stretched flame would be higher (lower) than that of an unstretched flame for positive (negative) values of Markstein length. At higher values of stretch, nonlinear effects become important, leading to more complex dependence of flame speed on flame stretch.

EXERCISES

6.1 A premixed laminar flame is stabilised in a shear layer where the horizontal velocity of the reactant mixture varies linearly from 900 m/s to 1400 m/s over a distance of 30 mm. If the flame speed for the mixture is 0.36 m/s, determine the flame shape.

6.2 A cylindrical tube of diameter D is filled with fuel-air mixture. The temperature gradient near the wall can be calculated by assuming linear increase from wall temperature to burned gas temperature over a distance D/b, where b is a constant. Derive an expression for the largest diameter of the tube that can prevent flashback in terms of planar flame speed, gas properties and b. Assume the flame to remain planar.

6.3 A cylindrical tube of diameter D and length L, closed at both ends, is filled with a stoichiometric mixture of methane and air. If the mixture is ignited at one end, derive an expression for the rate of pressure rise inside the tube as a function of the flame position. Assume that both burned and unburned gases undergo isentropic process and the mixture undergoes complete reaction without dissociation.

6.4 A laboratory experiment has to be designed in which stoichiometric methane-air mixture at 298 K and 1 bar has to be ignited by means of an electric spark. Calculate the minimum spark energy needed to ignite the mixture. Take thermophysical properties of the mixture equal to that of air at 298 K, 1 bar.

6.5 An annular space formed between two cylindrical surfaces of diameters D_i and D_o ($D_o > D_i$) is filled with fuel-air mixture. The inner surface of the annular space is insulated, and the temperature gradient near the outer wall can be calculated by assuming linear increase from wall temperature to burned gas temperature over a distance

D_o/b, where b is a constant. Derive an expression for the largest value of the ratio of the outer and inner diameters that can prevent flashback in terms of planar flame speed, gas properties, D_i and b. Assume the flame to remain planar. How would be the expression change for very high values of D_o and D_i such that $(D_o - D_i) << D_i$?

REFERENCES

1. Botha, J. P. and Spalding, D. B., The laminar flame speed of propane-air mixtures with heat extraction from the flame, *Proceedings of the Royal Society London* A, **225**, 71–96, 1954.
2. Mallard, E. and Le Chatelier, H. L., Thermal model for flame propagation, *Annals of Mines*, **4**, 379, 1883.
3. Spalding, D. B., A theory of inflammability limits and flame-quenching, *Proceedings of the Royal Society London A*, **240**, 83–100, 1957.
4. Turns, S. R. *An Introduction to Combustion*, New York: McGraw-Hill, 1996.
5. Leach, T. T., Cadou, C. P. and Jackson, G. S., Effect of structural conduction and heat loss on combustion in micro-channels, *Combustion Theory and Modelling*, **10**, 85–103, 2006.
6. Lieuwen, T., McDonnell, V., Santavicca, D. A. and Sattelmayer, T., Burner development and operability issues associated with steady flowing syngas fired combustors, *Combustion Science and Technology*, **180**, 1169–1192, 2008.
7. Law, C. K. *Combustion Physics*. Cambridge, CA: Cambridge University Press, 2010.
8. Kerstein A. R., Ashurst, W. T. and Williams, F. A., Field equation for interface propagation in an unsteady homogeneous field, *Physical Review A*, **37**, 2728–2731, 1988.
9. Peters, N., *Turbulent Combustion*, Cambridge, UK: Cambridge University Press, 2000.
10. Tien, J. H. and Matalon, M., On the burning velocity of stretched flames, *Combustion and Flame*, **84**, 238–248, 1991.
11. Matalon, M. and Matkowsky, B. J., Flames as gas dynamic discontinuities, *Journal of Fluid Mechanics*, **124**, 239, 1982.
12. Matalon, M. and Matkowsky, B. J., Flames in fluids: Their interaction and stability, *Combustion Science and Technology*, **34**, 295, 1983.

Chapter 7

Laminar Non-Premixed Flames

7.1 PHYSICAL DESCRIPTION

In the case of a premixed flame, we saw that the fuel and oxidiser were mixed before entering into the combustion zone/combustor. In another type of flame, fuel and oxidiser are fed to the combustor separately. Mixing of fuel and air occurs in the combustor. This type of flame is known as a non-premixed flame. As the mixing process is governed by diffusion of fuel and air, it is called a diffusion flame.

Diffusion flames can be achieved easily. As fuel and oxidiser come as separate streams, the safety of these flames is high. The industrial combustor flames are generally not a premixed one. There we may found two types of flames: diffusion and partially premixed. In the latter case, fuel is mixed with part of the oxidiser before reaching the combustion area. The rest of the oxidiser is supplied during combustion. So, it is neither a premixed nor a non-premixed one.

A very simple example of a non-premixed flame is jet flame. Here, the burner consists of two concentric tubes, with a vertical axis. Fuel comes out of the central tube while the oxidiser/air flows through the outer annulus. A non-premixed jet flame is shown in Figure 7.1. This type of flames is commonly known as co-flow flame. From the photograph, we see that initially the flame is diverging a little bit and then gradually converging on the centreline at some height.

To understand this shape, we need to think of a jet coming out of a vertical nozzle. We consider a non-reacting jet, as shown in Figure 7.2, where fuel is coming out of a nozzle of $2R$ diameter with an axial velocity V_e. After coming out of the nozzle, the flow decelerates and diverges. The initial jet momentum must be conserved in the flow field. As the jet moves into the surroundings, some of the momentum is transferred to air, decreasing the velocity of the jet. Along the jet-increasing quantities of air are entrained into the jet as it proceeds downstream. The edge of the fuel jet is shown in the figure, which is gradually increasing in diameter. At the nozzle mouth, we can find a conical zone. Within this zone, the jet velocity does not change from its original value. So, within this zone, velocity is constant at a value V_e. This zone is known as the potential core.

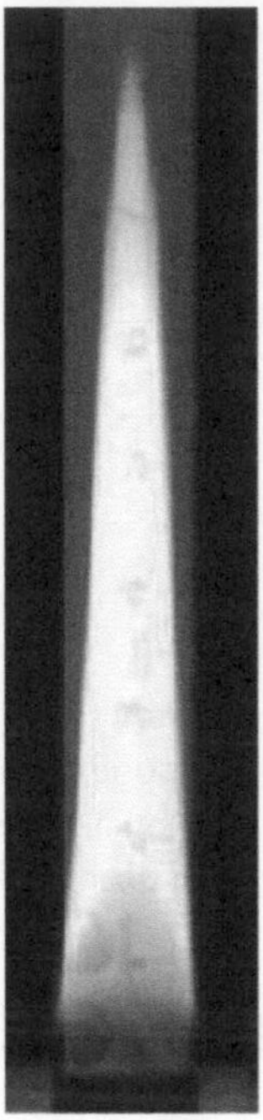

Figure 7.1 A non-premixed co-flow flame.

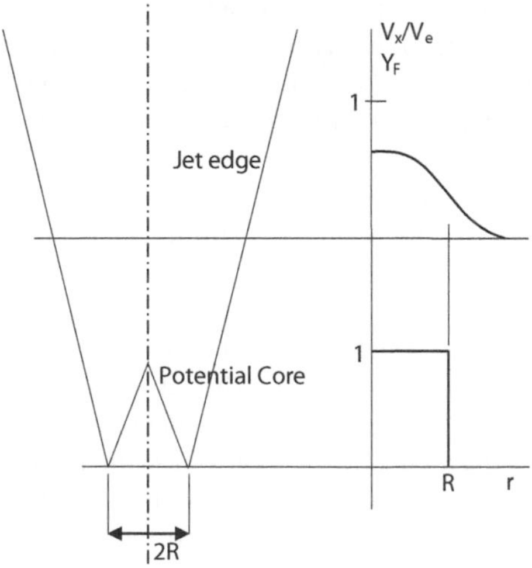

Figure 7.2 Non-reacting fuel jet.

In Figure 7.2, we can take two horizontal planes: one at the nozzle mouth and the other at a height as shown. At the nozzle mouth, there is a top hat velocity distribution. If the axial velocity is scaled in terms of exit velocity, we get a constant velocity of unity at all radius. If we plot the fuel mass fraction, the distribution will be similar to that of the velocity. In the region between the potential core and the jet edge, both the velocity and fuel concentration decrease monotonically to zero at the jet edge. Beyond the potential core, the viscous shear and diffusion effects are active across the whole field of the jet. Fuel molecules diffuse radially outward according to Fick's law. The effect of moving downstream is to increase time available for diffusion. At a higher location, the velocity distribution is spread up to the jet edge. So, as we are moving higher, the velocity as well as fuel mass fraction are reducing and spreading up to a larger radius. The locus of this point is the edge of the jet. Details of non-reacting jets may be found in [1,2].

From the above discussion, we can understand the initial diverging nature of the flame. Subsequently, as we move upward, the density of gases decreases due to high temperature. The buoyancy force increases and gas velocity also increases. As the flow is now accelerating, the jet now converges to give a tulip-like closed shape to the flame.

However, a flame may not take a closed tulip shape, depending on the supply of fuel and air. When the air supply is more than the stoichiometric requirement, the flame is called over-ventilated. In that case, the flame takes a closed shape. Figure 7.3 shows the flames established on a co-flow burner. The burner consists of two concentric tubes. Fuel flows through the inner one and outer annular area is supplying air. For over-ventilated flame, the flame is closed and of larger height (H_F). If a fuel jet is released in a quiescent atmosphere, the flame becomes an over-ventilated flame through a sufficient oxygen supply from the atmosphere and takes a closed shape. When the air supply is such that fuel does not get oxygen in stoichiometric proportion, it is called an under-ventilated flame. Here, the fuel diffuses more outward in search of oxygen and the flame takes an open shape with a smaller flame height.

The concentric burner shown in Figure 7.3 establishes an axisymmetric flame. This type of flame is less suitable for any property measurement, particularly in the case of optical or non-intrusive diagnosis. This may be eased out in case of slot burners, where there are three parallel slots for fuel and air flow [3]. The fuel may come out from the middle slot, and the slots, one on either side, are for air flow, as shown in Figure 7.4.

Fuel and air are mixed above the burner. The fuel-air ratio varies at different locations due to varied diffusion. We can assume that the flame has a luminous thin surface, where the fuel-air ratio is stoichiometric. We can consider the equivalence ratio (Φ) distribution in the whole domain, and the flame is an iso-Φ surface of value $\Phi = 1$.

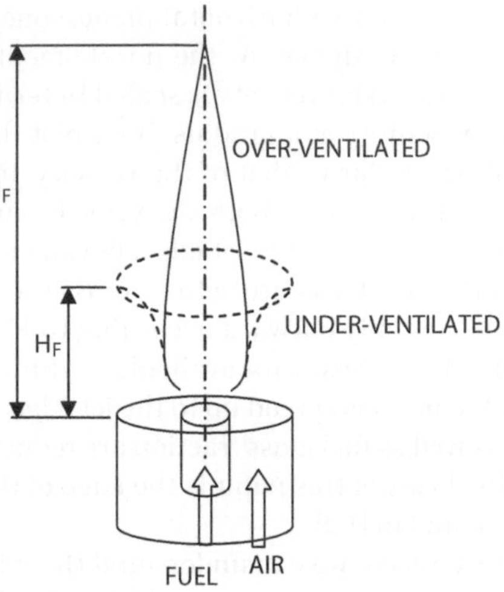

Figure 7.3 Over-ventilated and under-ventilated flames.

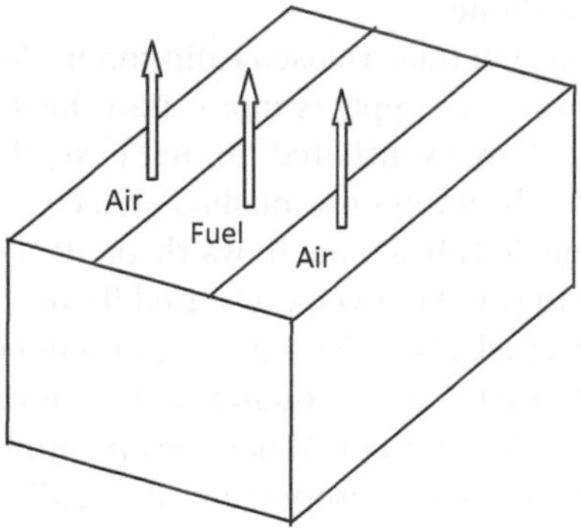

Figure 7.4 Schematic of a slot burner.

In a co-flow jet flame, the fuel and the oxidiser are coming out from the burner in the axial direction, which is the main flow direction. However, the molecular diffusion different species are occurring in the direction normal to the main flow. The fuel is diffusing radially outward, while oxygen and nitrogen (components of air) are diffusing in radially. There is a bulk motion of the species in a direction normal to main flow. This diffusion is responsible for the mixing of fuel and oxidiser, generating a distribution of equivalence ratio in the flow field.

On the flame, fuel and oxygen should be completely consumed. So, their mass fractions must be zero on flame surface, as shown in Figure 7.5.

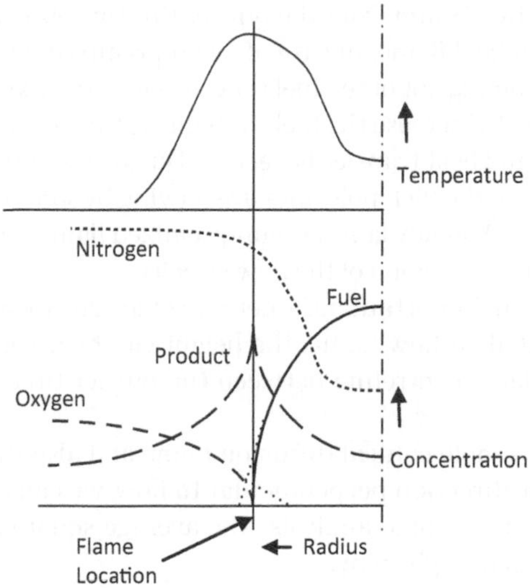

Figure 7.5 Temperature and species distribution in a co-flow flame.

Fuel concentration increases within the flame and maximises at flame centreline. Oxygen concentration increases outside the flame. Product formation will be highest at the flame. Product concentration decreases on either side of the flame as product diffusion takes place. As reaction and heat generation occur at flame surface, temperature is highest at that location. We can think of this flame as a closed surface on which combustion/reaction occurs. No reaction takes place on either side of the flame.

The flame is not actually a thin surface. A thin flame is the result of infinite reaction rate assumption. However, relaxation of infinite reaction rate makes the flame thicker. In case of a thick flame, the fuel and oxygen concentration do not come to zero at the same point. The fuel and oxygen concentrations, in the case of a thick flame, are shown with dotted lines, which shows fuel and oxygen concentration drops to zero at the outer and inner edges of the flame, respectively.

In the case of a thin flame, the diffusion rates of fuel and oxygen towards each other are in stoichiometric proportion. This can be inferred as fuel and oxygen are reacting at stoichiometric proportion and concentrations of both species are sinking to zero on the flame surface. The flame temperature should be very close to stoichiometric adiabatic flame temperature. If finite reaction rate is considered, the flame temperature is about 90% of the stoichiometric adiabatic flame temperature [4,5].

A close observation of Figure 7.1 reveals that the flame formed is not attached to the tip of the burner. A scale image is superimposed with the flame photograph in the figure to measure the flame height. From the scale, we can see that the flame is lifted by 4/5 mm from the burner tip. This observation differs from the same for a premixed flame. In case of a non-premixed flame, the fuel is mixing with air after coming out of the fuel tube. Along with mixing, associated heat transfer also occurs to increase the fuel/air temperature. Before reaction of a fuel molecule, mixing and heat transfer is needed. During the time required for mixing and heat transfer, the fuel molecules are moving by some distance before the start of the reaction. A steady laminar non-premixed flame is not attached to the burner tip. This lift is a function of the flow velocity.

Flame height is an important parameter to characterise a flame. In the case of an over-ventilated co-flow flame, the height can be ascertained as the distance along the flame centreline between the burner tip and the location of $\Phi = 1$.

If we assume a simple conical diffusion flame and also that the diffusion is present only in the direction perpendicular to flow, we can estimate the flame height through a very simple analysis. The average square displacement of a molecule (say, oxygen) is given by:

$$\bar{y}^2 = 2Dt \tag{7.1}$$

where, $\bar{y}$ is the average displacement, D is the diffusion coefficient and t is the required time. At flame height, the oxygen molecules are diffused up to the centreline. The displacement can be taken as the fuel nozzle radius, R. The time required for diffusion can be written as:

$$t = R^2 / 2D \tag{7.2}$$

The time can also be equated to:

$$t = L_F / v_e \tag{7.3}$$

where L_F is the flame height and v_e is the average velocity of fuel at nozzle exit. We can write the volume flow rate, Q, as:

$$Q = v_e \pi R^2 \tag{7.4}$$

and the flame height can be expressed as:

$$L_F = \frac{v_e R^2}{2D} = Q / \pi D \tag{7.5}$$

Although very crude, this approximate analysis permits us to predict flame height. The flame height is independent of the burner diameter and is proportional to the volume flow rate of fuel.

Non-premixed flames usually look yellow-orange. This appearance is quite different from the premixed flame where the colour was mostly violet-blue. The reason behind the colour of the non-premixed flame is radiation from soot particles within the flame. In non-premixed flame, soot particles are formed within the reaction zone. The particles flow through the reaction zone and burn within it. The soot particles at flame temperature emit a broadband emission starting from yellow to infrared. Thus the colour of the flame appears. Sometimes the burning soot particles escape from the reaction zone. Burning of these particles happens in atmosphere, taking oxygen from the surroundings. Those particles extend the visible flame height from $\Phi = 1$ contour. Note that all escaping particles do not burn. Some of them escape as particulate matter. This results in black smoke escaping from the flame. In such cases, the top of the flame appears to open near the tip location. A close observation of Figure 7.1 shows this type of opening near the flame tip, which is called soot wing of the flame.

The non-premixed flame shows a regular flickering with a particular frequency. This flame flicker comes from instability in the shear flow layers. When the two layers are shearing over each other, vortices are formed and impart a flicker on the flow. Details of the phenomenon have been studied by a number of researchers, including Sahu et al. [6]. This flicker frequency is of the order of 10–15 Hz.

When the co-flow jet flame is studied under zero-gravity situation, we find that the flame shape becomes nearly spherical. The tulip structure of the jet flame has already been discussed here, and it is understood that the shape is due to high buoyant force. However, in zero gravity, buoyancy is absent. Flow is not accelerated at the top portion of the flame. The fuel species diffuses outwards to give a wider shape of the flame. At the later part, dilatation of gas induces an inward motion. This phenomenon is pulling the flame again and the flame takes a near-spherical shape. It is also observed that at micro-gravity or zero gravity, the flame takes a blue colour in comparison to the orange-yellowish colour of the normal gravity flames.

A vast use of non-premixed flame is wick flame. The fuel for this type of flame may be solid, like a candle, or liquid, like a kerosene lamp or kerosene stove. In a candle, the fuel is solid paraffin. The top portion of the solid paraffin melts by taking heat from the flame. The liquefied fuel is then moving upwards through wick due to capillary action. Finally, it vaporises and burns in the flame, with the help of atmospheric oxygen. For liquid fuel, the liquefaction process is not needed.

Counter-flow flame is another type of non-premixed flame. Here, fuel and oxidiser flow in opposite directions. Those streams are interacting to form a planar flame. We shall discuss this flame in more detail later in this chapter.

7.2 SIMPLIFIED ANALYSIS OF DIFFUSION CONTROLLED SYSTEMS

Detailed analysis of a non-premixed flame is a complex process. But we can predict the shape and height of the flame on the basis of some simplified assumptions.

Burke and Schumann [7] first used a simplified model to assess flame shape. The model included axial convection and radial diffusion effects while ignoring axial diffusion and radial convection effects, along with many other assumptions. Using this model, flame length was predicted reasonably well for axisymmetric burners. Roper [8–10] assumed temperature and velocity variation in the axial direction. But the quantities are constant in the cross-sectional plane. Thus, one may account for buoyancy. This model also does not include axial diffusion like that of Burke and Schumann. Spalding [11] used the jet velocity profile, obtained from Schlichting [2] and assumed equal rates of diffusion of species and momentum. Axial diffusion was neglected in this model also. Buoyancy was not included in this model. As these models did not consider axial diffusion, those are not applicable to flames where axial diffusion effects may be important, like micro-scale flames. Chung and Law [12] extended the Burke-Schumann model for flame shape to include axial diffusion effects and presented the results for the slot burner geometry. However, the axial velocity is assumed constant in both these models. The summary of assumptions for these models is shown in Table 7.1. There are many more researchers who addressed this issue. A discussion may be found in [13].

TABLE 7.1 SUMMARY OF ASSUMPTIONS FOR SIMPLIFIED MODELS

Author/ References	Axial Diffusion	Axial Convection	Radial Diffusion	Radial Convection	Buoyancy Effects
Burke-Schumann	Neglected	Included	Included	Neglected	No
Roper	Neglected	Included	Included	Included	Yes
Spalding	Neglected	Included	Included	Included	No
Chung and Law	Included	Included	Included	Neglected	No

Before going into detail about the above-mentioned results, let us see the mathematical description and realise the difficulty to obtain the solution. First, we can assume certain features to simplify the flame. But we must be careful so that the assumptions maintain the physics of the problem. The assumptions are as follows:

- The flow is laminar, steady, axisymmetric and coming out of a simple nozzle.
- Fuel, oxidiser and products are considered as three species.
- Fuel and oxidiser react in stoichiometric proportion only with an infinitely fast chemistry.
- Species transport is according to binary diffusion governed by Fick's law.
- Thermal energy and species diffusivities are the same, leading to unity Lewis number ($Le = \propto / D$).
- Radiation heat transfer and axial diffusion are neglected.

The conservation equations now can be simplified using the above assumptions. For the details of a flame, we require velocity, temperature, species distributions. To know these, we need to solve the conservation equations like continuity, momentum, energy, species and so on. We can solve continuity, axial momentum, energy and two species equations to know about axial and radial velocities, temperature, concentration of fuel and air. As pressure is not changing, solution of radial momentum is not necessary. However, when we solve the equations, we need to specify temperature and concentration boundary conditions on the flame. But the flame location is not known a priori. To resolve this issue, we need to think of some variables, known as conserved scalars, which are conserved in the flow field. The governing equation can be developed for the conserved scalar. For solving the governing equation, boundary conditions can be specified at nozzle exit plane, axis or far field, that is, at the physical boundaries of the problem domain.

7.3 CONSERVED SCALAR FORMULATION

As discussed in the previous section, defining conserved scalar is required. Conserved scalar is a scalar property, which is conserved throughout the flow field. Conserved scalars are neither generated nor destroyed within the flow field. Mass can be identified as one such scalar which cannot be generated or destroyed in the flow field. There is no source or sink for mass.

Previously we have derived a number of conservation equations for different variables, like mass, momentum, species, energy and so on. Remember that the conservation equations take a generic form containing mainly four terms: temporal, advection, diffusion and source. If we think of a scalar property φ, the conservation equation can be written as:

$$\rho\frac{\partial\varphi}{\partial t}+\rho\boldsymbol{V}\cdot\nabla\varphi=\nabla\cdot\left[\rho\delta(\nabla\varphi)\right]+S \tag{7.6}$$

where δ is the type of diffusivity associated with the scalar φ. Here, the two terms in LHS are temporal and advective, respectively. The first term in RHS is the diffusion term and the last one is the source term. For conserved scalars, there should not be any source term in the conservation equation; as an example, the mass conservation equation does not contain a non-zero source term. For conserved scalars, the conservation equation will look as follows:

$$\rho\frac{\partial\varphi}{\partial t}+\rho\boldsymbol{V}\cdot\nabla\varphi-\nabla\cdot\left[\rho\delta(\nabla\varphi)\right]=0 \tag{7.7}$$

Conceptually we can identify some scalars like mass, mass of carbon, mass of oxygen and so on. We can also derive properties (by suitable combination of elementary properties) that obey the above condition or follow the type of conservation equation.

7.3.1 Mixture Fraction

Mixture fraction is a conserved scalar that is very commonly used for solution of non-premixed flames. Let us restrict our discussion to a system of a single inlet stream of pure fuel, as in [1]. Similarly, we assume a single stream of pure oxidiser. We also take all the products of combustion as a single entity. The mixture fraction, Z, can be defined as the ratio of mass of material having its origin in the fuel stream to the mass of the mixture.

For the three species system, we can write:

$$1\,\text{kg of fuel}+\gamma\,\text{kg of oxidizer}\rightarrow(\gamma+1)\text{kg of products} \tag{7.8}$$

In a small volume, where the content is a mixture of the above three species, if we want to find the mixture fraction, it will have contributions from all three species:

$$Z=(1)(Y_F)+(0)(Y_O)+\left(\frac{1}{\gamma+1}\right)(Y_{Pr}) \tag{7.9}$$

where Y_F, Y_O and Y_{Pr} are the mass fractions of fuel, oxidiser and product, respectively. In our system, the mixture fraction in the stream of pure fuel is 1 and it is 0 in the pure oxidiser stream. So the contribution from the mass fractions of fuel and oxidiser to mixture fraction should be 1 and 0, respectively. Similarly, the fraction of fuel stuff mass per unit mass of products is $\left(\frac{1}{\gamma+1}\right)$. So that should be the contribution from products mass fraction. The mixture fraction can be determined as follows:

$$Z = Y_F + \left(\frac{1}{\gamma+1}\right) Y_{Pr} \tag{7.10}$$

Another generalised discussion and details about mixture fraction may be found in [14].

7.3.2 Conservation Equation for Mixture Fraction

We can write the species conservation equations of fuel and oxidiser in one dimension as follows:

$$\dot{m}'' \frac{dY_F}{dx} - \frac{d}{dx}\left(\rho D \frac{dY_F}{dx}\right) = \dot{m}_F''' \tag{7.11}$$

$$\dot{m}'' \frac{dY_{\mathrm{Pr}}}{dx} - \frac{d}{dx}\left(\rho D \frac{dY_{\mathrm{Pr}}}{dx}\right) = \dot{m}_{\mathrm{Pr}}''' \tag{7.12}$$

Now dividing Equation (7.12) by $(\gamma+1)$ and adding Equation (7.11), we can write:

$$\dot{m}'' \frac{d}{dx}\left(Y_F + \frac{Y_{\mathrm{Pr}}}{\gamma+1}\right) - \frac{d}{dx}\left[\rho D \frac{d}{dx}\left(Y_F + \frac{Y_{\mathrm{Pr}}}{\gamma+1}\right)\right] = \dot{m}_F''' + \dot{m}_{\mathrm{Pr}}''' / (\gamma+1) \tag{7.13}$$

Here, $\dot{m}''$ is the total mass flux (kg/m^{2}s); $\dot{m}_F'''$ and $\dot{m}_{\mathrm{Pr}}'''$ are the mass production rate per unit volume for fuel and product, respectively. One is positive and the other is negative here, as fuel is consumed during the reaction and product is generating. As reaction occurs at the stoichiometric proportion, it can be written as follows:

$$\dot{m}_F''' = -\dot{m}_{\mathrm{Pr}}''' / (\gamma+1) \tag{7.14}$$

Substitution of Equation (7.14) in Equation (7.13) results in:

$$\dot{m}'' \frac{dZ}{dx} - \frac{d}{dx}\left[\rho D \frac{dZ}{dx}\right] = 0 \tag{7.15}$$

Equation (7.15) can be called a mixture fraction conservation equation. The equation can be also developed in two or three dimensions. Here, we have developed a steady conservation equation. One can try to develop the unsteady conservation equation also.

Example 7.1

For a methane-air non-premixed flame, the mole fractions of CH_4, O_2, N_2, CO_2 and H_2O are known at a location. Express the mixture fraction at the said location in terms of the known mole fractions. Assume any other species are not present.

Solution

First, we assume that air is a mixture of oxygen and nitrogen only. Carbon and hydrogen (the fuel staff in this case) is coming from the fuel stream only. Now, let us write the expression of mixture fraction in terms of mass, considering our definition:

$$Z = \frac{[m_C + m_H]_{mix}}{m_{mix}} = Y_{CH_4}\frac{M_C}{M_{CH_4}} + Y_{CO_2}\frac{M_C}{M_{CO_2}} + Y_{H_2O}\frac{M_{H_2}}{M_{H_2O}} \tag{7.16}$$

Here, M is the molecular weight. The first two terms are contributing to the ratio of mass of carbon to the mass of the mixture. The second term is the same ratio for hydrogen. The mass fraction is related to the mole fraction as follows:

$$Y_i = X_i \frac{M_i}{M_{mixture}} \tag{7.17}$$

The mixture fraction can be written as:

$$Z = X_{CH_4}\frac{M_{CH_4}}{M_{mixture}}\frac{M_C}{M_{CH_4}} + X_{CO_2}\frac{M_{CO_2}}{M_{mixture}}\frac{M_C}{M_{CO_2}} + X_{H_2O}\frac{M_{H_2O}}{M_{mixture}}\frac{M_{H_2}}{M_{H_2O}} \tag{7.18}$$

where

$$M_{mixture} = \sum X_i M_i = X_{CH_4}M_{CH_4} + X_{CO_2}M_{CO_2} + X_{H_2O}M_{H_2O} + X_{O_2}M_{O_2} + X_{N_2}M_{N_2}. \tag{7.19}$$

Example 7.2

The mole fractions in a methane-air flame are given for some species. The mole fractions are $X_{CH4} = 2000$ ppm, $X_{O2} = 0.019$, $X_{CO2} = 0.099$ and $X_{H2O} = 0.15$. The balance of the mixture is nitrogen. Find out the mixture fraction and fuel-air ratio.

Solution

The mole fraction of nitrogen can be found as:

$$X_{N_2} = 1 - \left(X_{CH_4} + X_{O_2} + X_{CO_2} + X_{H_2O}\right)$$
$$= 1 - (0.000002 + 0.019 + 0.099 + 0.15) = 0.7319$$

The molecular weight of the mixture is:

$$M_{mixture} = \sum X_i M_i$$
$$= (0.000002)(16) + (0.019)(32) + (0.099)(44) + (0.15)(18) + (0.7319)(28)$$
$$= 28.16$$

The mixture fraction can be found using Equation (7.18):

$$Z = \frac{1}{M_{mixture}}\left[M_C\left(X_{CH_4} + X_{CO_2}\right) + M_{H_2}\left(X_{H_2O}\right)\right]$$
$$= \left(\frac{1}{28.16}\right)\left[12(0.000002 + 0.099) + 2(0.15)\right] = 0.053$$

As the mixture fraction is a kind of fuel mass fraction, the fuel air ratio can be written as:

$$\frac{F}{A} = \frac{Z}{1 - Z} = \frac{0.053}{1 - 0.053} = 0.056$$

Example 7.3

Calculate the stoichiometric mixture fraction for methane and acetylene, when burning in air.

Solution

The stoichiometric mixture fraction is the fuel mass fraction in a stoichiometric mixture. For methane, we can write the stoichiometric equation:

$$\mathrm{CH_4} + 2(\mathrm{O_2} + 3.76\mathrm{N_2}) \rightarrow 2\mathrm{H_2O} + \mathrm{CO_2} + (2 \times 3.76)\mathrm{N_2}$$

The fuel mass fraction can be calculated as:

$$Y_{F_{St}} = Z_{St} = \frac{M_{\mathrm{CH_4}}}{M_{\mathrm{CH_4}} + 2M_{\mathrm{O_2}} + (2 \times 3.76)M_{\mathrm{N_2}}}$$

$$= \frac{16}{16 + (2 \times 32) + (2 \times 3.76 \times 28)} = 0.055$$

The stoichiometric equation can be written for acetylene also:

$$C_2H_2 + 2.5(O_2 + 3.76N_2) \rightarrow H_2O + 2CO_2 + (2.5 \times 3.76)N_2$$

$$Z_{St} = \frac{M_{C_2H_2}}{M_{C_2H_2} + 2.5M_{O_2} + (2.5 \times 3.76)M_{N_2}} = \frac{26}{26 + (2.5 \times 32) + (2,5 \times 3.76 \times 28)} = 0.07$$

Example 7.4

A non-premixed flame has an ethylene fuel stream. The oxidiser stream is an equimolar mixture of oxygen and carbon dioxide. How can you find the mixture fraction?

Solution

Here, all the carbon atoms are not coming from the fuel. The hydrogen atoms are generated in the fuel stream only. The mixture fraction must be proportional to hydrogen mass fraction. So Z can be defined as:

$$Z = \frac{\text{mass of fuel}}{\text{mass of hydrogen}} \frac{\text{mass of hydrogen}}{\text{mass of mixture}} = \frac{28}{4 \times 1} Y_{H_2} = 7Y_{H_2}$$

Knowing the species values, we can find the hydrogen mass fraction first and then the mixture fraction.

7.4 SHVAB-ZELDOVICH FORMULATION

We can recall the energy equation from Chapter 5: Equation (5.56). For a reacting system, we can neglect the first three terms in RHS. If there is no appreciable variation in pressure, the first term can be neglected. We can assume a negligible viscous dissipation. Also, there is no volumetric energy generation. Here, it should be noted that the energy generation due to combustion (heat of combustion) is not included in this term. That contribution to energy equation will come separately through heat of formation of products. The last term $\boldsymbol{q}$ represents the energy transfer across the boundary. In Chapter 5, we considered conduction heat transfer in this category. Here, we need to add the energy transfer due to species diffusion.

First, to facilitate understanding, we consider a steady one-dimensional system in Cartesian coordinates. The equation can be written as:

$$\rho v_x \frac{dh}{dx} = -\frac{dq_x}{dx} \tag{7.20}$$

A modification of Equation (7.20) is possible through suitable assumptions so that species mass fluxes and enthalpies are eliminated from the equation and replaced with terms having temperature. This modification process of the energy equation is known as the Shvab-Zeldovich formulation. One of the important assumptions takes the Lewis number (Le) equal to unity, which means $k = \rho c_P D$. Another one tells the validity of Fick's law for estimating diffusion flux. Now, the heat transfer can be written as:

$$q_x = -k\frac{dT}{dx} + \sum \dot{m}''_{i\,diff} h_i = -k\frac{dT}{dx} - \sum \rho D\frac{dY_i}{dx} h_i = -k\frac{dT}{dx} - \rho D\sum \frac{dY_i}{dx} h_i$$
$$= -k\frac{dT}{dx} - \rho D\frac{d\sum h_i Y_i}{dx} + \rho D\sum Y_i\frac{dh_i}{dx} \tag{7.21}$$

We can write $\sum Y_i \frac{dh_i}{dx} = \sum Y_i c_{pi} \frac{dT}{dx} = c_p \frac{dT}{dx}$. Substituting this relation into Equation (7.21), we get

$$\begin{aligned} q_x &= -k\frac{dT}{dx} - \rho D\frac{dh}{dx} + \rho D c_p \frac{dT}{dx} \\ &= -\rho\alpha c_p\frac{dT}{dx} - \rho D\frac{dh}{dx} + \rho D c_p\frac{dT}{dx} = -\rho D\frac{dh}{dx} \end{aligned} \tag{7.22}$$

Here, we invoked $D = \alpha$ from the Lewis number assumption. We can see three terms in the RHS, out of which two are cancelled out. The first term is the sensible enthalpy flux due to conduction, and the third one is the sensible enthalpy flux for species diffusion. The second term is the absolute enthalpy flux due to species diffusion. Substituting the relation obtained in Equation (7.22), we find:

$$\frac{d}{dx}\left(\rho D\frac{dh}{dx}\right) = \dot{m}''\frac{dh}{dx} \tag{7.23}$$

Equation (7.23) is a convection-diffusion equation of absolute enthalpy without any source term. This implies that absolute enthalpy, h, is a conserved scalar. Now, absolute enthalpy can be written in terms of the enthalpy of formation and sensible enthalpy as:

$$h = \sum Y_i h^0_{f\,i} + \int_{T_{ref}}^{T} c_p dT \tag{7.24}$$

Now, substituting Equation (7.24) into Equation (7.23), we get:

$$\frac{d}{dx}\left[\rho D\sum h^0_{f\,i}\left(\frac{dY_i}{dx}\right) + \rho D\frac{d}{dx}\int c_p dT\right] = \dot{m}''\sum h^0_{f\,i}\left(\frac{dY_i}{dx}\right) + \dot{m}''\frac{d}{dx}\int c_p dT \tag{7.25}$$

which can be rearranged as:

$$\dot{m}''\frac{d}{dx}\int c_p dT + \frac{d}{dx}\left[-\rho D\frac{d}{dx}\int c_p dT\right]$$
$$= -\frac{d}{dx}\sum (h^0_{f\,i}\dot{m}'') = -\sum (h^0_{f\,i}\dot{m}''') \quad (7.26)$$

Equation (7.26) is known as the Shvab-Zeldovich form of energy equation. The first term LHS is due to advection transport of sensible enthalpy. The second term denotes the sensible enthalpy transport due to diffusion. The RHS accounts for the enthalpy production due to chemical reaction. This signifies the amount of energy generation due to chemical reaction dissipated through sensible enthalpy advection and diffusion. The general form of this equation, which can be derived by the reader, is:

$$\nabla\cdot\left[\dot{m}''\int c_p dT - \rho D\nabla\int c_p dT\right] = -\sum (h^0_{f\,i}\dot{m}''') \quad (7.27)$$

7.5 ANALYSIS OF TYPICAL FLAME CONFIGURATIONS

Now, we will try to analyze two simple problems of non-premixed flame. The first one is a co-flow jet flame. And the other is the counterflow flame. For co-flow flame, we shall follow the analysis carried out by Burke and Schumann [7]; although simple, this will open up certain treatments we normally follow for solving non-premixed flame.

7.5.1 Burke-Schumann Flame

The original work of Burke and Schumann was on an axisymmetric burner. Here we shall consider a slot burner, which is a similar flame in the Cartesian system, for the sake of simplicity.

Consider a two-feed system, in Cartesian geometry. Fuel is flowing through a central slot of width $2L_1$. Oxidiser is flowing through a space of $(L_2 - L_1)$ on either side, as shown in Figure 7.6. A few assumptions are made to make the analysis simple:

1. The velocity of fuel and oxidiser in the region of flame is constant.
2. The coefficient of inter-diffusion of two gas streams is constant.
3. The diffusion is only in the radial direction.
4. The mixing of fuel and oxidiser is through diffusion only.
5. The flame is steady, laminar, and the density is same for all gases and is constant.
6. The flame is the surface where the mixture is stoichiometric.

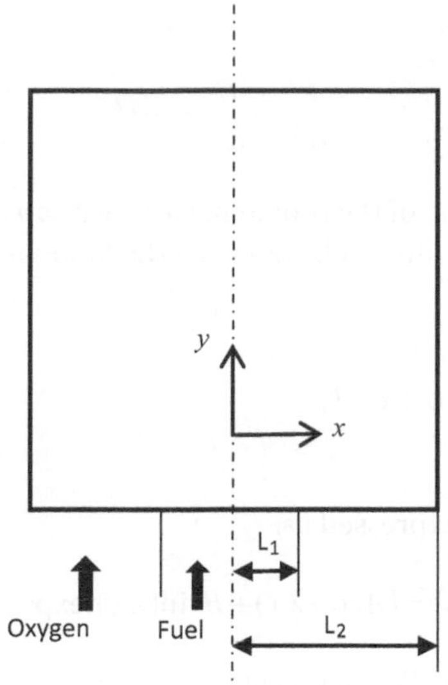

Figure 7.6 Geometry and flow of the flame.

Considering the above assumptions, we can write the conservation equation for mixture fraction:

$$\rho V \frac{\partial Z}{\partial y} = \frac{\partial}{\partial x}\left(\rho D \frac{\partial Z}{\partial x} \right) \tag{7.28}$$

where V is the flow velocity, D is the diffusion coefficient, Z is mixture fraction and ρ is the flow density. Considering $\alpha = D/V$, and because V and D are constants, we can write:

$$\frac{1}{\alpha}\frac{\partial Z}{\partial y} = \frac{\partial^2 Z}{\partial x^2} \tag{7.29}$$

Using the product solution, we can take the following:

$$Z(x,y) = F(x)G(y) \tag{7.30}$$

Equation (7.29) can written as:

$$\frac{1}{\alpha} G'F = GF''$$

or

$$\frac{F''}{\alpha F} = \frac{G'}{\alpha G} = -\lambda^2 \tag{7.31}$$

where λ is the eigenvalue of the problem. Here, we can apply the standard solution methodology of product solution with the following boundary conditions:

1. At $x = 0$, $\frac{\partial Z}{\partial x} = 0$.
2. At $x = L_2$, $Z = 0$.
3. At $y = 0$, $Z = 1$ for $0 \le x \le L_1$.

$Z = 0$ for $L_1 < x < L_2$

The function Z can be expressed as:

$$Z(x,y) = [A\cos(\lambda x) + B\sin(\lambda x)]\exp(-\lambda^2 \alpha y) \tag{7.32}$$

Differentiating Equation (7.32), we get:

$$\frac{\partial Z}{\partial x} = [-A\lambda \sin(\lambda x) + B\lambda \cos(\lambda x)]\exp(-\lambda^2 \alpha y) \tag{7.33}$$

From boundary condition (1), we get $B = 0$. From boundary condition (2), it is found $A\cos(\lambda L_2) = 0$. This leads us to $\cos(\lambda L_2) = 0$, as $A \neq 0$. Solving, we can find:

$$\lambda_n = \frac{1}{L_2}\frac{2n+1}{2}\pi, (n = 0,1,2....) \tag{7.34}$$

As the solutions are linear they can be superposed to give the solution. So,

$$Z = \sum_{n=0}^{\infty} A_n \cos\left(\frac{2n+1}{2L_2}\pi x\right)\exp(-\lambda_n^2 \alpha y) \tag{7.35}$$

From boundary condition (3) and applying orthogonality, we get:

$$\int_0^{L_2} Z(y=0)\cos\left(\frac{2n+1}{2L_2}\pi x\right)dx = \int_0^{L_2} A_n \cos^2\left(\frac{2n+1}{2L_2}\pi x\right)dx \tag{7.36}$$

From the above, we get:

$$A_n = \frac{4}{(2n+1)\pi}\sin\left(\frac{2n+1}{2L_2}\pi L_1\right) \tag{7.37}$$

Finally, the mixture fraction distribution can be expressed as:

$$Z = \sum_{n=0}^{\infty}\left[\frac{4}{(2n+1)\pi}\sin\left(\frac{2n+1}{2L_2}\pi L_1\right)\right]\cos\left(\frac{2n+1}{2L_2}\pi x\right)\exp\left(-\frac{1}{L_2}\frac{2n+1}{2}\pi\alpha y\right) \quad (7.38)$$

On the flame, (x_f, y_f), the value of the mixture fraction will be Z_{St}. Substituting this in Equation (7.38), the flame shape can be determined. For an over-ventilated flame, sufficient oxidiser is entering into flame zone. So, $L_2 \gg x$ can be assumed for that, while $L_2 \sim x$ (both the quantities are of same order) may be considered. The flame shapes predicted for cylindrical jet flame are shown in Figure 7.7.

Here, one should note that any other conserved scalar distribution can be similarly found. The appropriate conservation equation and boundary conditions should be found.

Our primary interest should be to find different properties like mass fraction, temperature, density and so on, at different location of the solution domain. After knowing the mixture fraction distribution, we should calculate the values of the primitive variables as a function on the mixture fraction (or any other conserved scalar). So, we need to know the state relationships to find out the variables of interest, for example:

$$Y_F = Y_F(Z) \quad (7.39)$$

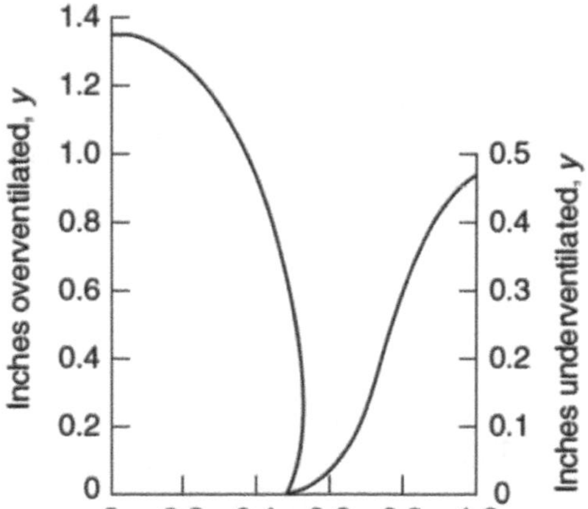

Figure 7.7 Flame shape predicted by Burke and Schumann [4] for axisymmetric jet flame. (Reprinted from Glassman, I., and Yetter, R. A., *Combustion*, 2008. With Permission.)

$$Y_{O_2} = Y_{O_2}(Z) \tag{7.40}$$

$$Y_{Pr} = Y_{Pr}(Z) \tag{7.41}$$

$$T = T(Z) \tag{7.42}$$

$$\rho = \rho(Z) \tag{7.43}$$

The mixture fraction equal to 0 is in the oxygen stream. At $Z = 0$, Y_{O_2} (oxygen mass fraction) will be one. Similarly, the fuel mass fraction, Y_F will be one at $Z = 1$ (at fuel stream). At the flame, where $Z = Z_{St}$, both fuel and oxygen are consumed and the species mass fractions become zero, while the product mass fraction is maximum. A linear variation of mass fractions with mixture fraction can be shown similar to Figure 7.8a. Now let us try to formulate the state relation based on the diagram:

$$Y_{O_2} = 1 - \frac{Z}{Z_{St}} \quad for\, 0 \le Z \le Z_{St}$$

$$= 0 \quad for\, Z_{St} < Z \le 1 \tag{7.44}$$

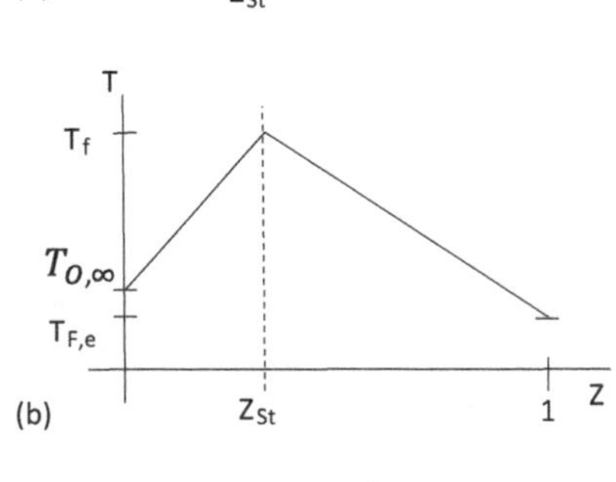

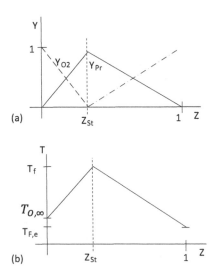

Figure 7.8 Simplified state relationship for (a) mass fractions and (b) temperature in mixture.

$$Y_F = \frac{Z - Z_{St}}{1 - Z_{St}} \quad for\, Z_{St} < Z \le 1$$

$$= 0 \quad for\, 0 \le Z \le Z_{St} \tag{7.45}$$

$$Y_{Pr} = \frac{1 - Z}{1 - Z_{St}} \quad for\, Z_{St} < Z \le 1$$

$$= \frac{Z}{Z_{St}} \quad for\, 0 \le Z \le Z_{St} \tag{7.46}$$

To determine the temperature in terms of the mixture fraction, we need the following assumptions:

- Specific heats of all the species; that is, fuel, oxygen, product, are equal: $c_{pF} = c_{pO2} = c_{pPr} = c_p$.
- Enthalpy of formation of oxygen and products are zero and for fuel, it is the heat of combustion: $h^0_{f,O2} = h^0_{f,Pr} = 0; h^0_{f,F} = \Delta h_c$.

Absolute enthalpy can be written as:

$$h = \sum Y_i h^0_{f,i} = Y_F \Delta h_c + c_p\left(T - T_{ref}\right) \tag{7.47}$$

We can consider equivalence between absolute enthalpy and mixture fraction, as the conservation equations are of same form, and write:

$$T = (Z - Y_F)\frac{\Delta h_c}{c_p} + Z(T_{F,e} - T_{O,\infty}) + T_{O,\infty} \tag{7.48}$$

Substituting the values of Y_F for different zones, we can write:

$$T = Z\left[\frac{\Delta h_c}{c_p} + \left(T_{F,e} - T_{O,\infty}\right)\right] - \frac{Z - Z_{St}}{1 - Z_{St}}\frac{\Delta h_c}{c_p} + T_{O,\infty} \quad for\, Z_{St} < Z \le 1 \tag{7.49}$$

$$T = Z\left(\frac{\Delta h_c}{c_p} + T_{F,e} - T_{O,\infty}\right) + T_{O,\infty} \quad for\, 0 \le Z \le Z_{St} \tag{7.50}$$

The temperature distribution based on the state relationship derived is shown in Figure 7.8b. Now, it is understood that we can formulate state relationships with different flow variables as a function of the mixture fraction (or any other conserved scalar). Here, our considerations were simple and thus the variation considered was linear. It is also understood that the flame temperature shown here at Z_{St}

is the adiabatic temperature. We can make our state relationships more complex and realistic, depending on our assumptions. It is clear, however, that by knowing the distribution of the conserved scalar, we can determine the flame properties. This approach to the solution is well accepted for non-premixed flames.

7.5.2 Counterflow Flame

A counterflow diffusion flame, where the flame can be established in the space of impingement of two opposed gaseous flows, can provide a detailed structure of a pure diffusion flame. Counterflow diffusion flames, which are generally referred to as pure diffusion flames, are mostly used for investigating the effects of different parameters on diffusion flames due their simplicity and one-dimensional nature [15]. The counterflow flame has been studied widely for last five decades for understanding extinction mechanism and complex chemical kinetics and transport processes in diffusion flames. A large number of theoretical and experimental works have been reported in the literature on the structure and extinction of flat laminar diffusion counterflow flames. Smooke and coworkers [16–21] have computationally and experimentally investigated chemical kinetics and transport processes in counterflow diffusion flame. The counterflow diffusion flames can be divided into two groups: (1) the counterflow diffusion flame between two opposed gaseous jets of fuel and oxidant, and (2) the counterflow diffusion flame established in the forward stagnation region of a porous burner. Different configurations of counterflow flames are presented in Figure 7.9 [22]. In recent times, [23] introduced an opposed jet tubular counterflow flame which can be used to study the effects of strain rate and curvature in diffusion flame. In opposed the tubular burner, a counterflow flame of cylindrical shape is established by issuing fuel from an inner porous tube towards an oxidiser that is issued from an outer porous tube, or vice versa. Reaction products exit the burner in the axial direction. Due to rotational symmetry, the tubular flame is one-dimensional in the radial direction. Well-defined boundary conditions allow controlled experimental conditions and facilitate comparisons to numerical results. Furthermore, this flame configuration allows flame curvature and strain rate to be varied independently.

The objective of studying laminar counterflow flame is not only to understand fundamental processes but also to model complex turbulent diffusion flame [24–26]. A turbulent diffusion flame can be imagined as a collection of many stretched and curved laminar flamelets. A flamelet library is built using laminar flame details which is useful for turbulent combustion simulation (Figure 7.10).

A schematic illustration of the counterflow flame experimental configuration is shown in Figure 7.11. Experiments are performed with fuel introduced from the bottom duct and oxidiser introduced from the top duct. An annular nitrogen gas curtain is provided to isolate the flow from the atmosphere.

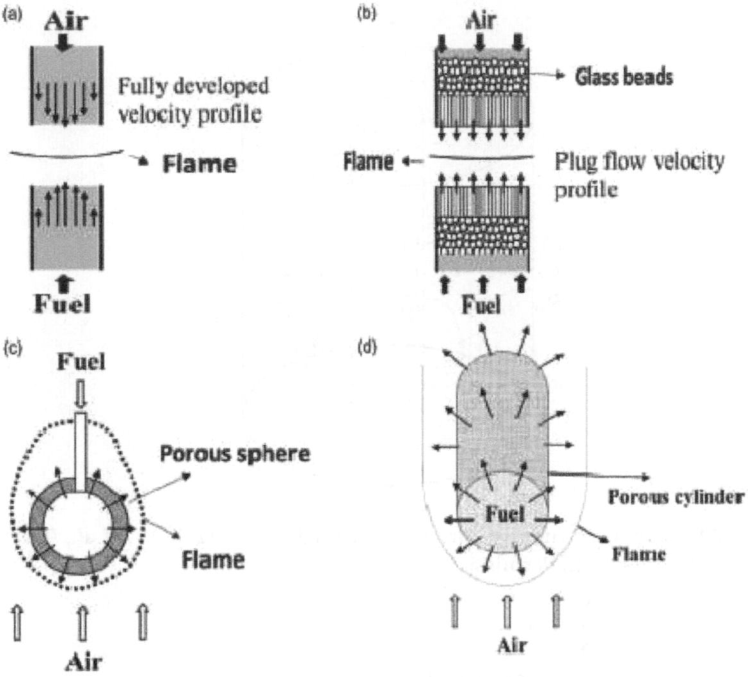

Figure 7.9 Schematic of various configurations of the counterflow burner: (a) cylindrical tubes without porous medium leading to fully developed velocity profile at the exit, (b) cylindrical tube with flow stabiliser leading to plug flow velocity profile, (c) hollow sphere with porous surface leading to spherically symmetric radially outward flow of fuel, (d) fuel flows radially outward from a cylindrical porous surface providing an axisymmetric flow field. (Reprinted from Ravikrishna, R. V., and Sahu, A.B., *Int. J. Spray Combust.*, 1756827717738168, 2017, With Permission.)

The opposing jets of oxidiser and fuel create a stagnation plane where axial velocity (v_x) is equal to zero. The relative magnitude of oxidiser and fuel jet initial momentum flux determines the location of the stagnation plane. The stagnation plane resides at the midpoint of the separation gap between the nozzles for equal momentum flux of both streams. The location of the flame is determined by the mixture fraction value. The flame establishes itself where mixture fraction attains stoichiometric conditions. For hydrocarbon fuel (methane, propane and so on) burning in air, flame lies in the oxidiser side as it requires more air than fuel at stoichiometry $(f_{\text{stoic}} = 0.06)$. The flame lies in the fuel side for hydrogen–air flame where more fuel is required than air in stoichiometric conditions.

Now, let us try to develop a mathematical description. Here, we shall assume a steady, axisymmetric laminar stagnation point flow in cylindrical

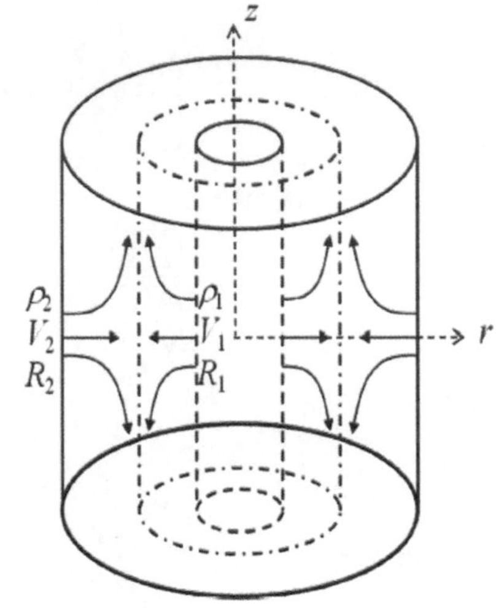

Figure 7.10 Opposed tubular burner. (Reprinted from Pitz, R. W., et al., *Prog. Energ. Combust.*, **42**, 2014. With Permission.)

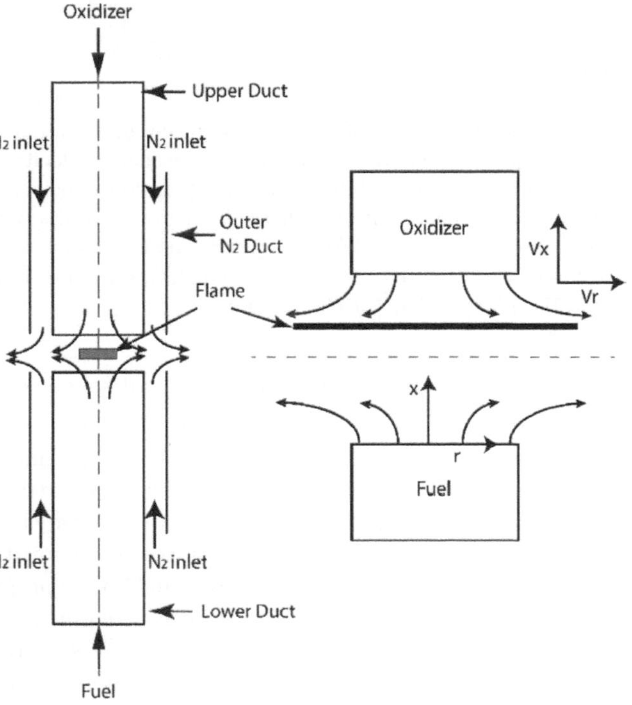

Figure 7.11 Schematic of counterflow diffusion flame.

coordinates. The conservation equation of mass, axial momentum, radial momentum, energy and specie equations is written in the the forms:

$$\frac{1}{r}\frac{\partial}{\partial r}(r\rho v_r)+\frac{\partial}{\partial x}(\rho v_x)=0 \tag{7.51}$$

$$\frac{\partial}{\partial x}(r\rho v_x v_x)+\frac{\partial}{\partial r}(r\rho v_r v_x)=\frac{\partial}{\partial r}(r\tau_{rx})+r\frac{\partial \tau_{xx}}{\partial x}-r\frac{\partial P}{\partial x} \tag{7.52}$$

$$\frac{\partial}{\partial x}(r\rho v_r v_x)+\frac{\partial}{\partial r}(r\rho v_r v_r)=\frac{\partial}{\partial r}(r\tau_{rr})+r\frac{\partial \tau_{rx}}{\partial x}-r\frac{\partial P}{\partial r} \tag{7.53}$$

$$\rho v_r\frac{\partial Y_k}{\partial r}+\rho v_x\frac{\partial Y_k}{\partial x}+\frac{\partial}{\partial x}(\rho Y_k V_k)-\left(\dot{w}_k W_k\right)=0,\quad k=1,2,\ldots,K, \tag{7.54}$$

$$\rho v_r c_p\frac{\partial T}{\partial r}+\rho v_x c_p\frac{\partial T}{\partial x}-\frac{\partial}{\partial x}\left(\lambda\frac{\partial T}{\partial x}\right)-\sum_{k=1}^{K}\rho Y_k V_k c_{pk}\frac{\partial T}{\partial x}+\sum_{k=1}^{K}\dot{w}_k W_k h_k=0 \tag{7.55}$$

The system is closed with the ideal gas law:

$$\rho=\frac{p\bar{W}}{RT} \tag{7.56}$$

These are the following constitutive relations:

$$\tau_{xx}=\mu\left[2\frac{dv_x}{dx}-\frac{2}{3}(\nabla.V)\right] \tag{7.57}$$

$$\tau_{rr}=\mu\left[2\frac{dv_r}{dr}-\frac{2}{3}(\nabla.V)\right]\tau_{rx}=\mu\left[2\frac{dv_x}{dr}+\frac{dv_r}{dx}\right] \tag{7.58}$$

where

$$\nabla.V=\frac{1}{r}\frac{\partial}{\partial r}(rv_r)+\frac{\partial v_x}{\partial x} \tag{7.59}$$

In these equations, T denotes the temperature; Y_k the mass fraction of the kth species; p the pressure; v_r and v_x the radial and the axial components of the velocity, respectively; ρ the mass density; W_k the molecular weight of the kth species; $\bar{W}$ the mean molecular weight of the mixture; R the universal gas constant; λ the thermal conductivity of the mixture; c_p the constant pressure heat capacity of the mixture; c_{pk} the constant pressure heat capacity of the kth species; $\dot{w}_k$ the molar rate of production of the kth species per unit volume; h_k, the specific enthalpy of the kth species; μ the viscosity of the mixture and V_k the diffusion velocity of the kth species in the x direction. K denotes the total number of species.

These sets of axisymmetric, coupled nonlinear governing partial differential equations need to be transformed into a coupled system of ordinary differential equations. We have adopted the following approach, which was originally developed by Kee et al. for premixed flame [27]. The OPPDIF code, which is used to simulate counterflow flame, also uses following formulation. This approach starts with the assumption of similarity variables. It is assumed that $\frac{v_r}{r}$ and v_x are function x only.

So we define the following:

$$G(x)=\frac{\rho v_r}{r} \tag{7.60}$$

$$F(x)=\frac{\rho v_x}{2} \tag{7.61}$$

For which the continuity Equation (7.1) reduces to:

$$G(x)=\frac{dF(x)}{dx} \tag{7.62}$$

We introduce the above assumptions in the axial and radial momentum equations. Then these partial differential equations reduce to third-order ordinary differential equations, which are stated as follows:

$$\frac{\partial P}{\partial x}=-4F\frac{d}{dx}\left(\frac{F}{\rho}\right)-2\mu\frac{d}{dx}\left(\frac{1}{\rho}G\right)+\frac{4}{3}\frac{d}{dx}\left[2\mu\frac{d}{dx}\left(\frac{F}{\rho}\right)+\frac{\mu}{\rho}G\right]=f_1(x) \tag{7.63}$$

$$\frac{1}{r}\frac{\partial P}{\partial r}=\frac{d}{dx}\left(\frac{2F}{\rho}G\right)-\frac{3}{\rho}G^2-\frac{d}{dx}\left[\mu\frac{d}{dx}\left(\frac{1}{\rho}G\right)\right]=f_2(x) \tag{7.64}$$

If we try to relate the left-hand side of Equations (7.63) and (7.64), we have:

$$\frac{\partial}{\partial x}\left(\frac{1}{r}\frac{\partial P}{\partial r}\right)=\frac{1}{r}\frac{\partial}{\partial x}\left(\frac{\partial P}{\partial r}\right)=\frac{1}{r}\frac{\partial}{\partial r}\left(\frac{\partial P}{\partial x}\right) \tag{7.65}$$

As $\frac{\partial P}{\partial x}$ and $\frac{1}{r}\frac{\partial P}{\partial r}$ are functions of x only, it follows that:

$$\frac{\partial}{\partial x}\left(\frac{1}{r}\frac{\partial P}{\partial r}\right)=\frac{1}{r}\frac{\partial}{\partial r}\left(\frac{\partial P}{\partial x}\right)=0 \tag{7.66}$$

and

$$\frac{1}{r}\frac{\partial P}{\partial r}=constant\equiv H \tag{7.67}$$

So the radial pressure gradient forms an eigenvalue problem and enters into the set of ordinary differential equations as:

$$\frac{dH}{dx}=0 \tag{7.68}$$

The radial momentum equation is:

$$\frac{d}{dx}\left[\mu\frac{d}{dx}\left(\frac{G}{\rho}\right)\right]-2\frac{d}{dx}\left(\frac{FG}{\rho}\right)+\frac{3}{\rho}G^2+H=0 \tag{7.69}$$

Similarly, energy and species equations are:

$$2Fc_p\frac{dT}{dx}-\frac{d}{dx}\left(\lambda\frac{dT}{dx}\right)+\sum_{k=1}^{K}\rho Y_k V_k c_{pk}\frac{\partial T}{\partial x}\sum_{k=1}^{K}\dot{w}_k W_k h_k=0 \tag{7.70}$$

$$2F\frac{dY_k}{dx}+\frac{d}{dx}(\rho Y_k V_k)-(\dot{w}_k W_k)=0 \tag{7.71}$$

The diffusion velocities are given by the following multicomponent formulation:

$$V_k=\frac{1}{X_k\bar{W}}\sum_{j\neq k}^{K}W_j D_{k,j}\frac{dX_j}{dx}-\frac{D_k^T}{\rho Y_k}\frac{1}{T}\frac{dT}{dx} \tag{7.72}$$

where X_k is the mole fraction, $D_{k,j}$ is the ordinary multicomponent diffusion coefficient, and D_k^T is the thermal diffusion coefficient.

The boundary conditions for the fuel (F) and oxidiser (O) streams at the nozzles are:

$$x=0: \quad F=\frac{\rho_F v_{e,F}}{2}, G=0,\ T=T_F,\ \rho v_x Y_k+\rho Y_k V_k=\left(\rho v_x Y_k\right)_F$$

$$x=L: \quad F=\frac{\rho_O v_O}{2}, G=0,\ T=T_O,\ \rho v_x Y_k+\rho Y_k V_k=\left(\rho v_x Y_k\right)_O$$

The above boundary conditions form a boundary value problem for the dependent variables F,G,$T$and Y_k. The four differential equations (Equations 7.61, 7.62, 7.69 and 7.70) and the eigenvalue H (Equation 7.68) completely describe the mathematical model of a counterflow flame.

Determination of the source term of the species conservation equation requires an extensive calculation. The laminar finite rate model is used to calculate the source term $\dot{w}_k$ as gas flows are laminar. Elementary reversible (or irreversible) reactions involving K chemical species are commonly represented in the following general form:

$$\sum_{k=1}^{K} v'_{ki}\chi_k \rightleftharpoons \sum_{k=1}^{K} v''_{ki}\chi_k \ \left(\mathrm{i}=1,2\ldots\ldots.\mathrm{I}\right) \tag{7.73}$$

The stoichiometric coefficients are v'_{ki} and are integer numbers, and χ is the chemical symbol for the kth species. The superscript $'$ indicates forward stoichiometric coefficients; $''$ denotes reverse stoichiometric coefficients.

The production rate $\dot{w}_k$ of the kth species can be expressed as a summation of the rate of progress variables for all reactions involving the kth species:

$$\dot{w}_k=\sum_{i=1}^{I} v_{ki}\, q_i \tag{7.74}$$

where

$$v_{ki}=\left(v''_{ki}-v'_{ki}\right) \tag{7.75}$$

The rate of progress variable q_i for the ith reaction is expressed by the difference of the forward and reverse rates as:

$$q_i=k_{fi}\prod_{k=1}^{K}[X]_k^{v'_{ki}}-k_{ri}\prod_{k=1}^{K}[X]_k^{v''_{ki}} \tag{7.76}$$

where X_k is the molar concentration of the kth species and k_{fi} and k_{ri} are the forward and reverse rate constants of the ith reaction.

It is assumed that the forward rate constants will follow the Arrhenius temperature dependence:

$$k_{fi} = A_i T^{\beta_i} \exp\left(\frac{E_i}{RT}\right) \tag{7.77}$$

where the pre-exponential factor A_i, the temperature exponent β_i, and the activation energy E_i are specified. These three parameters are required input to the gas-phase kinetics package for each reaction.

In thermal systems, the reverse rate constants $k_{r,i}$ are related to the forward rate constants through the equilibrium constants by:

$$k_{r,i} = {}^{k_{fi}}\!/_{K_c} \tag{7.78}$$

where K_c is the equilibrium constant of the ith reaction, which is calculated from enthalpy and entropy of the species evaluated at the respective temperature and pressure.

The steps for calculating $\dot{w}_k$ are:

1. A_i, β_i, and E_i will be taken from reaction mechanism chemistry (like Table 7.2 for methane-air) to calculate k_{fi}.
2. K_c will be calculated from the enthalpy and entropy of the species evaluated at the respective temperature and pressure (thermodynamic data).
3. k_{ri} will be calculated from Equation (7.78).
4. q_i will be calculated from Equation (7.76).
5. $\dot{w}_k$ will be calculated from Equation (7.74).

The above counterflow model is used to analyze the structure of a methane-air counterflow flame. It considers an opposed flow configuration with a separation distance between the duct as 1.4876 cm. The axial velocities of methane and air stream at the exit of the nozzles are 76.8 cm/s and 73.4 cm/s, respectively. Pt versus Pt 10% Rh thermocouples are used for temperature measurement. A reaction mechanism which consists of 78 reactions, including C_2 chemistry [17], is used for calculation (Table 7.2). The axial velocity profile is shown in Figure 7.12. The point 0.0 lies 6.0 mm from the bottom fuel duct. It is found from the velocity profile, and the strain rate for aforesaid condition is 54.67 s^{-1}. The stagnation plane is located 5.687 mm from the bottom fuel duct. The comparion of the temperature profile obtained from numerical calculations and experimental measurements is presented in Figure 7.13. The maximum temperatures obtained from numerical calculations and experimental measurements are 2010 K and 1950 K, respectively.

TABLE 7.2 METHANE-AIR REACTION MECHANISM

Serial No.	Reaction	A	β	E
1	$CH_3 + H \rightleftharpoons CH_4$	1.90E + 36	-7.000	9050
2	$CH_4 + O_2 \rightleftharpoons CH_3 + HO_2$	7.90E + 13	0.000	56000
3	$CH_4 + H \rightleftharpoons CH_3 + H_2$	2.20E + 04	3.000	8750
4	$CH_4 + O \rightleftharpoons CH_3 + OH$	1.60E + 06	2.360	7400
5	$CH_4 + OH \rightleftharpoons CH_3 + H_2O$	1.60E + 06	2.100	2460
6	$CH_2O + OH \rightleftharpoons HCO + H_2O$	7.53E + 12	0.000	167
7	$CH_2O + H \rightleftharpoons HCO + H_2$	3.31E + 14	0.000	10500
8	$CH_2O + M \rightleftharpoons HCO + H + M$	3.31E + 16	0.000	81000
9	$CH_2O + O \rightleftharpoons HCO + OH$	1.81E + 13	0.000	3082
10	$HCO + OH \rightleftharpoons CO + H_2O$	5.00E + 12	0.000	0
11	$HCO + M \rightleftharpoons H + CO + M$	1.60E + 14	0.000	14700
12	$HCO + H \rightleftharpoons CO + H_2$	4.00E + 13	0.000	0
13	$HCO + O \rightleftharpoons OH + CO$	1.00E + 13	0.000	0
14	$HCO + O_2 \rightleftharpoons HO_2 + CO$	3.00E + 12	0.000	0
15	$CO + O + M \rightleftharpoons CO_2 + M$	3.20E + 13	0.000	-4200
16	$CO + OH \rightleftharpoons CO_2 + H$	1.51E + 07	1.300	-758
17	$CO + O_2 \rightleftharpoons CO_2 + O$	1.60E + 13	0.000	41000
18	$CH_3 + O_2 \rightleftharpoons CH_3O + O$	7.00E + 12	0.000	25652
19	$CH_3O + M \rightleftharpoons CH_2O + H + M$	2.40E + 13	0.000	28812
20	$CH_3O + H \rightleftharpoons CH_2O + H_2$	2.00E + 13	0.000	0
21	$CH_3O + OH \rightleftharpoons CH_2O + H_2O$	1.00E + 13	0.000	0
22	$CH_3O + O \rightleftharpoons CH_2O + OH$	1.00E + 13	0.000	0
23	$CH_3O + O_2 \rightleftharpoons CH_2O + HO_2$	6.30E + 10	0.000	2600
24	$CH_3 + O_2 \rightleftharpoons CH_2O + OH$	5.20E + 13	0.000	34574
25	$CH_3 + O \rightleftharpoons CH_2O + H$	6.80E + 13	0.000	0
26	$CH_3 + OH \rightleftharpoons CH_2O + H_2$	7.50E + 12	0.000	0
27	$CH_2 + H \rightleftharpoons CH + H_2$	4.00E + 13	0.000	0
28	$CH_2 + O \rightleftharpoons CO + H + H$	5.00E + 13	0.000	0
29	$CH_2 + O_2 \rightleftharpoons CO_2 + H + H$	1.30E + 13	0.000	1500
30	$CH_2 + CH_3 \rightleftharpoons C_2H_4 + H$	4.00E + 13	0.000	0
31	$CH + O \rightleftharpoons CO + H$	4.00E + 13	0.000	0
32	$CH + O_2 \rightleftharpoons CO + OH$	2.00E + 13	0.000	0

(*Continued*)

TABLE 7.2 (*Continued*) METHANE-AIR REACTION MECHANISM

Serial No.	Reaction	A	β	E
33	$CH_3 + CH_3 \rightleftharpoons C_2H_6$	1.70E + 53	-12.000	19400
34	$CH_3 + CH_3 \rightleftharpoons C_2H_5 + H$	8.00E + 13	0.000	26500
35	$C_2H_6 + H \rightleftharpoons C_2H_5 + H$	5.40E + 02	3.500	5240
36	$C_2H_6 + O \rightleftharpoons C_2H_5 + OH$	3.00E + 07	2.000	5100
37	$C_2H_6 + OH \rightleftharpoons C_2H_5 + H_2O$	6.00E + 13	0.000	19400
38	$C_2H_5 + O_2 \rightleftharpoons C_2H_4 + HO_2$	2.00E + 13	0.000	5000
39	$C_2H_5 \rightleftharpoons C_2H_4 + H$	2.60E + 43	-9.250	52580
40	$C_2H_4 + O \rightleftharpoons HCO + CH_3$	1.60E + 09	1.200	740
41	$C_2H_4 + OH \rightleftharpoons C_2H_3 + H_2O$	4.80E + 12	0.000	1230
42	$C_2H_4 + H \rightleftharpoons C_2H_3 + H_2$	1.10E + 14	0.000	8500
43	$C_2H_3 + H \rightleftharpoons C_2H_2 + H_2$	6.00E + 12	0.000	0
44	$C_2H_3 + O_2 \rightleftharpoons C_2H_2 + HO_2$	1.58E + 13	0.000	10060
45	$C_2H_3 \rightleftharpoons C_2H_2 + H$	1.60E + 32	-5.500	46200
46	$C_2H_2 + O \rightleftharpoons CH_2 + CO$	2.20E + 10	1.000	2583
47	$C_2H_2 + OH \rightleftharpoons CH_2CO + H$	3.20E + 11	0.000	200
48	$CH_2CO + H \rightleftharpoons CH_3 + CO$	7.00E + 12	0.000	3000
49	$CH_2CO + O \rightleftharpoons HCO + HCO$	2.00E + 13	0.000	2300
50	$CH_2CO + OH \rightleftharpoons CH_2O + HCO$	1.00E + 13	0.000	0
51	$CH_2CO + M \rightleftharpoons CH_2 + CO + M$	1.000E + 16	0.000	59250
52	$C_2H_2 + O \rightleftharpoons HCCO + H$	3.56E + 04	2.700	1391
53	$HCCO + H \rightleftharpoons CH_2 + CO$	3.00E + 13	0.000	0
54	$HCCO + O \rightleftharpoons CO + CO + H$	1.20E + 13	0.000	0
55	$C_2H_2 + OH \rightleftharpoons C_2H + H_2O$	1.00E + 13	0.000	7000
56	$C_2H + O \rightleftharpoons CO + CH$	1.00E + 13	0.000	0
57	$C_2H + H_2 \rightleftharpoons C_2H_2 + H$	3.50E + 12	0.000	2100
58	$C_2H + O_2 \rightleftharpoons CO + HCO$	5.00E + 13	0.000	1500
59	$HO_2 + CO \rightleftharpoons CO_2 + OH$	5.80E + 13	0.000	22934
60	$H_2 + O_2 \rightleftharpoons 2OH$	1.70E + 13	0.000	47780
61	$OH + H_2 \rightleftharpoons H_2O + H$	1.17E + 09	1.300	3626
62	$H + O_2 \rightleftharpoons OH + O$	2.20E + 14	0.000	16800
63	$O + H_2 \rightleftharpoons OH + H$	1.80E + 10	1.000	8826
64	$H + O_2 + M \rightleftharpoons HO_2 + M$[a]	2.10E + 18	-1.000	0

(*Continued*)

TABLE 7.2 (*Continued*) METHANE-AIR REACTION MECHANISM

Serial No.	Reaction	A	β	E
65	$H + O_2 + O_2 \rightleftharpoons HO_2 + O_2$	6.70E + 19	-1.420	0
66	$H + O_2 + N2 \rightleftharpoons HO_2 + N_2$	6.70E + 19	-1.420	0
67	$OH + HO_2 \rightleftharpoons H_2O + O_2$	5.00E + 13	0.000	1000
68	$H + HO_2 \rightleftharpoons 2OH$	2.50E + 14	0.000	1900
69	$O + HO_2 \rightleftharpoons O_2 + OH$	4.80E + 13	0.000	1000
70	$2OH \rightleftharpoons O + H_2$	6.00E + 08	1.300	0
71	$H_2 + M \rightleftharpoons H + H + M^b$	2.23E + 12	0.500	92600
72	$O_2 + M \rightleftharpoons O + O + M$	1.85E + 11	0.500	95560
73	$H + OH + M \rightleftharpoons H_2O + M^c$	7.50E + 23	-2.600	0
74	$H + HO_2 \rightleftharpoons H_2 + O_2$	2.50E + 13	0.000	700
75	$HO_2 + HO_2 \rightleftharpoons H_2O_2 + O_2$	2.00E + 12	0.000	0
76	$H_2O_2 + M \rightleftharpoons OH + OH + M$	1.30E + 17	0.000	45500
77	$H_2O_2 + H \rightleftharpoons HO_2 + H_2$	1.60 + 12	0.000	3800
78	$H_2O_2 + OH \rightleftharpoons H_2O + HO_2$	1.00E + 13	0.000	1800

Source: Reprinted by permission of the publisher Taylor & Francis Group, http://www.tandfonline.com. (From Puri, I.K. et al., *Combust. Sci. Technol.*, **56**, 1–22, 1987.)

Note: Reaction mechanism rate coefficients are in the form $k_i = AT^{\beta}\exp\left(\frac{E_0}{RT}\right)$ Units are moles, cubic centimeters, seconds, Kelvins and calories/mole.

[a] Third body efficiencies:
$k_{64}(H_2O) = 21k_{64}(Ar), k_{64}(H_2) = 3.3k_{64}(Ar), k_{64}(N_2) = 3.3k_{64}(O_2) = 0$.

[b] Third body efficiencies:
$k_{71}(H_2O) = 6k_{71}(Ar), k_{71}(H) = 2k_{71}(Ar), k_{71}(H_2) = 3k_{71}(Ar)$.

[c] Third body efficiencies:
$k_{73}(H_2O) = 20k_{73}(Ar)$.

The maximum temperature location signifies the stoichiometric mixture fraction location. Errors assosiated the velocity measurement of the reactants at the exit of the ducts might be the cause of this devation. The experimental and computational profiles of the major stable species in the flame are compared in Figure 7.14. There is excellent agrement for CH_4, N_2 and O_2 conentration profiles between the experimental value and the numerical calculation. The concetration of CH_4 and O_2 are maximum at their respective inlets and start to deplete along the axis due to reaction. The profiles of minor species are shown in Figure 7.15. Due to the inclusion of detailed multistep chemisty, intermediate minor species also give an excellent agreement with the experimental mesurements.

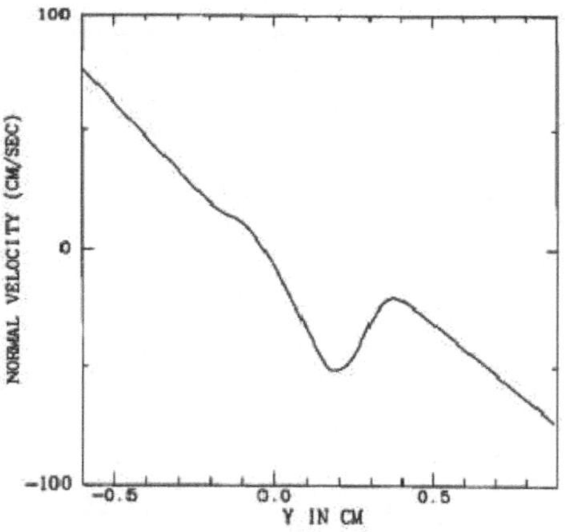

Figure 7.12 Calculated profile of the axial velocity component. (With permission from Taylor and Francis: *Combust. Sci. Technol.*, A comparison between numerical calculations and experimental measurements of the structure of a counterflow methane-air diffusion flame, 56, 1987, 1–22, Puri, I. K., et al.; http://www.tandfonline.com.)

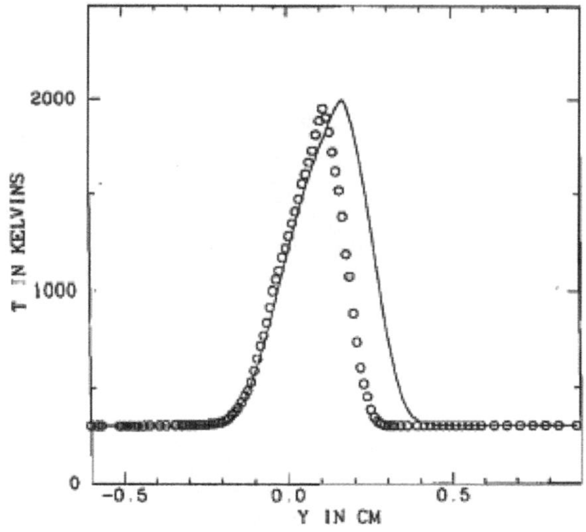

Figure 7.13 Comparison between measured (O) and calculated values (solid line) of the temperature profile. (With permission from Taylor and Francis: *Combust. Sci. Technol.*, A comparison between numerical calculations and experimental measurements of the structure of a counterflow methane-air diffusion flame, 56, 1987, 1–22, Puri, I. K., et al.; http://www.tandfonline.com.)

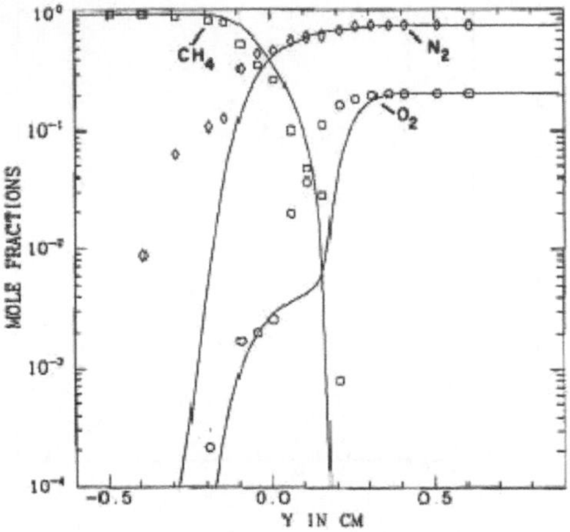

Figure 7.14 Comparison between measured CH_4 (□) O_2 (○) and N_2 (◊) profiles and corresponding calculated values (solid lines). (With permission from Taylor and Francis: *Combust. Sci. Technol.*, A comparison between numerical calculations and experimental measurements of the structure of a counterflow methane-air diffusion flame, 56, 1987, 1–22, Puri, I.K. et al.; http://www.tandfonline.com.)

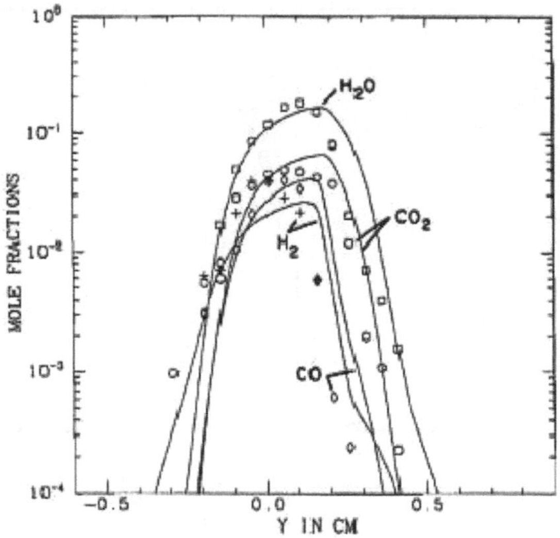

Figure 7.15 Comparison between measured H_2O (□), CO_2 (○). H, (+) and CO (◊) profiles and corresponding calculated values (solid lines). (With permission from Taylor and Francis: *Combust. Sci. Technol.*, A comparison between numerical calculations and experimental measurements of the structure of a counterflow methane-air diffusion flame, 56, 1987, 1–22, Puri, I. K., et al.; http://www.tandfonline.com.)

7.6 PARTIALLY PREMIXED FLAMES

Hybrid flames that contain multiple reaction zones can be called partially premixed flames (PPFs). As the name suggests, full premixing is not done here. The fuel can be mixed with a portion of the oxidiser, and the rest may be supplied separately from another stream. In a sense, these flames have features of both non-premixed as well as premixed flames. Thus, these flames have the advantages of both nonpremixed and premixed flames. The advantages can enhance flame safety with lower pollutant emission and flame stability. PPFs may be seen in many real-life applications. Gas-fired domestic burners, Bunsen burners, industrial furnaces, and so on are examples of PPFs. In turbulent combustion, due to local extinction and reignition processes, we can see the effect of partial premixing. In practical spray systems, when local fuel vapor-rich zones are found, the features of partial premixing can be seen.

Schematic representations of non-premixed, premixed and PPFs are shown in Figure 7.16. In a non-premixed flame, shown in Figure 7.16a, the flame surface separates the fuel and air stream. In a premixed flame, the flame surface is basically an interface of the fuel-air mixture and product gases. The equivalence ratio of the mixture is 1 or less, which means the mixture is either stoichiometric or lean. In case of a PPF, the fuel side is supplied with some amount of air. The mixture of fuel and air becomes rich in this case. The other side of the flame can be only air or that also can be a lean mixture of fuel and air.

As fuel and oxidiser in premixed combustion are mixed with each other beforehand, more complete combustion can be ensured compared to that in

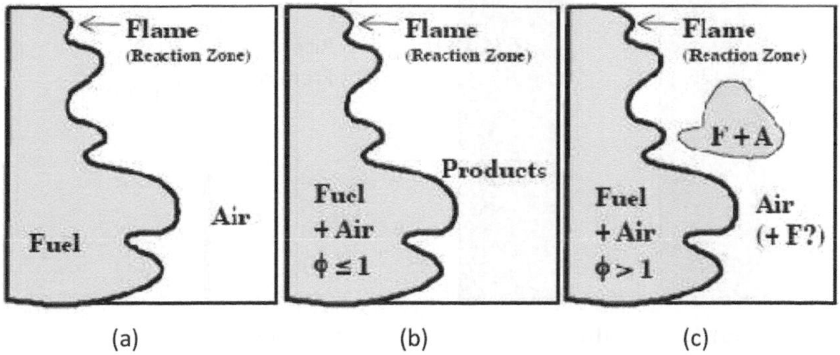

Figure 7.16 Schematic representations of (a) non-premixed, (b) premixed, and (c) partially premixed flames.

non-premixed flames. Still, for real-life applications, non-premixed flames are mostly used. In the case of premixed flames, the possibility of flashback cannot ensure safety. As a compromise, a portion of air/oxygen can be mixed with the fuel and the rest can be supplied as a separate stream. Usually, it is called primary and secondary air. This arrangement can develop a PPF. The fraction of air, used as primary, should be such that the fuel-air ratio is well above flammability limit. Thus, flashback can be avoided. On the other hand, the advantages of premixed combustion can be availed partly.

Let us consider a co-flow burner. The inner passage is used for a mixture of fuel and primary air. Secondary air is supplied through the outer slot. The flame shape will be like that of Figure 7.17a, where half of the flame is shown. There is a bell-shaped flame, called rich-premixed flame, at the centre. The fuel-rich flow, after coming out from the passage, burns at the mouth of the burner. Surrounding that we can see the surface of the non-premixed flame. The unburned fuel in the rich premixed flame is burning with atmospheric air. The structure schematic can be seen in Figure 7.17c. This is called a double flame, due to its two flame surfaces.

If we supply a fuel lean mixture through the outer slot, we can see a third flame surrounding the two flames. The third flame is called the lean premixed flame, and the three-flame structure as a whole is called triple flame (Figure 7.17b). The shape of these three flames depends on a number of factors, like flow rate, equivalence ratio and so on. The detailed discussion is beyond the scope of this book and may be found in the works by Aggarwal, Puri and coworkers [28–30].

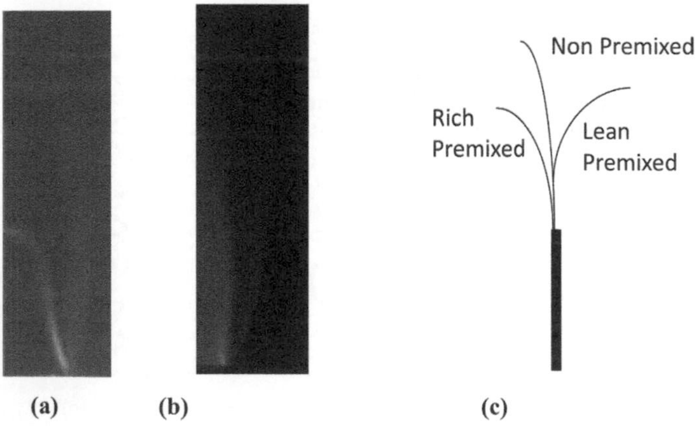

Figure 7.17 (a) Co-flow double flame, (b) co-flow triple flame, (c) schematic of flame structure.

7.7 SOOT GENERATION

Non-premixed flames usually look yellow-orange. This is mainly due to radiation coming out of the soot particles, which are generated within the flame. Soot particles emit energy, and they are formed by combination of hydrocarbon molecules and due to the dearth of oxygen. Normally soot particles are formed at the richer side of the flame, where sufficient oxygen is absent. PPFs also generate soot at the non-premixed part of the flame. These particles are considered an environmental pollutant and result in flame heat loss through radiation.

Soot particles are mainly carbonaceous particles of an amorphous nature. Close observation shows the presence of hydrogen atoms also [1]. The typical size of a soot particle is 20–50 nm. The particles are agglomerated together to form a cluster, as shown in Figure 7.18. The size of the primary particles reduces with partial premixing. However, the agglomeration size is more or less constant [31].

The formation mechanism of soot particles is quite complex and not yet completely understood. Four broad steps have been identified:

1. Precursor formation
2. Particle inception
3. Growth and agglomeration
4. Oxidation

Soot particles start forming in the nucleation stage from gaseous soot precursors. Lin et al. [32] claimed acetylene as a soot precursor, while polycyclic aromatic hydrocarbons (PAHs) are taken as soot precursors by Dobbins et al. [33]. They showed PAHs of mass range 202–472 amu act as soot precursors. The nucleation starts at a temperature as low as 1200 K and may continue up to

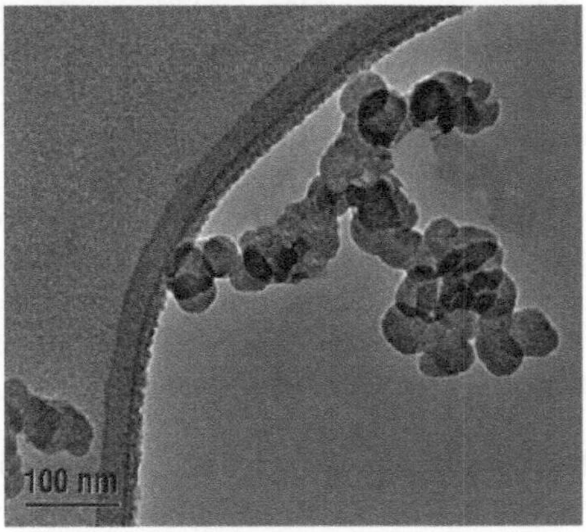

Figure 7.18 TEM image of soot particle agglomeration.

1800 K until the equivalence ratio comes below 1.14. The first step of nucleation starts with the formation of an aromatic ring from aliphatic species, mainly acetylene. The hydrogen abstracted from the chain of hydrocarbon forming radicals is due to bond breaking. Subsequently the radicals join together to form aromatic compounds. The mechanism is termed a hydrogen abstraction carbon addition (HACA) mechanism [34].

When the PAHs become larger, they form soot precursor nanoparticles, or nano-organic carbon (NOC), or young soot through graphitisation [35]. The nascent soot particles formed can grow through a surface growth mechanism. Carbon is added to the surface of the soot particle from gas phase PAHs to increase its size. In this stage, the number density of soot particle does not change normally. Only increment of size occurs.

The grown particles collide and coalesce with each other to form a large aggregate. The aggregates are normally fractal shape in nature, as shown in Figure 7.18. The particles may initially be of a viscous nature and thus coalesce with each other. The agglomerates may form because of Van der Waals forces of attraction between very small entities. It is also true that there is no sharp demarcation between growth and agglomeration. In fact, the two processes occur simultaneously within the flame. The aggregation of particles may hinder the surface growth also.

Oxidation of soot particles occurs when the soot aggregates come in contact with oxygen. Carbon monoxide and carbon dioxide are the expected products of soot oxidation. In jet flames, this stage occurs at the upper zone of the flame. The particle size gradually decreases, even vanishes, due to oxidation. If soot particles are coming out across the flame tip, they are visible as a black smoke, and the flame can be termed a sooting flame. In the case of a non-sooting flame, no soot escapes through the tip, but the above four stages are present inside the flame. All soot particles are consumed by oxidation inside the flame.

Soot is normally quantified as soot volume fraction, that is, volume of soot per unit volume of gas. We can define this in the form of mass concentration – mass of soot per unit volume of gas. Number density (number of particles per unit volume) and particle size can be other ways of quantification.

Soot generation in non-premixed and PPF depends on the type of fuel used. The soot propensity of a fuel can be quantified in terms of the smoke point of a fuel. If we increase the fuel flow in a laminar non-premixed co-flow jet flame without changing the air, we find that the flame height also increases. The fuel flow rate can be taken as the smoke point when soot starts coming out across the flame tip [36].

It is observed that the soot volume fraction increases initially with a small amount of partial premixing of air (in a zone of equivalence ratio of infinity to 24). As the primary air proportion increases, the soot volume decreases gradually [31]. These observations were made in a co-flow burner. We can also observe that the soot volume fraction increases with height above the burner first, reaches a maximum value and then it decreases. This feature is similar for all equivalence ratios. This implies that soot inception occurs near the burner mouth. It grows as it flows up. After that, oxidation starts, which reduces soot volume fraction as we move towards the flame tip.

7.8 EFFECT OF EXIT VELOCITY ON JET FLAMES

When the exit velocity increases for a jet flame, the flame height also increases. This happens as additional amount of fuel requires more volume to burn. This feature is true for laminar flames. After that as velocity increases at the nozzle outlet, the nature of the flame changes. Figure 7.19 shows schematically what happens as the velocity increases [4].

With entry in the turbulent regime, we find that flame distortion takes place. The eddy formation due to turbulence interacts with the flame structure. The smooth tulip-like structure of the laminar flame gets distorted. At the initial level, the distortion is observed at the tip area. The flow which comes out of the burner has some small eddies, which grow with length and are distorted at the top level of the flame. As exit velocity increases, the distortion comes down at the lower level. With more velocity, it comes out of the nozzle with turbulence.

The height of the flame in the turbulent area is more or less constant. In the laminar flow regime, diffusivity is a gas property and is more or less a constant for a fuel. The additional fuel requires additional space to burn. However, in the turbulent zone, the molecular diffusivity is no longer a determining parameter for fuel-air mixing. Rather, eddy diffusivity takes that place and better mixing is the outcome. This makes the flame height nearly constant.

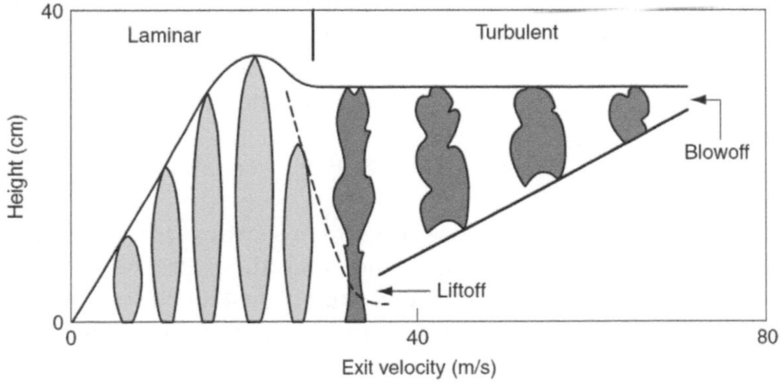

Figure 7.19 Exit velocity effect on non-premixed jet flame structure.

With an increase in exit velocity, the flame gets lifted from the burner tip. Flickering is also evident. But this flickering is different from the laminar flame flickering. The laminar flame flickering is a periodic phenomenon with a specific frequency. In the turbulent zone, we cannot observe any periodicity. The flicker is random in nature. The modelling of these phenomena will be explored in detail in Chapter 9.

EXERCISES

7.1 Develop a transient mixture fraction conservation equation in one dimension.

7.2 Derive a two-dimensional, steady mixture fraction conservation equation.

7.3 Develop a three-dimensional unsteady mixture fraction conservation equation.

7.4 Derive the mixture fraction conservation equation in an axisymmetric coordinate.

7.5 Find the stoichiometric mixture fraction for acetylene while burning with an oxidiser, an equimolar mixture of oxygen and nitrogen.

7.6 Find the stoichiometric mixture fraction for ethylene while burning with an oxidiser, an equimolar mixture of oxygen and carbon dioxide.

7.7 C_2H_6 is supplied to a non-premixed flame. The oxidiser is a mixture of oxygen and carbon dioxide in a 1:2 molar ratio. At a location within the flame, mass fractions are known for all species, namely, C_2H_6, CO_2, CO, O_2, H_2, OH and H_2O. Find the mixture fraction at that location.

7.8 Consider an ethylene air flame, where mole fractions of the following species are measured as $X_{CO} = 950$ ppm, $X_{H2O} = 0.148$, $X_{CO2} = 0.089$, $X_{O2} = 0.0188$, $X_{H2} = 345$ ppm, $X_{OH} = 1290$ ppm. The balance amount is of nitrogen. Find the mixture fraction. Also find the equivalence ratio.

7.9 Derive the conserved scalar enthalpy equation in an axisymmetric system.

7.10 Derive the conserved scalar enthalpy equation in spherical coordinates.

7.11 State relations relate various mixture properties with the mixture fraction or any conserved scalar, and the state relations are frequently used for non-premixed flame. Construct a state relation, relating adiabatic flame temperature and mixture density with mixture fraction.

7.12 Find the flame shapes for over-ventilated and under-ventilated flames on a slot burner following the analysis of Burke and Schumann.

7.13 Derive the expression for the mixture fraction for an axisymmetric flame following Burke-Schumann analysis. Also find out the shapes of under-ventilated and over-ventilated flames.

7.14 Try to formulate the governing equations if the buoyancy effect is considered in addition to Burke-Schumann assumptions for a co-flow flame.

REFERENCES

1. Turns, S. R., *An Introduction to Combustion: Concepts and Applications*, McGraw-Hill, 2000.
2. Schlichting, H., *Boundary Layer Theory*, McGraw-Hill, 1979.
3. Smyth, K. C., NO production and destruction in a methane/air diffusion flame, *Combustion Science and Technology*, **115**, 151–176, 1996.
4. Glassman, I. and Yetter, R. A., *Combustion*, Academic Press, 2008.
5. Williams, F. A., *Combustion Theory, 2nd Ed.*, Benjamin-Cummins, Menlo Park, CA, 1985.
6. Sahu, K. B., Kundu, A., Ganguly, R. and Datta, A., Effects of fuel type and equivalence ratios on the flickering of triple flames, *Combustion and Flame*, **156**, 484–493, 2009.
7. Burke, S. P. and Schumann, T. E. W. Diffusion flames, *Industrail and Engineering Chemistry*, **29**, 998–1004, 1928.
8. Roper, F. G., The prediction of laminar jet diffusion flame sizes: Part I. Theoretical model, *Combustion and Flame*, **29**, 219–226, 1977.
9. Roper, F. G., Smith, C. and Cunningham, A.C., The prediction of laminar jet diffusion flame sizes: Part II. Experimental verification, *Combustion and Flame*, **29**, 227–234, 1977.
10. Roper, F. G., Laminar diffusion flame sizes for curved slot burners giving fan-shaped flames, *Combustion and Flame*, **31**, 251–259, 1978.
11. Spalding, D. B., *Combustion and Mass Transfer*, Pergamon Press, 1979.
12. Chung, S.H. and Law, C.K., Burke-Schumann flame with streamwise and preferential diffusion, *Combustion Science and Technology*, **37**, 21–46, 1984.
13. Krishnan, S. S., Abshire, J. M., Sunderland, P. B., Yuan, Z.-G. and Gore, J. P., Analytical predictions of shapes of laminar diffusion flames in microgravity and earth gravity, *Combustion Theory and Modelling*, **12**, 605–620, 2008.
14. Peters, N., *Fifteen Lectures on Laminar and Turbulent Combustion, Ercoftac Summer School*, Aachen, Germany, 1992. https://www.itv.rwth-aachen.de/fileadmin/Downloads/Summerschools/SummerSchool.pdf.
15. Tsuji, H., Counterflow diffusion flames, *Progress in Energy and Combustion Science*, **8**, 93–119, 1982.
16. Smooke, M. D., Puri, I. K. and Seshadri, K., A comparison between numerical calculations and experimental measurements of the structure of a counterflow diffusion flame burning diluted methane in diluted air, *Symposium (International) on Combustion*, **21**, 1783–1792, 1988.

17. Puri, I. K., Seshadri, K, Smooke, M. D. and Keyes, D., A comparison between numerical calculations and experimental measurements of the structure of a counterflow methane-air diffusion flame, *Combustion Science and Technology*, **56**, 1–22, 1987.
18. Seshadri, K., Trevino, C. and Smooke, M. D., Analysis of the structure and mechanisms of extinction of a counterflow methanol-air diffusion flame, *Combustion and Flame*, **76**, 111–132, 1989.
19. Smooke, M. D., Yetter, R. A., Parr, T. P., Hanson-Parr, D. M., Tanoff, M. A., Colket, M. B. and Hall, R. J., Computational and experimental study of ammonium perchlorate/ethylene counterflow diffusion flames, *Proceedings of the Combustion Institute*, **28**, 2013–2020, 2000.
20. Amantini, G., Frank, J. H., Smooke, M. D. and Gomez, A., Computational and experimental study of standing methane edge flames in the two-dimensional axisymmetric counterflow geometry, *Combustion and Flame*, **147**, 133–149, 2006.
21. Amantini, G., Frank, J.H., Smooke, M. D. and Gomez, A., Computational and experimental study of steady axisymmetric non-premixed methane counterflow flames, *Combustion Theory and Modelling*, **11**, 47–72. 2007.
22. Ravikrishna, R. V. and Sahu, A.B., Advances in understanding combustion phenomena using non-premixed and partially premixed counterflow flames: A review, *International Journal of Spray and Combustion Dynamics*, 1756827717738168, 2017.
23. Pitz, R. W., Hu, S. and Wang, P., Tubular premixed and diffusion flames: Effect of stretch and curvature, *Progress in Energy and Combustion Science*, **42**, 2014.
24. Peters, N., Laminar diffusion flamelet models in non-premixed turbulent combustion, *Progress in Energy and Combustion Science*, **10**, 319–339, 1984.
25. Peters, N., Laminar flamelet concepts in turbulent combustion, *Symposium (International) on Combustion*, **21**, 1231–1250, 1988.
26. Benim, A. C., and Syed, K. J., Laminar flamelet modelling of turbulent premixed combustion, *Applied Mathematical Modelling*, **22**, 113–136, 1998.
27. Kee, R. J., Miller, J. A., Evans, G. H. and Dixon-Lewis, G., A computational model of the structure and extinction of strained, opposed flow, premixed methane-air flames, *Symposium (International) on Combustion*, **22**, 1479–1494, 1989.
28. Shu, Z., Aggarwal, S. K., Katta, V. and Puri, I. K., Flame-vortex dynamics in an inverse partially premixed combustor: The froude number effects, *Combustion and Flame*, **111**, 276–295, 1997.
29. Aggarwal, S. K., Extinction of laminar partially premixed flames, *Programming Energy Combinatorial Science*, **35**, 528–528. 2009.
30. Aggarwal, S. K. and Puri, I. K., Flame structure interactions and state relationships in an unsteady partially premixed flame, *AIAA Journal*, **36**, 1190–1199, 1998.
31. Arana, C. P., Pontoni, M., Sen, S. and Puri, I. K., Field measurements of soot volume fractions in laminar partially premixed coflow ethylene/air flames, *Combustion and Flame*, **134**, 362–372, 2004.
32. Lin, K. C., Sunderland, P. B. and Faeth, G. M., Soot nucleation and growth in acetylene air laminar coflowing jet diffusion flame, *Combustion and Flame*, **104**, 369–375, 1996.

33. Dobbins, R. A., Fletcher, R. A. and Chang, H. C., The evolution of soot precursor particles in a diffusion flame, *Combustion and Flame*, **115**, 285–298, 1998.
34. Frenklach, M., Reaction mechanism of soot formation in flames, *Physical Chemistry Chemical Physics*, **4**, 2028–2037, 2002.
35. D'Anna, A., Combustion formed nanoparticles, *Proceeding Combustion Institute*, **32**, 593–613, 2009.
36. Kent, J. H., A quantitative relationship between soot yield and smoke point measurements, *Combustion and Flame*, **63**, 349–358, 1986.

Chapter 8

Droplet and Spray Combustion

A majority of combustion-powered devices use liquid fuels. Liquid fuels possess several advantages, like high energy content per unit volume, compared to gaseous fuels due to the high density of liquids. Liquid fuels are used in diverse applications like automobiles, airplanes, spacecraft, oil-fired furnaces and boilers. In most of these applications, liquid is introduced into the combustion chamber in the form of a spray of tiny droplets. These droplets evaporate and the fuel vapour reacts with air. The behaviour of individual droplets inside the combustor strongly influences the emission and combustion characteristics. Hence it is important to study the behaviour of individual droplets inside the combustor.

8.1 SPRAY FORMATION

Liquid fuel is generally introduced in the form of a jet or a sheet. On introduction into the combustion chamber, the jet or sheet is broken up into a spray of fine droplets of diverse sizes and velocities. As the liquid comes out of the orifice into the surrounding gaseous medium, the liquid sheet or jet first develops some ripples or perturbations. Under conditions favourable for spray formation, these perturbations grow, leading to stretching and/or thinning of the jet or sheet till the continuous liquid stream disintegrates into smaller segments called ligaments. These ligaments undergo further disintegration, forming droplets. The breakup of a liquid jet or sheet into droplets is called **primary atomisation**. Droplets formed during primary atomisation may undergo further disintegration into smaller droplets due to various forces like vibration and shear. This process of formation of smaller droplets by breakup of larger droplets is called **secondary atomisation**. Sometimes smaller droplets can collide with one another in regions with a high density of droplets and coalesce into larger droplets.

Spray can be formed in many ways. The most common ways of spray formation are through pressurised jet atomisation and air-blast atomisation. In pressurised jet atomisation, liquid at high pressure is forced through a small orifice at a high velocity. Often the liquid is admitted tangentially to impart a swirling motion, which further aids in breakup. In air-blast atomisation, the momentum of the high-velocity air jet is used to disintegrate the liquid sheet. Since the energy needed for atomisation comes from the air stream, a relatively low velocity of the liquid jet or sheet is sufficient.

The process of spray formation includes many stochastic elements. Hence the size and velocity of droplets formed are described by a probability distribution function. This distribution is represented generally in the form of a histogram. Droplet size distribution is generally represented in terms of the fraction ΔN_i of the droplets whose diameter lies in the diameter interval ΔD_i, with D_i as the middle of the interval. Droplets can also be characterised in terms of cumulative number fraction and cumulative volume fraction. **Cumulative number fraction**, which denotes the fraction of droplets smaller than a given size D_i, is defined as:

$$CNF_k = \frac{\sum_{i=1}^{k} D_i \Delta N_i}{\sum_{i=1}^{\infty} D_i \Delta N_i}$$

Similarly, one can define **cumulative volume fraction** as the volume fraction of all droplets smaller than a given diameter D_i as:

$$CVF_k = \frac{\sum_{i=1}^{k} D_i^3 \Delta N_i}{\sum_{i=1}^{\infty} D_i^3 \Delta N_i}$$

The size distribution is characterised in terms of different average velocities. Some commonly used mean diameters are:

$$\text{Number mean diameter } D_{10} = \frac{\sum_{i=1}^{\infty} D_i \Delta N_i}{\sum_{i=1}^{\infty} \Delta N_i} = \sum_{i=1}^{\infty} D_i \Delta N_i \text{ since } \sum_{i=1}^{\infty} \Delta N_i = 1$$

$$\text{Volume (or mass) mean diameter } D_{30} = \left(\frac{\sum_{i=1}^{\infty} D_i^3 \Delta N_i}{\sum_{i=1}^{\infty} \Delta N_i} \right)^{1/3} = \left(\sum_{i=1}^{\infty} D_i^3 \Delta N_i \right)^{1/3}$$

$$\text{Sauter mean diameter } D_{32} = \frac{\sum_{i=1}^{\infty} D_i^3 \Delta N_i}{\sum_{i=1}^{\infty} D_i^2 \Delta N_i}$$

One of these mean diameters is often taken as a representative single size of the droplet for spray calculations.

Example 8.1

A fuel spray has the following size distribution 25% of the droplets have a 20-micron diameter, 50% of the droplets have a 40-micron diameter, 25% of the droplets have a 60-micron diameter. Calculate each of the following: (a) number mean diameter, (b) Sauter mean diameter, (c) volume mean diameter, (d) CNF and (e) CVF.

Solution

a. Number mean diameter

$$D_{10} = \sum_{i-1}^{\infty} D_i \Delta N_i = (0.25 \times 20 + 0.5 \times 40 + 0.25 \times 60)\text{microns} = 40 \text{ microns}$$

b. Sauter mean diameter

$$D_{32} = \frac{\sum_{i=1}^{\infty} D_i^3 \Delta N_i}{\sum_{i=1}^{\infty} D_i^2 \Delta N_i} = \frac{0.25 \times 20^3 + 0.5 \times 40^3 + 0.25 \times 60^3}{0.25 \times 20^2 + 0.5 \times 40^2 + 0.25 \times 60^2} \text{ microns} = 48.89 \text{ microns}$$

c. Volume mean diameter

$$D_{30} = \left(\sum_{i=1}^{\infty} D_i^3 \Delta N_i \right)^{1/3} = \left(0.25 \times 20^3 + 0.5 \times 40^3 + 0.25 \times 60^3\right)^{1/3} \text{ microns} = 44.48 \text{ microns}$$

d. CNF

D_i (microns)	$CNF_k = \dfrac{\sum_{i=1}^{k} D_i \Delta N_i}{\sum_{i=1}^{3} D_i \Delta N_i}$
20	0.125
40	0.625
60	1

e. CVF

D_i (microns)	$CVF_k = \dfrac{\sum_{i=1}^{k} D_i^3 \Delta N_i}{\sum_{i=1}^{3} D_i^3 \Delta N_i}$
20	0.02
40	0.39
60	1

8.2 EVAPORATION OF A SINGLE DROPLET

In most situations, the droplets are too densely packed and there is not enough oxygen supply in their vicinity to sustain distinct flames around individual droplets. Instead, in most situations, individual droplets evaporate and contribute to a common pool of fuel vapour. Thus, it is important to analyze the evaporation of a single droplet in a hot environment. The major assumptions in the analysis of evaporation of a single droplet are:

1. Relative motion between the droplets and the carrier gas phase, inevitable in a practical situation, affects heat and mass transfer from the droplet, leading to modification in the evaporation rate. However, very small-sized droplets tend to acquire the velocity of the gas phase very early, and hence relative velocity between the two becomes negligible. Hence, relative motion between the droplet and the surrounding gas phase can be neglected. Thus, the droplet can be assumed to be evaporating in a quiescent medium, considerably simplifying the analysis.

2. It can also be shown that, due to much higher inertia of the liquid phase, the vapour phase processes have a much shorter timescale than their liquid phase counterpart. This observation can lead to a simplification that the vapour phase responds instantly to changes in boundary conditions and hence the processes can be assumed to be quasi-steady.
3. The droplet is assumed to consist of a single component. It is also assumed that the gases in the vicinity of the droplet do not dissolve in the droplet. This latter assumption becomes invalid at very high pressures (e.g., at near-critical or super-critical pressures) when gases like nitrogen dissolve in the droplet. This assumption also breaks down in case of alcohol droplets in which moisture from the air may condense and be dissolved in the droplet.
4. The droplet temperature is assumed to be uniform. This assumption, which eliminates the need to solve the transport equations in the liquid phase, is strictly valid for very small droplets but can be reasonably accurate for larger droplets.
5. The initial heat-up period of the droplet is neglected. Instead, the droplet is assumed to be at the boiling point of the liquid.
6. Binary diffusion is assumed for mass diffusion, with the process being described by Fick's law.

With the above assumptions, the transport process can be described by a steady, one-dimensional, spherically symmetric model. Although the gas phase is considered quiescent, convective effects arise due to the motion generated by evaporation of the liquid (Stefan flow). With the above assumptions, the transport equations in the spherical coordinate system can be written as follows.

Mass:

$$\frac{d}{dr}\left(\rho r^2 V_r\right) = 0 \Rightarrow \rho r^2 V_r = \frac{\dot{m}}{4\pi} = const \tag{8.1}$$

Species:

$$\frac{d}{dr}\left(\rho r^2 V_r y_F - r^2 \rho_g D_g \frac{dy_F}{dr}\right) = \frac{d}{dr}\left(\dot{m} y_F - 4\pi r^2 \rho_g D_g \frac{dy_F}{dr}\right) = 0 \tag{8.2}$$

Integrating Equation (8.2) and noting that the total mass flux in the vapour phase is equal to that of the fuel vapour, one obtains:

$$\dot{m} y_F - 4\pi r^2 \rho_g D_g \frac{dy_F}{dr} = const = \dot{m}$$

The above can be rearranged as:

$$\frac{dy_F}{1-y_F}=-\frac{\dot{m}}{4\pi\rho_g D_g}\frac{dr}{r^2} \tag{8.3}$$

Integrating Equation (8.3) from $r=r_d$ (where r_d is the droplet radius) to $r\to\infty$ and using the boundary conditions $y_F=y_{F,s}$ at $r=r_d$ and $y_F\to y_{F,\infty}$ as $r\to\infty$, the spatial distribution of fuel vapour is obtained as:

$$\ln\frac{1-y_{F,s}}{1-y_{F,\infty}}=\frac{dy_F}{1-y_F}=-\frac{\dot{m}}{4\pi\rho_g D_g r_d}$$

Assuming $y_{F,\infty}=0$, the mass flux is obtained as:

$$\dot{m}=-4\pi\rho_g D_g r_d\ln(1-y_{F,s})=4\pi\rho_g D_g r_d\ln\left(\frac{1}{1-y_{F,s}}\right)=4\pi\rho_g D_g r_d\ln(1+B_Y) \tag{8.4}$$

Here, $B_Y=\frac{y_{F,s}}{1-y_{F,s}}$ is known as the **Spalding transfer number** (based on mass transfer).

The mass of the droplet is given by:

$$m_d=\frac{4}{3}\pi r_d^3\rho_l \tag{8.5}$$

Differentiating Equation (8.5) and noting that $\dot{m}=-\frac{dm_d}{dt}$, one can obtain:

$$\dot{m}=4\pi\rho_g D_g r_d\ln(1+B_Y)=-4\pi\rho_l r_d^2\frac{dr_d}{dt}=-4\pi\rho_l\frac{d}{dt}\left(\frac{r_d^2}{2}\right)$$

Rearranging the above, one can write:

$$\frac{d}{dt}(D^2)=-8\frac{\rho_g}{\rho_l}D_g\ln(1+B_Y)=-K=const \tag{8.6}$$

where $D=2r_d$ is the droplet diameter. K is a constant for a given value of fuel concentration at the droplet surface and is known as the **evaporation constant.**

Equation (8.6) gives the rate of evaporation of an isolated stationary droplet in a stationary medium. It shows that the surface area of the droplet decreases as a linear function of time, and the relation is referred to as **D²-law**. For a droplet with initial diameter D_0, the droplet lifetime is estimated as $t_d=\frac{D_0^2}{K}$.

The use of this expression requires knowledge of the fuel concentration at the droplet surface.

An alternative expression for the evaporation constant can be analogously derived from the energy equation. The energy equation can be calculated as:

$$\frac{d}{dr}\left(\rho r^2 V_r C_p T - r^2 k \frac{dT}{dr}\right) = \frac{d}{dr}\left(\dot{m} C_p T - 4\pi r^2 k \frac{dT}{dr}\right) = 0 \tag{8.7}$$

Equation (8.7) implies that:

$$\dot{m} C_p T - 4\pi r^2 k \frac{dT}{dr} = const = \left[\dot{m} C_p T - 4\pi r^2 k \frac{dT}{dr}\right]_{r=r_d} = \dot{m} C_p T_s - 4\pi r_d{}^2 k \left.\frac{dT}{dr}\right|_{r=r_d} \tag{8.8}$$

where T_s denotes the temperature at the droplet surface. Since the initial heat-up period is neglected, all the heat supplied from the gas phase to the droplet is used to provide the latent heat of evaporation. Thus, $4\pi r_d{}^2 k \frac{dT}{dr}\big|_{r=r_d} = \dot{m} h_{fg}$. Substituting this into Equation (8.8), one obtains:

$$\dot{m}\left[C_p (T - T_s) + h_{fg}\right] = 4\pi r^2 k \frac{dT}{dr}$$

After some algebraic steps, the above equation leads to the following expression for the evaporation rate of the droplet:

$$\dot{m} = 4\pi k r_d \ln\left[\frac{C_p (T_\infty - T_s) + h_{fg}}{h_{fg}}\right] = 4\pi k r_d \ln(1 + B_q) \tag{8.9}$$

Here, $B_q = \frac{C_p (T_\infty - T_s)}{h_{fg}}$ is also known as Spalding transfer number (based on heat transfer).

Following similar steps as in the previous case, one can obtain the evaporation rate as:

$$K = -\frac{d}{dt}\left(D^2\right) = 8 \frac{k}{\rho_l C_p} \ln\left(1 + B_q\right) \tag{8.10}$$

From Equation (8.10), the evaporation rate can be calculated, provided the temperature at the droplet surface is known. Often, a simplifying assumption is made that at the quasi–steady state of evaporation, the droplet is at the boiling point of the liquid. This assumption follows from a simplified model that the droplet initially gets heated to its boiling point with negligible evaporation and then undergoes evaporation after reaching the boiling point with no further increase in temperature. This approximation is not justifiable, however, under all conditions. A more involved analysis, shown below, allows one to relax this assumption and determine the temperature and fuel vapour concentration at the droplet surface as part of the solution.

Comparing Equations (8.6) and (8.10), one can write:

$$\frac{k}{\rho_l C_p}\ln\left(1+B_q\right)=\frac{\rho_g}{\rho_l}D_g\ln\left(1+B_Y\right)$$

Assuming the Lewis number for vapour phase, $Le=\frac{k}{\rho_g D_g C_p}=1$, one can write $B_Y=B_q$, which leads to:

$$y_{F,s}=\frac{C_p\left(T_\infty-T_s\right)}{h_{fg}+C_p\left(T_\infty-T_s\right)} \tag{8.11}$$

The surface temperature and fuel vapour concentration can also be related through the phase equilibrium relation. The phase equilibrium can be expressed in terms of the Clapeyron equation as:

$$\frac{dp_{sat}}{dT}=\frac{h_{fg}}{Tv_{fg}} \tag{8.12}$$

For ideal gases at sufficiently low pressures, where $v_g >> v_f$, and using the equation of state $p_{sat}v_g = RT$, one can simplify the Clapeyron equation to obtain the Clausius–Calpeyron equation:

$$\frac{dp_{sat}}{dT}=p_{sat}\frac{h_{fg}}{RT^2} \tag{8.13}$$

Assuming h_{fg} to be constant and integrating Equation (8.13), one can obtain:

$$\ln\frac{p_{sat}}{p_{atm}}=-\frac{h_{fg}}{R}\left(\frac{1}{T_s}-\frac{1}{T_{nb}}\right) \tag{8.14}$$

where T_{nb} is the normal boiling point (i.e., the boiling point at atmospheric pressure). From the definition of partial pressure, one can write $p_{sat}=x_{F,s}P$, where P is the total pressure and the mole fraction of the fuel vapour at the droplet surface, $x_{F,s}$, is related to the corresponding mass fraction at the droplet surface, $y_{F,s}$, as follows:

$$y_{F,s}=\frac{x_{F,s}M_F}{x_{F,s}M_F+x_{NC,s}M_{NC}}=\frac{x_{F,s}M_F}{x_{F,s}M_F+\left(1-x_{F,s}\right)M_{NC}} \tag{8.15}$$

Substituting Equation (8.15) into Equation (8.14), the relation between $y_{F,s}$ and T_s is obtained as follows:

$$y_{F,s} = \frac{M_F \dfrac{P}{p_{atm}} \exp\left[-\dfrac{h_{fg}}{R}\left(\dfrac{1}{T_s} - \dfrac{1}{T_{nb}}\right)\right]}{\left(M_F - M_{NC}\right) \dfrac{P}{p_{atm}} \exp\left[-\dfrac{h_{fg}}{R}\left(\dfrac{1}{T_s} - \dfrac{1}{T_{nb}}\right)\right] + M_{NC}} \tag{8.16}$$

From Equations (8.11) and (8.16), one can write:

$$\frac{M_F \dfrac{P}{p_{atm}} \exp\left[-\dfrac{h_{fg}}{R}\left(\dfrac{1}{T_s} - \dfrac{1}{T_{nb}}\right)\right]}{\left(M_F - M_{NC}\right) \dfrac{P}{p_{atm}} \exp\left[-\dfrac{h_{fg}}{R}\left(\dfrac{1}{T_s} - \dfrac{1}{T_{nb}}\right)\right] + M_{NC}} = \frac{C_p\left(T_\infty - T_s\right)}{h_{fg} + C_p\left(T_\infty - T_s\right)} \tag{8.17}$$

The solution of Equation (8.17) gives the temperature at the droplet surface. This temperature is often referred to as the **wet bulb temperature** and is somewhat lower than the normal boiling point. Once the temperature at the droplet surface is known, the evaporation rate can be calculated using Equation (8.10).

Example 8.2

Calculate the evaporation rate and lifetime of a 70-micron heptane droplet evaporating in an ambient at 1000 K temperature. Assume the droplet to be at its normal boiling point of 371 K. Take the density and enthalpy of vaporisation for heptane as 684 kg/m³ and 316 kJ/kg, respectively. The specific heat and thermal conductivity of the gas phase are 1.06 kJ/kgK and 0.039W/mK, respectively.

Solution

$$\text{Transfer number } B_q = \frac{C_p\left(T_\infty - T_s\right)}{h_{fg}} = 2.1$$

$$\text{Evaporation rate } K = 8\frac{k}{\rho_l C_p}\ln\left(1 + B_q\right) = 4.88 \times 10^{-7}\,\mathrm{m^2/s}$$

$$\text{Droplet lifetime } t_d = \frac{D_0^2}{K} = 100 \text{ microseconds}$$

8.3 COMBUSTION OF A SINGLE DROPLET

Contrary to the scenario described in the previous section, sometimes flames form around individual droplets. In these cases, the fuel vapour originating from an individual droplet and oxygen from the immediate surroundings of the droplet approach each other and a flame is formed at the location where the fuel and the oxidiser are present in stoichiometric proportions. For simplicity, an infinite reaction rate is assumed. This assumption implies an infinitesimally thin flame, separating the fuel and the oxidiser sides. Infinite reaction rate also implies that both the fuel and oxidiser are completely consumed at the flame. Thus, no fuel is present on the oxidiser side, and vice versa. It also implies that both fuel and oxidiser concentrations become zero at the flame. With these assumptions, the model for combustion of an individual stationary droplet in a quiescent infinite oxidiser medium can be developed as follows. The governing equations for the fuel side transport are similar to that developed in the previous section. Only the boundary at infinity has to be replaced by the boundary at the flame. The transport equations on the fuel side are given by the following.

Species equation:

$$\frac{d}{dr}\left(\dot{m}y_F - 4\pi r^2 \rho_g D_g \frac{dy_F}{dr}\right) = 0 \tag{8.18}$$

The boundary conditions are given by:

$$r = r_d : y_F = y_{F,s};\ r = r_f : y_F = 0 \tag{8.19}$$

Using the boundary conditions given by Equation (8.19) and following a procedure identical to that in the previous section, one obtains:

$$\ln\left(1 - y_{F,s}\right) = -\frac{\dot{m}}{4\pi\rho_g D_g}\left(\frac{1}{r_d} - \frac{1}{r_f}\right) \tag{8.20}$$

Energy equation:

$$\frac{d}{dr}\left(\dot{m}C_p T^- - 4\pi r^2 k \frac{dT^-}{dr}\right) = 0 \tag{8.21}$$

The minus sign in the superscript denotes temperature on the fuel side; the plus sign in the superscript denotes the temperature on the oxidiser side.

The boundary conditions are given by:

$$r = r_d : T^- = T_s;\ r = r_f : T^- = T_f$$

Proceeding in a similar manner as in the previous section, the solution is obtained as follows:

$$\ln\left(\frac{T_f - T_s + \frac{h_{fg}}{C_p}}{\frac{h_{fg}}{C_p}}\right) = -\frac{\dot{m}C_p}{4\pi k}\left(\frac{1}{r_d} - \frac{1}{r_f}\right) \tag{8.22}$$

The conservation equations and the boundary conditions on the oxidiser side are as follows.

Species:

$$\frac{d}{dr}\left(\dot{m}y_o - 4\pi r^2 \rho_g D_g \frac{dy_o}{dr}\right) = 0 \tag{8.23}$$

The boundary conditions are given by:

$$r = r_f : y_o = 0; \quad r \to \infty : y_o \to y_{o,\infty} \tag{8.24}$$

Energy:

$$\frac{d}{dr}\left(\dot{m}C_p T^+ - 4\pi r^2 k \frac{dT^+}{dr}\right) = 0 \tag{8.25}$$

The boundary conditions are given by:

$$r = r_f : T^+ = T_f; \quad r \to \infty : T^+ \to T_\infty \tag{8.26}$$

Integrating Equation (8.23) and noting that the oxidiser and the fuel mass fluxes must be in stoichiometric proportions, one obtains:

$$\dot{m}y_o - 4\pi r^2 \rho_g D_g \frac{dy_o}{dr} = const = \dot{m}_o = -\nu \dot{m}_F = -\nu \dot{m} \tag{8.27}$$

Integrating the above equation and using the boundary conditions from Equation (8.24), one obtains:

$$\ln \frac{y_{o,\infty} + \nu}{\nu} = \frac{\dot{m}}{4\pi \rho_g D_g r_f} \tag{8.28}$$

Integrating Equation (8.25), one obtains:

$$\dot{m}C_p T^+ - 4\pi r^2 k \frac{dT^+}{dr} = const = \dot{m}C_p T_f - 4\pi r^2 k \left.\frac{dT^+}{dr}\right|_{r_f} \tag{8.29}$$

The temperature gradient at the flame $\left.\frac{dT^+}{dr}\right|_{r_f}$ is evaluated from the energy balance at the flame. The energy balance at the flame is obtained as follows:

$$-4\pi r_f^2 k \left.\frac{dT^+}{dr}\right|_{r_f} + 4\pi r_f^2 k \left.\frac{dT^-}{dr}\right|_{r_f} = \dot{m}_F \Delta h_C = \dot{m}\Delta h_C \tag{8.30}$$

From Equation (8.21) and using the relation $4\pi r_d^2 k \frac{dT^-}{dr}\Big|_{r=r_d} = \dot{m}h_{fg}$, one can write:

$$\left[\dot{m}C_p T^- - 4\pi r^2 k \frac{dT^-}{dr}\right]_{r_f} = \left[\dot{m}C_p T^- - 4\pi r^2 k \frac{dT^-}{dr}\right]_{r_d} \Rightarrow 4\pi r_f^2 k \frac{dT^-}{dr}\Big|_{r_f}$$

$$= \dot{m}\left[C_p(T_f - T_s) + h_{fg}\right]$$

Hence, from Equation (8.30), one can write:

$$4\pi r_f^2 k \frac{dT^+}{dr}\Big|_{r_f} = 4\pi r_f^2 k \frac{dT^-}{dr}\Big|_{r_f} - \dot{m}\Delta h_C = \dot{m}\left[C_p\left(T_f - T_s\right) + h_{fg} - \Delta h_C\right] \quad (8.31)$$

Substituting Equation (8.31) into Equation (8.29), one obtains:

$$\dot{m}\left[C_p\left(T^+ - T_s\right) + h_{fg}\right] - 4\pi r^2 k \frac{dT^+}{dr} = \dot{m}\Delta h_C \quad (8.32)$$

Multiplying Equation (8.27) by Δh_C, one obtains:

$$\dot{m}\left(\frac{y_O}{\nu}\Delta h_C\right) - 4\pi r^2 \rho_g D_g \frac{d}{dr}\left(\frac{y_O}{\nu}\Delta h_C\right) - \dot{m}\Delta h_C \quad (8.33)$$

Adding Equations (8.32) and (8.33), one obtains:

$$\dot{m}\left[C_p\left(T^+ - T_s\right) + h_{fg} + \frac{y_O}{\nu}\Delta h_C\right] - 4\pi r^2 \frac{k}{C_p}\frac{d}{dr}\left[C_p\left(T^+ - T_s\right) + h_{fg}\right]$$

$$-4\pi r^2 \rho_g D_g \frac{d}{dr}\left(\frac{y_O}{\nu}\Delta h_C\right) = 0$$

Assuming that the gas phase Lewis number $Le = \frac{k}{\rho_g D_g C_p} = 1$, the above equation simplifies to:

$$\dot{m}\left[C_p(T^+ - T_s) + h_{fg} + \frac{y_O}{\nu}\Delta h_C\right] - 4\pi r^2 \rho_g D_g \frac{d}{dr}\left[C_p(T^+ - T_s) + h_{fg} + \frac{y_O}{\nu}\Delta h_C\right] = 0 \quad (8.34)$$

Integrating Equation (8.34) and using the boundary conditions in Equations (8.24) and (8.26), one obtains:

$$\ln \frac{C_p\left(T_\infty - T_s\right) + h_{fg} + \frac{y_{O,\infty}}{\nu}\Delta h_C}{C_p\left(T_f - T_s\right) + h_{fg}} = \frac{\dot{m}}{4\pi \rho_g D_g}\frac{1}{r_f} \quad (8.35)$$

Using Equations (8.22) and (8.35), one obtains:

$$\frac{\dot{m}}{4\pi\rho_g D_g} = r_d \ln \frac{C_p\left(T_\infty - T_s\right) + h_{fg} + \frac{y_{o,\infty}}{\nu}\Delta h_C}{h_{fg}} = r_d \ln\left(1 + B_q\right)$$

Here, $B_q = \frac{C_p\left(T_\infty - T_s\right) + \frac{y_{o,\infty}}{\nu}\Delta h_C}{h_{fg}}$ is the transfer number for a burning droplet.

From Equations (8.28) and (8.35), one obtains the **flame radius** as:

$$r_f = \frac{\dot{m}}{4\pi\rho_g D_g \ln \frac{y_{o,\infty} + \nu}{\nu}} = \frac{r_d \ln\left(1 + B_q\right)}{\ln\left(1 + \frac{y_{o,\infty}}{\nu}\right)} \tag{8.36}$$

Another quantity of interest is the **flame stand-off ratio**, which is defined as:

$$\frac{r_f}{r_d} = \frac{\ln\left(1 + B_q\right)}{\ln\left(1 + \frac{y_{o,\infty}}{\nu}\right)} \tag{8.37}$$

Using Equations (8.28) and (8.35), one can obtain the **flame temperature** as:

$$T_f = T_s + \frac{h_{fg}}{C_p}\frac{\left(\nu B_q - y_{o,\infty}\right)}{\left(\nu + y_{o,\infty}\right)} \tag{8.38}$$

The above results show that all the important parameters can be determined once the droplet surface temperature is known. Following an approach similar to the one in the last section and using Equations (8.20) and (8.22), one can write:

$$-\frac{\dot{m}}{4\pi\rho_g D_g}\left(\frac{1}{r_d} - \frac{1}{r_f}\right) = \ln\left(1 - y_{F,s}\right) = \ln\left(\frac{T_f - T_s + \frac{h_{fg}}{C_p}}{\frac{h_{fg}}{C_p}}\right)$$

This can be simplified to:

$$y_{F,s} = \frac{B_q - \frac{y_{o,\infty}}{\nu}}{B_q + 1} \tag{8.39}$$

Equation (8.39), along with Equation (8.17), provides the two relations from which $y_{F,s}$ and T_s can be determined.

8.4 MULTICOMPONENT, HIGH-PRESSURE CONVECTIVE EFFECTS

The analyses presented in the two previous sections are valid for isolated stationary droplets of pure liquid fuels in a quiescent environment. In reality, however, fuels like gasoline, diesel and kerosene that are used in different applications are almost invariably multicomponent in nature. For multicomponent droplets, the components with lower boiling points evaporate from the surface at a faster rate. The supply of those components from the droplet core to the surface is largely governed by the mass diffusivity. Generally mass diffusivity in the liquid phase is much slower than the evaporation rate. This leads to a depletion of the volatile components near the surface and high concentration at the core, leading to a concentration gradient. This calls for solution of transport equations within the liquid phase also, unless the evaporation rate is very slow compared to the mass diffusivity. In case of multicomponent fuels, the steady state analysis leading to D^2-law is not ideally valid as the droplets always undergo both heating and significant evaporation throughout their lifetime. This is especially true for bicomponent droplets having components with widely varying boiling points. For such cases, several stages, several stages exist like the initial heating period, the phase dominated by significant evaporation of the lower boiling point component, the second heat-up period after near-complete depletion of the low boiling point component from the surface and a second evaporation phase dominated by evaporation of the high boiling point component. For multicomponent liquids or for bicomponent liquids with small differences in boiling points, these stages are generally not very prominent.

Another complexity neglected in the earlier phases of analysis is the effect of relative motion between the droplets and the gas phase. Although this effect was neglected in earlier sections because of the small size of the droplets, such effects can become important for large droplets. Even for small droplets, the high velocity of the gas stream can cause significant convective effects. The presence of convection increases the rate of heat and mass transfer between the droplet and the gas phase. The shearing action of the gas flow also induces circulation inside the droplet, leading to improved mixing of heat and species within the droplet.

Many of the assumptions made in earlier sections are valid only for low pressure. In most practical applications, however, sprays are introduced at very high pressures and temperatures. In some cases, the pressure exceeds the critical pressure of the liquid also. At these near critical, trans-critical and super-critical conditions, several assumptions break down. First, in the super-critical state, there is no distinction between the liquid and gas phases, and latent heat ceases to be meaningful. Also, at high pressures, the ideal gas equation of state needs to be replaced by real gas equations like the Redlich–Kwong equation or the Peng–Robinson equation. Ambient gases like nitrogen have appreciable solubility in hydrocarbon liquids at high pressure. Thus, even nominally single-component droplets acquire a multicomponent character.

8.5 PHYSICAL DESCRIPTION OF SPRAY COMBUSTION

Liquid fuel is introduced into the combustor in the form of a spray of tiny droplets. When these droplets are injected into a hot environment, which is typically encountered in a combustor, the droplets evaporate, creating a fuel-rich region in the vicinity of the droplets. Conversely, the region away from the droplet cluster is rich in air (oxidiser). Fuel and air diffuse towards each other, leading to formation of a predominantly non-premixed flame. As mentioned in Sections 8.2 and 8.3, two limiting situations arise. At one limit, oxygen penetrates in sufficient quantity close to the droplet surface such that envelope flames form around individual droplets. This is observed generally for very dilute sprays with large spacing between the droplets or when there is a very strong entrainment of air into the spray. At the other limit, when the evaporation rate is very fast or the droplets are closely spaced and thus form a dense spray, there is very little oxygen penetration into the interior of the spray. In such situations, the fuel evaporating from individual droplets creates a common vapour pool which reacts with air to create a common flame envelope around the entire cluster of droplets. Chiu and co-workers [1,2] identified the different regimes of spray combustion in terms of a **group combustion number:**

$$G = n\frac{D_\infty R_\infty r_d}{U_\infty}\left[4\pi\left(1+0.276\,\mathrm{Re}^{1/2}\,Sc^{1/3}\right)Le\right] \tag{8.40}$$

Here, n denotes the droplet number density (i.e., the number of droplets per unit volume); D_∞ and U_∞ are the mass diffusivity and velocity of the gas phase in the free stream, respectively; r_d denotes the mean droplet radius and R_∞ is the radius of the spray cluster while Re, Sc and Le represent the Reynolds number, Schmidt number and Lewis number, respectively.

At very high values of droplet number density, G also has very high values ($G > 100$). For these conditions, the inner core of the droplet is so densely packed with droplets that there is very little evaporation of the droplets. Outside the non-evaporating core, an annular region exists where evaporation takes place from the droplets. However, penetration of oxygen is much less, even in the outer evaporating region. Hence, no flames are formed around individual droplets. The fuel evaporating from individual droplets diffuses outwards and forms an external flame surrounding the entire cluster. This regime of spray combustion is known as **external group combustion with sheath evaporation.**

As the droplet number density decreases slightly, the non-evaporating core decreases in size. When $G \approx 1$, evaporation extends to the entire cluster. The flame still occurs as a common envelope around the entire cluster. This mode of combustion is known as **external group combustion with stand-off flame.**

With an additional decrease in droplet number density in the range of $0.01 < G < 1$, the main flame exists inside the cluster. The droplets at the interior of the cluster evaporate, and the resulting fuel vapour generates an envelope flame around the inner core of droplets. Outside the main flame, however, there is enough oxygen penetration in the vicinity of the droplets at the outer periphery of the cluster to form envelope flames around individual droplets. This regime is known as **internal group combustion.**

Finally, for very dilute sprays, when $G < 0.01$, oxygen can penetrate up to the centre of the droplet cluster, and all the droplets burn with individual flames. This is known as an **individual droplet combustion regime.**

8.6 SIMPLIFIED ANALYSIS OF SPRAY COMBUSTION

In this section, a simplified one-dimensional analysis of spray combustion is presented. A steady state, steady flow, one-dimensional combustor is considered. The analysis involves two sets of equations: (1) gas phase equations and (2) droplet equations. In addition, another set of equations for liquid phase may also be solved for modelling the transport processes inside the droplet. In many situations, however, this third set of equations, for obtaining the scalar fields inside the droplet, is not solved as the droplets are considered lumped parameters. The transport equations for the gas phase are identical to that described in the case of reacting flows with gaseous fuels, except for source terms in the equations representing a contribution to the gas phase transport from the droplets. A summary of the one-dimensional model is given below.

8.6.1 Gas Phase

Mass:

$$\frac{\partial \rho}{\partial t} + \frac{\partial}{\partial x}(\rho u) = n\dot{m} \tag{8.41}$$

Species:

$$\frac{\partial(\rho y_i)}{\partial t}+\frac{\partial}{\partial x}(\rho u y_i)=\frac{\partial}{\partial x}\left(\rho D\frac{\partial y_i}{\partial x}\right)+n\dot{m}\delta_{iF}+\omega_i \tag{8.42}$$

Energy:

$$\rho C_p\frac{\partial T}{\partial t}+\rho C_p u\frac{\partial T}{\partial x}=\frac{\partial}{\partial x}\left(k\frac{\partial T}{\partial x}\right)+n\dot{m}\left[h_{fg}+C_p(T-T_s)\right]-\omega_F\Delta h_c \tag{8.43}$$

Here n and $\dot{m}$ denote droplet number density and evaporation rate from a single droplet, respectively. The subscript i stands for F (fuel), o (oxidiser) and p (product); δ_{iF} denotes Kronecker delta, which is defined as follows:

$$\begin{aligned}\delta_{iF} &= 1, i = F\\ &0,\ i \neq \mathrm{F}\end{aligned} \tag{8.44}$$

8.6.2 Droplet Equations

The droplets are tracked in a Lagrangian manner, with each droplet considered a point mass. From knowledge of the initial position and velocity of the droplets and the force acting on each droplet, it is possible to calculate the instantaneous position of the droplet. The source terms in the gas phase equations for conservation of mass, species and energy at different locations are evaluated on the basis of the positions of the droplets. Since closed form solutions are generally not possible, the system of equations has to be solved numerically. The solutions for the gas phase variables from the grid points for the gas phase are interpolated to the droplet locations for calculating the evaporation rate and the drag force on the corresponding droplet. Similarly, the contribution from different droplets to the gas phase source terms are assigned to neighbouring grid points for the gas phase solution. The droplet equations are listed below.

Droplet position:

$$\frac{dX_k}{dt}=V_k \tag{8.45}$$

Droplet velocity:

$$\frac{dV_k}{dt}=\frac{1}{m_k}\left[\frac{\pi}{8}C_D\rho_g D_k^2|u-V_k|(u_k-V_k)\right] \tag{8.46}$$

Droplet temperature:

$$\frac{dT_k}{dt}=\frac{1}{m_kC_p}\left[\pi NukD_k\left(T-T_k\right)-\frac{dm_k}{dt}h_{fg}\right] \tag{8.47}$$

The subscript k signifies kth droplet. The evaporation rate in a convective environment is obtained as follows:

$$\frac{dm_k}{dt}=\pi\rho ShD_gD\left(y_{F,s}-y_{F,\infty}\right) \tag{8.48}$$

The drag coefficient C_D, Nusselt number Nu and Sherwood number Sh are determined from appropriate correlations.

Although the above model is very simple, it has the necessary framework for a typical spray combustion analysis. For practical combustion systems, the flow is almost invariably turbulent, so appropriate turbulence models and models for the turbulence-chemistry interaction are needed in the gas phase model.

EXERCISES

8.1 Consider the following distribution of droplets: one-sixth of the droplets have a diameter of 10 microns, one-third have a diameter of 20 microns, one-third have a diameter of 30 microns and one-sixth have a diameter of 40 microns. Calculate the SMD, CNF and CVF.

8.2 Determine the evaporation rate of a heptane droplet evaporating in nitrogen gas for an ambient temperature in the range 500 K–1200 K. Compare the evaporation rates at different ambient temperatures with thermal diffusivity of heptane. Assume that the droplet surface temperature is equal to its normal boiling point.

8.3 Determine the wet bulb temperature for a heptane droplet evaporating in nitrogen gas. The ambient conditions are 1000 K and 1 bar. Using this temperature as the droplet surface temperature, determine the evaporation rate. Compare the result with the evaporation rate determined in Example 8.2. Use liquid and gas phase property values given in Example 8.2.

8.4 Consider a stationary hexane droplet of diameter 70 μm burning in air. The ambient conditions are 500 K and 1 bar. Assume the droplet to be at its boiling point, and neglect droplet heating. Calculate the droplet lifetime, flame temperature and flame stand-off ratio. Evaluate the gas phase properties as that of nitrogen at 1200 K. Assume the droplet to have reached steady state and unity Lewis number.

REFERENCES

1. Chiu, H. H. and Liu, T. M., Group combustion of liquid droplets, *Combustion Science and Technology*, **17**, 127–142, 1977.
2. Chiu, H. H., Kim, H. Y. and Croke, E. I., Internal group combustion of liquid droplets, *Nineteenth Symposium (International) on Combustion*, **19**, 971–980, 1983.

Chapter 9

Modelling of Turbulent Combustion

9.1 TURBULENT FLOWS

The study of turbulent fluid flows – as in any other scientific studies – requires a tractable quantitative theory or a model which can be utilised to evaluate quantities that are of interest and have practical relevance. The inherent characteristics of turbulent flows make it difficult, however, to develop an accurate theory or model.

The velocity field is three-dimensional, time-dependent as well as random. Scale of the largest turbulent motions are comparable to the characteristic length of the flow, and hence they are significantly affected by the flow geometry. In contrast, the smallest length scales are comparable to the molecular dimensions. The associated time scales are also of such varied nature. Difficulties also arise due to the non-linear nature of the convective term and the pressure-gradient term in the Navier–Stokes equation. Any theory or model dealing with turbulence must take into account this wide range of length and time scales and should also be able to handle the non-linear character of the governing equations.

All turbulence modelling approaches which have been developed to date are based on solving the Navier–Stokes equation for determining the flow velocity and the associated properties. The differences between these approaches are due to the resolution required for the associated length and time scales. Direct numerical simulation (DNS) of turbulent flows involves solving the Navier–Stokes equations while resolving all the scales of motion. Conceptually, DNS is the simplest of all turbulence modelling approaches since it does not require any additional modelling and assumptions. It has the highest accuracy and produces the most detailed level of description of the flow. The only constraint is the extremely high computational cost associated with DNS. The cost increases rapidly with Reynolds number (cost is proportional to Re^3); as such, this approach is limited to flows with low and moderate Reynolds numbers. This hinders the widespread use of this approach. The modelling resolution needs to be compromised in order to bring down the computational cost.

This is precisely what is followed in the large eddy simulation (LES) approach, where the equations are solved for a filtered velocity field that is representative of the large-scale turbulent motions. The influence of the smaller-scale motions are accounted for with additional models. In contrast to these approaches, the Reynolds-averaged Navier–Stokes (RANS) equations solve for the mean velocity field while the effects of the fluctuating component of velocity are modelled. Several models are available for determining these effects. The resolution achieved using this approach is much lower and, as such, the computational cost for solving the RANS equations is much less compared to the LES equations.

The RANS and LES approaches are discussed in further detail in the following sections.

9.2 MODELLING APPROACHES FOR TURBULENCE

9.2.1 Function Decomposition Techniques

All turbulence models utilise some form of decomposition of the dependent variables representing the physical situation. Several techniques are available for carrying out such decomposition. Among these, the most generally used are the Reynolds decomposition and the Favre decomposition techniques. These are discussed below.

Reynolds decomposition technique: Using this technique, any parameter (ϕ) can be split into a mean component and a fluctuating component as follows [1,2]:

$$\phi = \overline{\phi} + \phi' \tag{9.1}$$

In the above equation, $\overline{\phi}$ represent the mean component, while ϕ' represent the fluctuating component. The mean quantity is defined in the following manner:

$$\overline{\phi} = \lim_{\tau \to \infty} \frac{1}{\tau} \int_0^{\tau} \phi(t)\,dt \tag{9.2}$$

This implies that the mean component is time-independent; hence, these must be computed at steady state only. Some useful relations derived from the Reynolds decomposition are summarised below:

$$\overline{\overline{\phi}} = \overline{\phi} \tag{9.3}$$

$$\overline{\phi'} = 0 \tag{9.4}$$

$$\overline{\bar{\phi}\phi'} = \overline{\phi'\bar{\phi}} = 0 \tag{9.5}$$

Favre decomposition technique: The Favre decomposition technique decomposes any parameter (ϕ) into its mean and fluctuating components as follows [3]:

$$\phi = \tilde{\phi} + \phi'' \tag{9.6}$$

In the above equation, $\tilde{\phi}$ represent the Favre-mean component, while ϕ'' represent the Favre-fluctuating component. The Favre-mean quantity is defined in the following manner:

$$\tilde{\phi} = \frac{\overline{\rho\phi}}{\bar{\rho}} \tag{9.7}$$

The following properties of Favre decomposition find extensive use:

$$\tilde{\tilde{\phi}} = \tilde{\phi} \tag{9.8}$$

$$\tilde{\phi''} = 0 \tag{9.9}$$

It can also be shown that the Reynolds-mean ($\bar{\phi}$) and the Favre-mean ($\tilde{\phi}$) components are not equivalent except in incompressible situations.

9.2.2 Length and Time Scales in Turbulence

One of the basic characteristics of any turbulent flow is the existence of a wide range of length and time scales [2]. This feature is essentially what presents the difficulty in numerical modelling of turbulence. Thus, it is especially important to understand the various length and time scales associated with turbulent flow.

There are, in general, four main sets of scales in a turbulent flow. They are described in the following subsections.

9.2.2.1 Large Scale

This is the largest of all the scales and is based on the problem domain geometry. The domain size (L) is considered to be the characteristic length, and the mean flow velocity (u) is assumed to be the characteristic velocity. The Reynolds number is then defined as:

$$Re_L = \frac{uL}{\nu} \tag{9.10}$$

where ν denotes the kinematic viscosity. The time scale of flow can then be expressed as:

$$\tau_c = \frac{L}{u} \tag{9.11}$$

τ_c is referred to as the convective time scale. In contrast, the diffusive time scale (τ_d) is defined as:

$$\tau_d = \frac{L^2}{\nu} \tag{9.12}$$

The ratio of the diffusive to the convective time scale gives:

$$\frac{\tau_d}{\tau_c} = \frac{L^2/\nu}{L/u} = Re_L \tag{9.13}$$

indicating that the Reynolds number can also be used to compare the rates of molecular diffusion to the macroscopic convection.

9.2.2.2 Integral Scale

The integral scale typically considers the largest eddies occurring within the flow [2]. These account for the major transport of momentum and energy in the flow. The integral length scale (l) is statistically determined using the fluctuating component of velocity (u'), as follows:

$$l = \frac{1}{\left\| u'^2 \right\|} \int_{-\infty}^{\infty} u'(x,t)u'(x+r,t)dr \tag{9.14}$$

The corresponding time scale and Reynolds number is defined as:

$$\tau_t = \frac{l}{u'} \tag{9.15}$$

$$Re_l = \frac{|u'|l}{\nu} \tag{9.16}$$

τ_t is also often referred to as the integral eddy turnover time. The fluctuating velocity component (u') is defined in terms of the turbulent kinetic energy (k) and the integral length scale (l) as follows:

$$u' = \sqrt{\frac{2k}{3}} \tag{9.17}$$

$$\varepsilon \sim \frac{k^{\frac{3}{2}}}{l} \tag{9.18}$$

$$l \sim \frac{u'^3}{\varepsilon} \tag{9.19}$$

$$\tau_t \sim \frac{k}{\varepsilon} \tag{9.20}$$

9.2.2.3 Taylor Micro-Scale

The Taylor micro-scale falls in the intermediate regime between the integral scale (which considers the largest eddies) and the Kolmogorov scale (which considers the smallest eddies). Fluid viscosity starts influencing the kinetic energy dissipation within the flow typically starting from the Taylor micro-scale [2]. The Taylor micro-scale length is defined as:

$$\lambda = \sqrt{\frac{\nu u'^2}{\varepsilon}} \tag{9.21}$$

Here, ε denotes the dissipation rate of turbulent kinetic energy. The corresponding time scale and Reynolds number is defined as:

$$\tau_\lambda = \frac{\lambda}{u'} \tag{9.22}$$

$$Re_\lambda = \frac{|u'|\lambda}{\nu} \tag{9.23}$$

9.2.2.4 Kolmogorov Scale

The Kolmogorov scale deals with the smallest eddies occurring within the flow [2]. It is assumed that at this small scale, only viscous effects would dominate in the energy dissipation process. The Kolmogorov length scale (η) is defined as:

$$\eta = \left(\frac{\nu^3}{\varepsilon} \right)^{\frac{1}{4}} \tag{9.24}$$

The characteristic velocity in the Kolmogorov range can be scaled as:

$$v_\eta \sim (\nu\varepsilon)^{1/4} \sim (\eta\varepsilon)^{1/3} \sim \frac{u'}{Re_\lambda^{\frac{1}{4}}} \tag{9.25}$$

The corresponding time scale is defined as:

$$\tau_\eta = \frac{\eta}{v_\eta} = \left(\frac{\nu}{\varepsilon} \right)^{\frac{1}{2}} \tag{9.26}$$

A ratio of the integral time scale and the Kolmogorov time scale yields:

$$\frac{\tau_t}{\tau_\eta} \sim (Re_l)^{\frac{1}{2}} \tag{9.27}$$

9.2.3 RANS Equations

The RANS equations are obtained by applying the Reynolds decomposition technique to the Navier–Stokes equations. The mass, momentum, and energy conservation equations in the case of incompressible flows are usually written in tensorial notation as follows:

$$\frac{\partial u_i}{\partial x_j} = 0 \tag{9.28}$$

$$\rho \left[\frac{\partial u_i}{\partial t} + u_j \frac{\partial u_i}{\partial x_j} \right] = -\frac{\partial p}{\partial x_j} + \frac{\partial}{\partial x_j} \left[\mu \frac{\partial u_i}{\partial x_j} \right] \tag{9.29}$$

$$\rho \left[\frac{\partial h}{\partial t} + u_j \frac{\partial h}{\partial x_j} \right] = \frac{\partial}{\partial x_j} \left[k \frac{\partial T}{\partial x_j} \right] - Q\dot{w}_f \tag{9.30}$$

On carrying out Reynolds decomposition of the above stated momentum balance equation in terms of u_i, the transport equation can be re-written as [2]:

$$\rho\left[\frac{\partial\left(\bar{u}_l+u_i'\right)}{\partial t}+\frac{\partial}{\partial x_k}\left(\overline{u_i u_j}+u_i' u_j'+\bar{u}_i u_j'+\bar{u}_j u_i'\right)\right]$$
$$=-\frac{\partial\left(\bar{p}+p'\right)}{\partial x_i}+\frac{\partial}{\partial x_k}\left[\mu\frac{\partial\left(\bar{u}_l+u_i'\right)}{\partial x_k}\right] \tag{9.31}$$

Similar decomposition of the energy balance equation can be carried out in terms of h. Taking a Reynolds average of the resulting expressions gives us the following averaged transport equations [2–5]:

$$\frac{\partial\bar{u}_i}{\partial x_j}=0 \tag{9.32}$$

$$\rho\left[\frac{\partial\bar{u}_i}{\partial t}+\bar{u}_j\frac{\partial\bar{u}_i}{\partial x_j}\right]=-\frac{\partial\bar{p}}{\partial x_j}+\frac{\partial}{\partial x_j}\left[\mu\frac{\partial\bar{u}_i}{\partial x_j}\right]-\frac{\partial\left(\overline{\rho u_i' u_j'}\right)}{\partial x_j} \tag{9.33}$$

$$\rho\left[\frac{\partial\bar{h}}{\partial t}+\bar{u}_j\frac{\partial\bar{h}}{\partial x_j}\right]=\frac{\partial}{\partial x_j}\left[k\frac{\partial\bar{T}}{\partial x_j}\right]-\frac{\partial\left(\overline{\rho u_i' h'}\right)}{\partial x_j}-\overline{Q\dot{w}_f} \tag{9.34}$$

The above equations are referred to as the unsteady Reynolds-averaged Navier–Stokes (URANS) equations. In the classical RANS formulation, the time derivative is neglected since the Reynolds-averaged quantity is independent of time by definition.

It can be observed that certain additional terms (involving $\overline{\rho u_i' u_j'}$) are present in the averaged equations as a result of the decomposition. This term is referred to as the Reynolds stress term and requires additional modelling in order to ensure closure of the RANS equations [2]. These unclosed terms can be determined either using algebraic models (Boussinesq hypothesis, mixing length approach etc.) or by solving additional transport equations [2]. These equations are discussed below.

9.2.3.1 Reynolds Stress Closure

9.2.3.1.1 Algebraic Closures

9.2.3.1.1.1 Boussinesq Hypothesis

The Boussinesq hypothesis relates the turbulent shear stress to the mean flow strain rate as follows [2]:

$$\overline{\rho u_i' u_j'} \sim \left(\frac{\partial \overline{u_i}}{\partial x_j} + \frac{\partial \overline{u_j}}{\partial x_i} \right) \tag{9.35}$$

Introducing a proportionality constant in the above scaling, we can write:

$$-\overline{\rho u_i' u_j'} = \mu_T \left(\frac{\partial \overline{u_i}}{\partial x_j} + \frac{\partial \overline{u_j}}{\partial x_i} \right) \tag{9.36}$$

$$\Rightarrow \nu_T = -\frac{\overline{u_i' u_j'}}{\left(\frac{\partial \overline{u_i}}{\partial x_j} + \frac{\partial \overline{u_j}}{\partial x_i} \right)} \tag{9.37}$$

Here, ν_T represents the turbulent eddy viscosity. Note that ν_T is not a physical property as opposed to that of fluid viscosity.

This hypothesis provides the most convenient closure to the RANS equations which then would have the same form as the Navier–Stokes equations. However, this hypothesis does not hold well in many situations [2].

9.2.3.1.1.2 Mixing Length Model

The mixing length theory, proposed by Prandtl in 1925, is based on the Boussinesq hypothesis and attempts to quantify the turbulent eddy viscosity as a function of position and velocity. Prandtl hypothesised that the fluid flow can be visualised as a collection of fluid parcels moving randomly with some characteristic speed (ν_{mix}) and would retain their momentum over some characteristic length (l_{mix}). Prandtl related these two characteristics as follows [2]:

$$\nu_{mix} \sim l_{mix} \left| \frac{\partial \overline{u_i}}{\partial x_j} \right| \tag{9.38}$$

From a dimensional standpoint, the turbulent eddy viscosity also scales with the mixing parameters as:

$$\nu_T \sim \nu_{mix} l_{mix} \tag{9.39}$$

$$\Rightarrow \nu_T \sim l_{mix}^{\;2} \left| \frac{\partial \bar{u}_l}{\partial x_j} \right| \tag{9.40}$$

In this manner, the turbulent eddy viscosity and hence, the Reynolds stress terms can be estimated. However, the characteristic mixing length still needs to be determined, and it depends on the flow geometry. Also, this model suggests that the flow velocity becomes zero where the velocity gradient is zero. Several situations are contrary, however, to this implication.

The expression determining the mixing length has undergone several modifications over the years and can be expressed as [3]:

$$l_{mix} = \kappa y \left[1 - e^{-\frac{y^+}{A_0^+}} \right] \tag{9.41}$$

Here, $y^+ = u_\tau y / \nu$ and $A_0^{\;+} = 26$. u_τ is defined as $\sqrt{\tau_{wall}/\rho}$.

9.2.3.1.2 Transport Equation Based Closures

9.2.3.1.2.1 Reynolds-Stress Model

An alternative method of achieving closure of the RANS equations is by determination of the Reynolds stress terms. The transport equation for the Reynolds stress term $(\overline{\rho u_i' u_j'})$ can be obtained starting from the transport equation of the fluctuating component of velocity (u_i'). The latter can be derived by subtracting the Reynolds-averaged momentum equation (Equation 9.33) from the Reynolds-decomposed form of the momentum equation (Equation 9.31). This is expressed as [2]:

$$\rho \left[\frac{\partial u_i'}{\partial t} + \frac{\partial}{\partial x_k} \left(u_i' u_k' + \bar{u}_i u_k' + \bar{u}_k u_i' \right) \right] = -\frac{\partial p'}{\partial x_i} + \frac{\partial}{\partial x_k} \left[\mu \frac{\partial u_i'}{\partial x_k} \right] + \frac{\partial \left(\overline{\rho u_i' u_k'} \right)}{\partial x_k} \tag{9.42}$$

Interchanging the indices, we can write:

$$\rho \left[\frac{\partial u_j'}{\partial t} + \frac{\partial}{\partial x_k} \left(u_j' u_k' + \bar{u}_j u_k' + \bar{u}_k u_j' \right) \right] = -\frac{\partial p'}{\partial x_j} + \frac{\partial}{\partial x_k} \left[\mu \frac{\partial u_j'}{\partial x_k} \right] + \frac{\partial \left(\overline{\rho u_j' u_k'} \right)}{\partial x_k} \tag{9.43}$$

Multiplying Equation (9.42) by u_j' and Equation (9.43) by u_i' and adding the resultant expressions, and then performing Reynolds averaging on the resulting equation, we obtain the transport equation for Reynolds stresses as follows [2]:

$$\rho\left[\frac{\partial \overline{u_i' u_j'}}{\partial t} + \frac{\partial}{\partial x_k}\left(\overline{u_i' u_j' \overline{u}_k}\right)\right] = -\overline{u_i' \frac{\partial p'}{\partial x_j}} - \overline{u_j' u_j' \frac{\partial p'}{\partial x_i}} - \frac{\partial\left(\overline{u_i' u_j' u_k'}\right)}{\partial x_k} + \frac{\partial}{\partial x_k}\left[\mu \frac{\overline{\partial u_i' u_j'}}{\partial x_k}\right]$$

(9.44)

$$-\rho \overline{u_i' u_k'} \frac{\partial \overline{u}_j}{\partial x_k} - \rho \overline{u_j' u_k'} \frac{\partial \overline{u}_i}{\partial x_k} - 2\mu \overline{\frac{\partial u_i'}{\partial x_k} \frac{\partial u_j'}{\partial x_k}}$$

9.2.3.1.2.2 Spalart–Allmaras Model
The Spalart–Allmaras model is a one-equation model which solves the transport equation for the turbulent viscosity (v_T) instead of the Reynolds stress terms. The model equation is expressed as [2]:

$$\left[\frac{\partial(\rho v_T)}{\partial t} + \frac{\partial}{\partial x_k}\left(\rho \overline{u}_k v_T\right)\right] = \frac{\partial}{\partial x_k}\left[\frac{\mu_T}{\sigma_v} \frac{\partial v_T}{\partial x_k}\right] + S_{v_T}$$

(9.45)

The term S_{v_T} represents the source term and depends on the laminar as well as turbulent viscosities.

9.2.3.1.2.3 Kinetic Energy Models
The turbulent kinetic energy equation can be obtained by putting $i = j$ in the Reynolds-stress transport equation [Equation (9.44)] and multiplying the resulting equation by one-half. It is expressed as follows [2]:

$$\rho\left[\frac{\partial k}{\partial t} + \frac{\partial}{\partial x_k}\left(\overline{u}_k k\right)\right] = \left[-\overline{u_i' \frac{\partial p'}{\partial x_j}} - \frac{\partial\left(\overline{u_i' u_j' u_k'}/2\right)}{\partial x_k}\right] + \frac{\partial}{\partial x_k}\left[\mu \frac{\partial k}{\partial x_k}\right]$$

$$-\rho \overline{u_i' u_k'} \frac{\partial \overline{u}_i}{\partial x_k} - \mu \overline{\frac{\partial u_i'}{\partial x_k} \frac{\partial u_i'}{\partial x_k}}$$

(9.46)

The terms in the square bracket in the RHS of Equation (9.46) can be modelled using the turbulent eddy viscosity as:

$$-\overline{u' \frac{\partial p'}{\partial x_j}} - \frac{\partial\left(\overline{u_i' u_j' u_k'}/2\right)}{\partial x_k} = \frac{\partial}{\partial x_k}\left[\frac{\mu_T}{\sigma_\varepsilon} \frac{\partial k}{\partial x_k}\right]$$

where $\mu_T = \rho \nu_T$. The turbulent kinetic energy expression can then be expressed as follows:

$$\rho\left[\frac{\partial k}{\partial t}+\frac{\partial}{\partial x_k}\left(\overline{u_k}k\right)\right]=\frac{\partial}{\partial x_k}\left[\left(\mu+\frac{\mu_T}{\sigma_k}\right)\frac{\partial k}{\partial x_k}\right]-\rho\overline{u_i'u_k'}\frac{\partial \overline{u_i}}{\partial x_k}-\mu\overline{\frac{\partial u_i'}{\partial x_k}\frac{\partial u_i'}{\partial x_k}} \tag{9.48}$$

The transport of the mean kinetic energy $(\rho\overline{u_i u_i}/2)$ can be obtained by multiplying the Reynolds-averaged momentum equation with u_i as follows [2]:

$$\rho\left[\frac{\partial\left(\overline{u_i}\overline{u_i}/2\right)}{\partial t}+\overline{u_j}\frac{\partial\left(\overline{u_i}\overline{u_i}/2\right)}{\partial x_j}\right]=-\overline{u_i}\frac{\partial \overline{p}}{\partial x_j}+\overline{u_i}\frac{\partial}{\partial x_j}\left[\mu\frac{\partial \overline{u_i}}{\partial x_j}\right]-\overline{u_i}\frac{\partial\left(\overline{\rho u_i' u_j'}\right)}{\partial x_j} \tag{9.49}$$

This can be also reorganised in the following manner using Boussinesq's hypothesis for turbulent viscosity:

$$\rho\left[\frac{\partial\left(\overline{u_i}\overline{u_i}/2\right)}{\partial t}+\overline{u_j}\frac{\partial\left(\overline{u_i}\overline{u_i}/2\right)}{\partial x_j}\right]=-\frac{\partial}{\partial x_j}\left[\overline{u_i}\overline{p}+\overline{\rho u_i' u_j'}\,\overline{u_i}-\mu\frac{\partial\left(\overline{u_i}\overline{u_i}/2\right)}{\partial x_j}\right]$$
$$-\mu_T\left(\frac{\partial \overline{u_i}}{\partial x_j}+\frac{\partial \overline{u_j}}{\partial x_i}\right)\frac{\partial \overline{u_i}}{\partial x_j}-\mu\frac{\partial \overline{u_i}}{\partial x_j}\frac{\partial \overline{u_i}}{\partial x_j} \tag{9.50}$$

A comparison of the mean kinetic energy equation [Equation (9.50)] and the turbulent kinetic energy equation [Equation (9.48)] reveals that the sink term in the former acts as the source term in the latter. Physically, this signifies that energy transfer in turbulence takes place from the mean flow to the fluctuating flow. This is responsible for turbulence generation and is in turn transformed into thermal energy as a result of dissipative action of viscosity. This dissipation is signified by the last term in Equation (9.48). The rate of dissipation of turbulent kinetic energy (ε) is expressed as [3]:

$$\varepsilon=\frac{\mu}{\rho}\overline{\frac{\partial u_i'}{\partial x_k}\frac{\partial u_i'}{\partial x_k}} \tag{9.51}$$

At local equilibrium in a quasi-steady state, the rate of turbulent kinetic energy generation should be balanced by its dissipation rate, that is:

$$\rho G=-\rho\overline{u_i'u_k'}\frac{\partial \overline{u_i}}{\partial x_k}=\rho\varepsilon \tag{9.52}$$

Utilising the integral length and time scales under the local equilibrium assumption, the dissipation rate can be scaled as follows [3]:

$$\rho G = \rho\varepsilon = -\rho\overline{u_i' u_k'}\frac{\partial \overline{u_i}}{\partial x_k} \sim \rho\frac{u'^3}{l} \tag{9.53}$$

$$\Rightarrow \varepsilon \sim \frac{u'^3}{l} \tag{9.54}$$

$$\Rightarrow \varepsilon \sim \frac{k^{3/2}}{l} \tag{9.55}$$

This signifies that the dissipation rate in the integral scale becomes independent of viscosity under local equilibrium conditions. The breakdown of the larger eddies into relatively smaller eddies is due solely to fluid dynamic instabilities, and the energy of such larger eddies are distributed among the smaller eddies.

The kinematic turbulent viscosity can also be subjected to a similar scaling, which gives us [2]:

$$\nu_T = \frac{-\overline{u_i' u_k'}}{\left(\frac{\partial \overline{u_i}}{\partial x_j} + \frac{\partial \overline{u_j}}{\partial x_i}\right)} \sim \frac{u'^2}{u'/l} \sim u'l \tag{9.56}$$

$$\Rightarrow \nu_T \sim k^{1/2}\frac{k^{3/2}}{\varepsilon} \sim \frac{k^2}{\varepsilon} \tag{9.57}$$

This scaling is utilised in determining the turbulent dissipation rate and the turbulent viscosity in one-equation turbulent kinetic energy models as follows [2]:

$$\varepsilon = C_D\frac{k^{\frac{3}{2}}}{l} \tag{9.58}$$

$$\nu_T = C_\mu\frac{k^2}{\varepsilon} \tag{9.59}$$

where C_D and C_μ are the model constants in the respective equations.

Two-equation models solve an additional transport equation (in terms of dissipation rate, etc.) in addition to the turbulent kinetic energy transport equation.

The $k-\varepsilon$ model involves solution of the transport equations of k and ε. The modelled transport equations in the $k-\varepsilon$ model are as follows [2]:

$$\left[\frac{\partial(\rho k)}{\partial t}+\frac{\partial}{\partial x_k}\left(\rho\overline{u_k}k\right)\right]=\frac{\partial}{\partial x_k}\left[\left(\mu+\frac{\mu_T}{\sigma_k}\right)\frac{\partial k}{\partial x_k}\right]+\mu_t\left(\frac{\partial\overline{u_i}}{\partial x_j}+\frac{\partial\overline{u_j}}{\partial x_i}\right)\frac{\partial\overline{u_i}}{\partial x_j}-C_\mu\rho\varepsilon \quad (9.60)$$

$$\left[\frac{\partial(\rho\varepsilon)}{\partial t}+\frac{\partial}{\partial x_k}\left(\rho\overline{u_k}\varepsilon\right)\right]=\frac{\partial}{\partial x_k}\left[\left(\mu+\frac{\mu_T}{\sigma_\varepsilon}\right)\frac{\partial\varepsilon}{\partial x_k}\right]+C_{\varepsilon1}\mu_t\left(\frac{\partial\overline{u_i}}{\partial x_j}+\frac{\partial\overline{u_j}}{\partial x_i}\right)\frac{\partial\overline{u_i}}{\partial x_j}\frac{\varepsilon}{k}-C_{\varepsilon2}\rho\frac{\varepsilon^2}{k} \quad (9.61)$$

In the above equations, $\sigma_k, \sigma_\varepsilon, C_\mu, C_{\varepsilon1}$ and $C_{\varepsilon2}$ are constants with magnitudes 1, 1.3, 0.09, 1.44 and 1.92, respectively.

The $k-\omega$ model involves solution of the transport equations of k and ω (specific rate of dissipation of k). The modelled transport equations in the $k-\omega$ model are as follows [2]:

$$\left[\frac{\partial(\rho k)}{\partial t}+\frac{\partial}{\partial x_k}\left(\rho\overline{u_k}k\right)\right]=\frac{\partial}{\partial x_k}\left[\left(\mu+\frac{\mu_T}{\sigma_k}\right)\frac{\partial k}{\partial x_k}\right]+\mu_t\left(\frac{\partial\overline{u_i}}{\partial x_j}+\frac{\partial\overline{u_j}}{\partial x_i}\right)\frac{\partial\overline{u_i}}{\partial x_j}-C_\mu\rho\varepsilon \quad (9.62)$$

$$\left[\frac{\partial(\rho\omega)}{\partial t}+\frac{\partial}{\partial x_k}\left(\rho\overline{u_k}\omega\right)\right]=\frac{\partial}{\partial x_k}\left[\left(\mu+\frac{\mu_T}{\sigma_\omega}\right)\frac{\partial\omega}{\partial x_k}\right]+C_{\varepsilon1}\mu_t\left(\frac{\partial\overline{u_i}}{\partial x_j}+\frac{\partial\overline{u_j}}{\partial x_i}\right)\frac{\partial\overline{u_i}}{\partial x_j}\omega-C_{\varepsilon2}\rho\omega^2 k \quad (9.63)$$

In the above equations, $\sigma_k, \sigma_\omega, C_\mu, C_{\varepsilon1}$ and $C_{\varepsilon2}$ are constants with magnitudes 2, 2, 0.09, 5/9 and 3/40, respectively.

9.2.3.2 Reynolds Scalar Flux Closure

The species transport equation can be written in a similar manner as Equations (9.28) to (9.30), as follows [4]:

$$\frac{\partial(\rho Y_\alpha)}{\partial t}+\frac{\partial(\rho u_j Y_\alpha)}{\partial x_j}=\frac{\partial}{\partial x_j}\left[\rho D\frac{\partial Y_\alpha}{\partial x_j}\right]+\dot{w}_\alpha \quad (9.64)$$

The Favre-averaged form of the species transport equation is expressed as [4]:

$$\frac{\partial\left(\bar{\rho}\tilde{Y}_\alpha\right)}{\partial t}+\frac{\partial\left(\bar{\rho}\tilde{u}_j\tilde{Y}_\alpha\right)}{\partial x_j}=\frac{\partial}{\partial x_j}\left[\overline{\rho D\frac{\partial Y_\alpha}{\partial x_j}}\right]-\frac{\partial\left(\bar{\rho}u_i{}''\tilde{Y}_\alpha{}''\right)}{\partial x_j}+\bar{\dot{w}}_\alpha \tag{9.65}$$

Similar to the closure of the Reynolds-stress terms, the unclosed Reynolds-scalar flux terms in Equation (9.65) ($\bar{\rho}u_i{}''\tilde{Y}_\alpha{}''$) or in Equation (9.34) ($\overline{\rho u_i' h'}$) are determined either from some algebraic models (such as the gradient hypothesis expressed in [Equation (9.65)]) or by solving the transport equation for the corresponding Reynolds-scalar flux.

Using the gradient hypothesis, the Reynolds-scalar flux in Equation (9.65) can be determined as follows [4]:

$$\bar{\rho}u_i{}''\widetilde{Y}_\alpha{}''=-\frac{\mu_t}{\sigma_t}\frac{\partial\widetilde{Y}_\alpha}{\partial x_j} \tag{9.66}$$

In Equation (9.69), the quantity σ_t denotes the turbulent Schmidt number. The transport equation for the Reynolds-scalar flux can be derived as [4]:

$$\frac{\partial\left(\bar{\rho}\widetilde{u_i{}''Y_\alpha{}''}\right)}{\partial t}+\frac{\partial\left(\bar{\rho}\tilde{u}_j\widetilde{u_i{}''Y_\alpha{}''}\right)}{\partial x_j}=-\frac{\partial\left(\overline{\rho u_j{}''u_i{}''Y_\alpha{}''}\right)}{\partial x_j}-\overline{\rho u_j{}''u_i{}''}\frac{\partial\widetilde{Y}_\alpha}{\partial x_j}-\overline{\rho u_j{}''Y_\alpha{}''}\frac{\partial\tilde{u}_i}{\partial x_j}$$

$$-\overline{Y_\alpha''}\frac{\partial\bar{p}}{\partial x_j}-\overline{Y_\alpha''\frac{\partial p'}{\partial x_j}}+\left[\overline{u_i\frac{\partial}{\partial x_j}\left[\rho D\frac{\partial Y_\alpha}{\partial x_j}\right]}-\tilde{u}_i\overline{\frac{\partial}{\partial x_j}\left[\rho D\frac{\partial Y_\alpha}{\partial x_j}\right]}\right]$$

$$+\left[\overline{Y_\alpha\frac{\partial\tau}{\partial x_j}}-\widetilde{Y}_\alpha\overline{\frac{\partial\tau}{\partial x_j}}\right]+\left[\overline{u_i\dot{w}_\alpha}-\tilde{u}_i\overline{\dot{w}_\alpha}\right] \tag{9.67}$$

Several of these terms are in an unclosed form and need further modelling. A similar approach may be adopted for the scalar flux in Equation (9.34).

9.2.4 LES Modelling

The LES modelling of the Navier–Stokes equations is based on the concept that the smaller scales of turbulence must be modelled in order to get a desired reduction of the computational requirements such that the numerical simulation becomes feasible [2]. This is achieved by directly simulating the larger three-dimensional turbulent motions while the effects of the smaller scales are modelled.

The LES approach can be conceptually divided into the following steps [2]:

1. Filtering: A filtering operation is carried out to decompose the velocity into the sum of a filtered (or resolved) component and a residual (or sub-grid scale [SGS]) component. The resolved component represents the motion of the larger eddies; the SGS component accounts for the smaller eddies.
2. Equations describing the evolution of the filtered velocity field are derived from the Navier–Stokes equations. These equations contain the SGS–stress tensor term (similar to the Reynolds-stress term) which accounts for the SGS motion.
3. Closure modelling of the SGS-stress tensor.
4. Numerical solution of the filtered velocity field utilising the closure models.

The filtering operation is formally expressed as:

$$\bar{u} = \int u(x-r,t)G(r,x)dr \tag{9.68}$$

Here, $G(r)$ is the filter kernel such that $\int G(r,x)dr = 1$.

The residual (or SGS) velocity field is defined as:

$$u' = u - \bar{u} \tag{9.69}$$

The velocity field is therefore represented as:

$$u = \bar{u} + u' \tag{9.70}$$

The scale at which the filtering is carried out is characterised by the filter width (Δ). Detailed discussions about the filters and the SGS models used in LES beyond the scope of this text.

Example 9.1

Consider a flow situation in a domain of length 1 m with velocity 10 m/s and kinematic viscosity and dissipation rate of 8.927×10^{-5} m^2/s and 100 m^2/s^3, respectively. Calculate the Kolmogorov length scale for this situation.

Solution

The Kolmogorov length scale can be determined using the kinematic viscosity and the dissipation rate as follows:

$$\eta = \left(\frac{\nu^3}{\varepsilon}\right)^{\frac{1}{4}}$$

$$= \left(\frac{\left(8.927\times10^{-5}\right)^3}{100}\right)^{\frac{1}{4}}$$

$$= \left(\frac{\left(7.114\times10^{-13}\right)}{100}\right)^{\frac{1}{4}}$$

$$= 0.00029\,m$$

9.3 NEED FOR COMBUSTION MODELLING

Apart from the achieving the closure of the Reynolds-scalar flux terms in the species transport equation [Equation (9.65)], it is also required to properly close the term $\overline{\dot{w}_\alpha}$ in a reactive system [1,4]. This serves as the motivation for combustion modelling.

Consider an irreversible reaction of the following form:

$$Fuel + sOxidiser \rightarrow (1+s)Products$$

The mean reaction rate ($\overline{\dot{w}_F}$) can be obtained from the Arrhenius law and by using Taylor's series:

$$\overline{\dot{w}_F} = A\tilde{T}^{\beta}\bar{\rho}^2 exp\left[-\frac{E_{ac}}{\bar{R}\tilde{T}}\right]\left[1+\frac{\widetilde{Y_F''Y_O''}}{\widetilde{Y_F}\widetilde{Y_O}}+\ldots\right] \tag{9.71}$$

This again requires additional closure of the averaged terms through algebraic expressions or transport equations. Also, large truncation errors occur when only the first few terms of the Taylor's series are utilised. In addition, realistic chemical schemes cannot be modelled using this approximation. As such, additional modelling of the mean reaction rate is required.

9.4 TURBULENT PREMIXED COMBUSTION

In premixed combustion, the fuel and the oxidiser are completely mixed before combustion is allowed to take place. Typical applications of premixed combustion can be found in spark-ignition engines, lean-burn gas turbines, industrial and domestic burners, and so on. In all these applications, the fuel and the air are mixed before being injected into the combustion chamber.

9.4.1 Structure of a Premixed Flame

The typical structure of a laminar premixed flame is represented in Figure 9.1. Fresh gases (fuel and oxidiser mixed at the molecular level) and burnt gases (combustion products) are separated by a thin reaction zone. The typical thermal flame thicknesses vary between 0.1 and 1 mm. A strong temperature gradient is observed within this flame thickness. The flame is described in

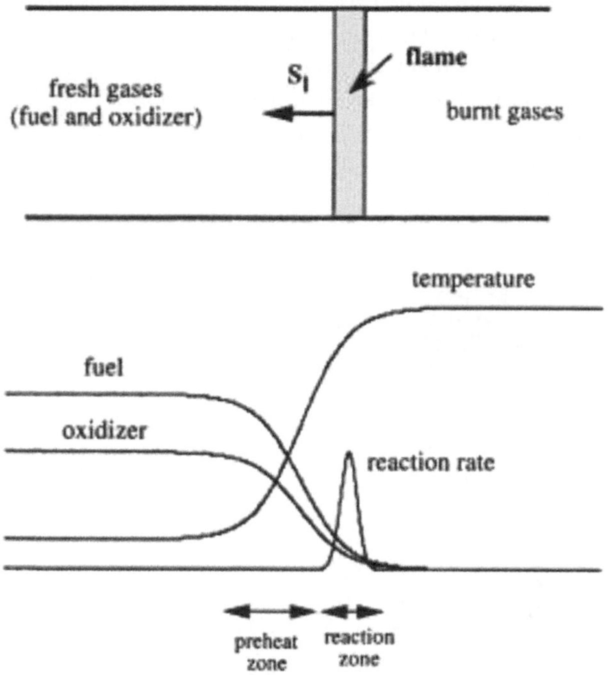

Figure 9.1 Structure of a laminar premixed flame. (From Veynante, D. and Vervisch, L., Turbulent combustion modelling, *Prog. Energ. Combust.*, **28**, 193–266, 2002. With Permission.)

terms of a reaction progress variable (c), such that $c = 0$ in the fresh gases and $c = 1$ in the fully burnt ones. This is defined in terms of reduced temperature or a reduced mass fraction as follows [1]:

$$c = \frac{T - T_u}{T_b - T_u} \quad \text{or} \quad c = \frac{Y_F - Y_F^{\,u}}{Y_F^{\,b} - Y_F^{\,u}} \tag{9.72}$$

where T, T_u and T_b are the local, the unburnt gas and the burnt gas temperatures, respectively. Y_F, $Y_F^{\,u}$ and $Y_F^{\,b}$ are the local, unburnt gas and burnt gas fuel mass fractions, respectively. The transport equation for the reaction progress variable is stated as follows [1]:

$$\frac{\partial(\rho c)}{\partial t} + \nabla.(\rho u c) = \nabla.(\rho D \nabla c) + \dot{\omega} \tag{9.73}$$

Visualisations of turbulent premixed flames indicate that the effect of turbulence is limited to wrinkling and distorting an essentially laminar flame front. Figure 9.2 represents the typical structure of a turbulent premixed flame. The instantaneous flame front has been experimentally observed to be highly convoluted with a very thin reaction zone. This is similar to the flame thickness observed in laminar premixed flames and hence is sometimes termed laminar flamelets. The time-averaged view of the flame gives the appearance, however, of a rather thick reaction zone (Figure 9.2b). This is generally referred to as the turbulent flame brush.

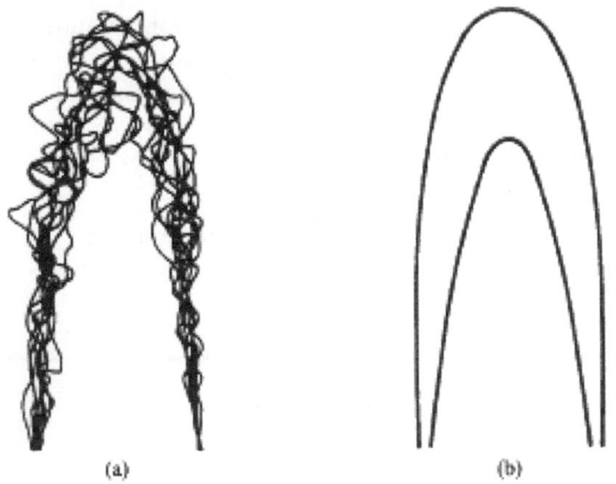

Figure 9.2 Structure of a turbulent premixed flame: (a) instantaneous position of reaction fronts and (b) time-averaged view. (From Turns, S. R., *An Introduction to Combustion: Concepts and Applications*, McGraw-Hill, 2000. With Permission.)

9.4.2 Laminar and Turbulent Burning Velocities

Another characteristic feature of a premixed flame is its ability to propagate towards the fresh gases. Because of the temperature gradient and the corresponding thermal fluxes, fresh gases are preheated, which then start burning. This local imbalance between the diffusion of heat and chemical consumption results in propagation of the flame front [1]. The velocity with which the flame front propagates normal to itself and relative to flow into the unburnt gas mixture is termed the laminar burning velocity, or the flame speed (S_L). This depends on various factors such as the compositions of fuel and oxidiser, the temperature of fresh gases, pressure, and so on. The thermal flame thickness (δ_L), the laminar flame speed (S_L) and the kinematic viscosity of the fresh gases (ν) are related by means of the flame Reynolds number, defined as follows [1]:

$$Re_f = \frac{\delta_L S_L}{\nu} \tag{9.74}$$

Turbulent premixed flames may be described in terms of a global turbulent flame speed (S_T). The following expression has been proposed for estimating S_T:

$$\frac{S_T}{S_L} = 1 + \alpha \left(\frac{u'}{S_L} \right)^n \tag{9.75}$$

where α and n are two model constants of the order of unity. The turbulent flame speed is not a very well-defined quantity, however, and the experimental data available in this respect exhibit a large scatter due to its dependence on a large number of parameters.

9.4.3 Regimes of Premixed Turbulent Combustion

The basic structure of a turbulent premixed flame is governed by the relationship of the Kolmogorov length scale (η) and the integral length scale (l) to the flame thickness (δ). Based on these length scales, turbulent premixed combustion is subdivided into the following three regimes:

1. Wrinkled laminar flamelets ($\delta \leq \eta$): When the flame thickness is thinner than the Kolmogorov length scale, the turbulent motion is only able to wrinkle or distort the thin reaction zone. This regime is characterised by fast chemistry ($Da >> 1$).

2. Flamelets in eddies, or thickened wrinkled flames ($l > \delta > \eta$): The laminar flame thickness remains in between the Kolmogorov length scale and the integral length scales in this regime.
3. Distributed reactions, or thickened flames ($\delta > l$): If the flame thickness is greater than all the turbulent length scales, then transport within the reaction zone is influenced by the turbulent nature of flow.

Two dimensionless parameters useful in demarcating the regimes of turbulent premixed flames are the Damköhler number (Da) and the Karlovitz number (Ka). The Damköhler number (Da) is defined as the ratio of the integral time scale (τ_t) and the chemical time scale (τ_c) as:

$$Da = \frac{\tau_t}{\tau_c} = \frac{l/u'}{\delta/S_L} \tag{9.76}$$

It is evident from this expression that, at very high values of Da, that is, $Da >> 1$, the chemical time scale is much smaller than the turbulent time scale. The inner structure of the flame therefore remains unaffected by turbulence motions. As such, the flame front remains thin and is only distorted at its outer surfaces. This corresponds to the wrinkled flamelet regime. In contrast, a low $Da\,(Da < 1)$ corresponds to the distributed reactions regime; an intermediate Da indicates the flamelet in eddies regime.

The dimensionless Karlovitz number $(Ka)$is defined as the ratio of the chemical time scale (τ_c) and the Kolmogorov time scale (τ_η) and is expressed as follows:

$$Ka = \frac{\tau_c}{\tau_\eta} = \frac{\delta/S_L}{\eta/v_\eta} \tag{9.77}$$

The Karlovitz number is used to quantify the Klimov-Williams criterion (corresponding to $Ka = 1$) which delineates the combustion regimes. A magnitude of $Ka < 1$ indicates the wrinkled flamelet regime; $Ka > 100$ indicates the distributed reaction regime. The intermediate regime is indicated by $1 < Ka < 100$ (Figure 9.3).

9.4.4 Modelling Approaches

9.4.4.1 Eddy-Breakup Model

The eddy-breakup model (EBU) is based on an assumption of very high Reynolds and Damköhler numbers, that is, $Re \gg 1$ and $Da \gg 1$. The reaction zone is considered to be a collection of fresh and burnt gas pockets such that the turbulent motion can break down the fresh gas pockets due to the cascading effect.

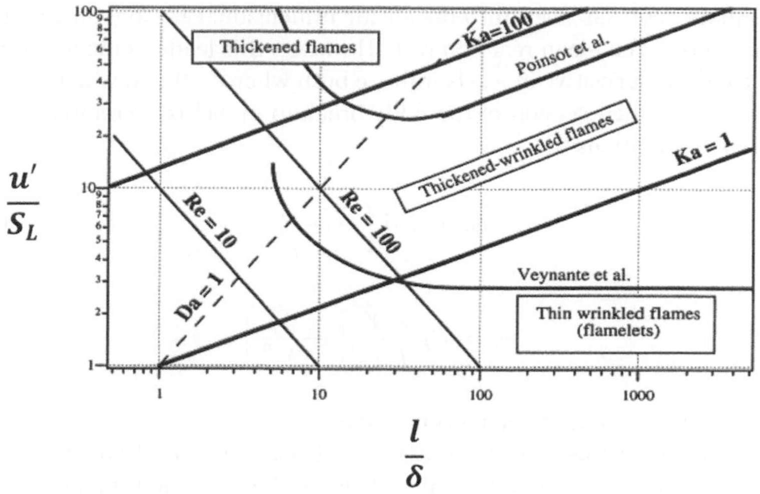

Figure 9.3 Turbulent premixed combustion regimes. (From Veynante, D. and Vervisch, L., Turbulent combustion modelling, *Prog. Energ. Combust.*, **28**, 193–266, 2002. With Permission.)

The mean reaction rate is therefore mainly governed by the turbulent mixing time (τ_t). The mean reaction rate is estimated as [6]:

$$\overline{\dot{\omega}_F} = -C\bar{\rho}\frac{\sqrt{\widetilde{c''^2}}}{\tau_t} \tag{9.78}$$

where Y_F'' denotes the fuel mass fraction fluctuations and C is a model constant of the order of unity. This expression of the mean reaction rate can be recast in terms of the reaction progress variable (c) as follows:

$$\overline{\dot{\omega}_F} = -C\bar{\rho}\frac{\sqrt{\widetilde{c''^2}}}{\tau_t} \tag{9.79}$$

It is evident from the above expressions that the fluctuations of the fuel mass fraction or the reaction progress variable must be modelled in order to determine the mean reaction rate. This is usually achieved by solving a scalar transport equation in any one of the two quantities. It can also be estimated by assuming an infinitely thin flame front, which gives us:

$$\bar{\rho}\widetilde{c''^2} = \overline{\rho(c-\tilde{c})^2} = \bar{\rho}\left(\widetilde{c^2} - \tilde{c}^2\right) = \bar{\rho}\tilde{c}(1-\tilde{c}) \tag{9.80}$$

The square root has been introduced for dimensional reasons in the above expressions of the mean reaction rate. However, this leads to inconsistencies because the $\tilde{c}$ derivative of $\overline{\dot{\omega}_F}$, is infinite both when $\tilde{c}=0$ and when $\tilde{c}=1$. The following modified version of the eddy-breakup model is therefore, used for practical simulations [1]:

$$\overline{\dot{\omega}_F} = -C\bar{\rho}\frac{\varepsilon}{k}\tilde{c}(1-\tilde{c}) \tag{9.81}$$

$$\overline{\dot{\omega}_F} = -C\bar{\rho}\frac{\varepsilon}{k}\frac{\widetilde{Y_F}}{Y_F^{\,0}}\left(1-\frac{\widetilde{Y_F}}{Y_F^{\,0}}\right) \tag{9.82}$$

where $Y_F^{\,0}$ is the initial mass fraction of fuel.

The EBU model was found to be attractive because the reaction rate is simply expressed as a function of known quantities without any additional transport equation. However, the modelled reaction rate does not depend on the chemical characteristics and assumes homogeneous and isotropic turbulence. Also, the EBU model tends to overestimate the reaction rate, especially in highly strained regions, where the ratio ε / k is large (flame-holder wakes, walls, etc.).

9.4.4.2 Bray-Moss-Libby Model

The Bray-Moss-Libby model combines a statistical approach, using probability density functions, with physical analysis to obtain the mean reaction rate. Considering a one-step, irreversible chemical reaction between two reacting species and assuming perfect gases, incompressible flows, constant chemical properties, unity Lewis numbers, and so on, the presumed probability density function of the reaction progress variable (c) is expressed as a sum of contributions from fresh, fully burnt and burning gases as follows [7]:

$$P(c:x,t)=\alpha(x,t)\delta(c)+\beta(x,t)\delta(1-c)+\gamma(x,t)f(c:x,t) \tag{9.83}$$

where α,β and γ denote the probability to have, at location (x,t), fresh gases, burnt gases and a reacting mixture, respectively. The variables $\delta(c)$ and $\delta(1-c)$ are the Dirac delta functions corresponding to fresh gases ($c=0$) and fully burnt ones ($c=1$), respectively.

The mean reaction rate is evaluated using the probability density function as follows:

$$\overline{\dot{\omega}} = \int_0^1 \dot{\omega}(c)P(c:x,t)dc \tag{9.84}$$

Note that, if the reaction zone is infinitely thin, that is, $Da >> 1$, the contribution from the reacting mixture becomes significantly less than the fresh and the fully burnt gases (i.e., $\alpha >> \gamma$ and $\beta >> \gamma$. The effect of the reacting mixture can be neglected in such a situation.

9.4.4.3 Flame Surface Density Model

Closure modelling of the mean reaction rate based on flame surface density can be achieved using algebraic expressions as well as transport equations.

9.4.4.3.1 Algebraic Closures

Bray, Moss, Libby and their co-workers proposed the following model based on flame surface density [8]:

$$\Sigma = \frac{g}{\sigma_y L_y} \frac{1+\tau}{(1+\tau\tilde{c})^2} \tilde{c}(1-\tilde{c}) \tag{9.85}$$

In this expression, g is a constant and has a magnitude of unity. The variable σ_y is a flamelet orientation factor and is assumed to be a universal model constant ($\sigma_y = 0.5$). L_y represents the flame front wrinkling length scale and is usually determined by assuming that it is proportional to the integral length scale, as follows:

$$L_y = C_l l \left(\frac{S_L}{u'} \right)^n \tag{9.86}$$

where C_l and n are two constants of the order of unity.

The flame surface density can also be derived from fractal theories leading to the following expression [9]

$$\Sigma = \frac{1}{L_{outer}} \left(\frac{L_{outer}}{L_{inner}} \right)^{D-2} \tag{9.87}$$

where L_{outer} and L_{inner} refer to the outer and inner cut-off length scales, respectively, and D is the fractal dimension of the flame surface.

The mean reaction rate can thus be determined using the above estimations of flame surface density as follows [1]:

$$\overline{\dot{\omega}} = \dot{\Omega}_c \Sigma = (\rho_0 S_L)\Sigma \tag{9.88}$$

9.4.4.3.2 Transport Equation Closures

The flame surface density can also be determined by solving an additional transport equation in terms of the flame surface density. However, a separate closure modelling is required for determining the turbulent flux of the flame surface density. Some of the main closure models are detailed below:

1. The *Cant-Pope-Bray (CPB) model* [10] is derived from the exact transport equation for the flame surface density. The strain rate due to the turbulent fluctuations is estimated from the time scale of the Kolmogorov structures. The turbulent strain rate is probably overestimated.
2. The *coherent flame model (CFM)* was developed by Candel and his co-workers following the initial work of Marble and Broadwell [11–13]. Different versions of this model are available depending on the mode of determination of the strain rate. In the original version, the strain rate is estimated from the characteristic time of the integral length scale. In the two succeeding formulations, the expression of the turbulent strain rate acting on the flame front were improved using results from direct numerical simulations and multi-fractal analysis.
3. The *Mantel-Borghi (MB) model* is based on an exact equation for the scalar dissipation rate [14].
4. The *Cheng-Diringer (CD) model* is similar to the original version of the coherent flame model. An additional term is proposed to take into account flame extinction under excessively high strain rates [15].
5. The *Choi-Huh (CH) model* has been devised for spark-ignited engines to recover experimental data obtained in a closed vessel [16].

9.5 TURBULENT NON-PREMIXED COMBUSTION

In many applications, the fuel and the oxidiser are injected separately into the combustion chamber, thereby eliminating any possibility of mixing prior to combustion. This is termed non-premixed combustion. Turbulent non-premixed combustion is utilised in the majority of the combustion systems due to the ease with which the flames can be controlled. Important applications include combustion in furnaces, diesel engines, gas turbines, and so on.

9.5.1 Structure of a Non-premixed Flame

In non-premixed combustion, the fuel and oxidiser are on both sides of a reaction zone where the heat is released (Figure 9.4). The burning rate is controlled by the molecular diffusion of the reactants towards the reaction zone. The structure of a steady diffusion flame in non-premixed combustion depends

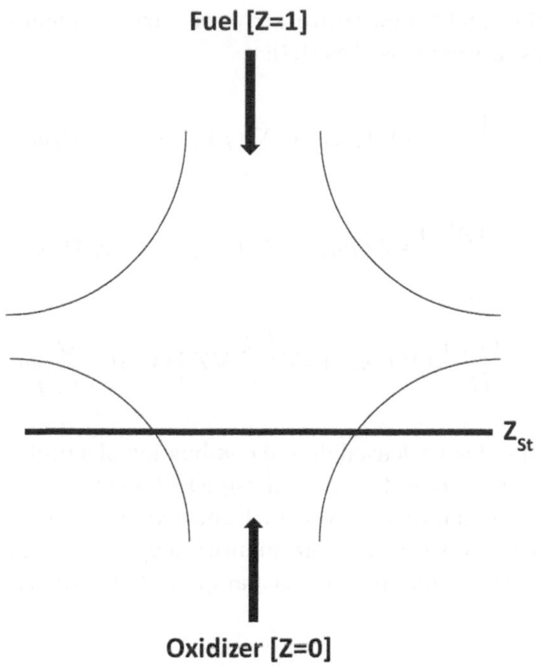

Figure 9.4 A typical laminar non-premixed (diffusion) flame.

on ratios between characteristic time scales representative of molecular diffusion and chemistry. The thicknesses of the mixing zone and that of the reaction zone vary with these characteristic time scales.

Consider the following irreversible. single-step chemical reaction between fuel and oxidiser:

$$F + sO \rightarrow (1+s)P$$

where s represents the mass stoichiometric coefficient. This reaction may be re-written in terms of mass fraction as follows:

$$\nu_F Y_F + \nu_O Y_O = \nu_P Y_P \tag{9.89}$$

In this expression, Y_F, Y_O and Y_P represents the fuel, oxidiser and product mass fractions, respectively. The quantities ν_F, ν_O and ν_P represent the corresponding stoichiometric molar coefficients of the reaction. The transport equations for mass fraction and temperature are necessary for identifying the flame properties. These are expressed as [1,4]:

$$\frac{\partial(\rho Y_F)}{\partial t}+\nabla.(\rho u Y_F)=\nabla.(\rho D_F \nabla Y_F)-\nu_F M_F \dot{\omega} \tag{9.90}$$

$$\frac{\partial(\rho Y_O)}{\partial t}+\nabla.(\rho u Y_O)=\nabla.(\rho D_O \nabla Y_O)-\nu_O M_O \dot{\omega} \tag{9.91}$$

$$\frac{\partial(\rho T)}{\partial t}+\nabla.(\rho u T)=\nabla.\left(\frac{k}{c_p}\nabla T\right)+\nu_F M_F\left(\frac{Q}{c_p}\right)\dot{\omega} \tag{9.92}$$

Q is the amount of heat released due to combustion of a unit mass of fuel. The molecular diffusion can be determined using Fick's law.

The transport equations for the fuel and oxidiser mass fraction can be combined to give a conserved scalar quantity φ by assuming equal molecular diffusivities in all the equations. The scalar quantity is expressed as [1]:

$$\varphi = Y_F - \frac{Y_O}{s} \tag{9.93}$$

The mixture fraction (Z) is defined by normalising φ using the mass fractions such that Z evolves through the diffusive layer from zero (oxidiser) to unity (fuel). Mathematically this is stated as [1]:

$$Z = \frac{\phi \frac{Y_F}{Y_{Fo}} - \frac{Y_O}{Y_{Oo}} + 1}{\phi + 1} \tag{9.94}$$

Here, Y_{Fo} and Y_{Oo} represent the mass fraction in the fuel feeding stream and oxidiser stream, respectively. The variable ϕ represents the chemical equivalence ratio and is expressed as:

$$\phi = \frac{s Y_{Fo}}{Y_{Oo}} \tag{9.95}$$

The mass stoichiometric coefficient is expressed as:

$$s = \frac{\nu_O M_O}{\nu_F M_F} \tag{9.96}$$

The transport equation in terms of the mixture fraction can be expressed as [1]:

$$\frac{\partial(\rho Z)}{\partial t}+\nabla.(\rho u Z)=\nabla.(\rho D \nabla Y_F) \tag{9.97}$$

In situations when the molecular diffusivities differ from each other, an additional definition of mixture fraction is required which would satisfy the mixture fraction transport equation. The modified mixture fraction is defined as [1]:

$$Z_L=\frac{\Phi\frac{Y_F}{Y_{Fo}}-\frac{Y_O}{Y_{Oo}}+1}{\Phi+1} \tag{9.98}$$

The mixture fraction transport equation is then expressed as:

$$\frac{\partial(\rho Z)}{\partial t}+\nabla.(\rho u Z)=\frac{1}{L}\nabla.(\rho D \nabla Y_F) \tag{9.99}$$

The terms L and Φ are defined as [1]:

$$\Phi=\frac{Le_O}{Le_F}\phi \tag{9.100}$$

$$L=Le_O\frac{\phi+1}{\Phi+1} \tag{9.101}$$

where Le denotes the Lewis number of the respective species.

The scalar dissipation rate of the mixture fraction is defined as [1]:

$$\chi=D|\nabla Z|^2 \tag{9.102}$$

Characterisation of turbulence requires the specification of a length scale as well as a velocity scale. The absence of any intrinsic characteristic velocity in non-premixed combustion makes it difficult to identify turbulent non-premixed combustion. Also, the thickness of the flame depends on the aerodynamics controlling the thickness of the local mixing layers developing between the fuel and the oxidiser. Hence, no fixed reference length scale can be easily determined for non-premixed flames.

Turbulent non-premixed combustion may be classified as follows:

- The turbulent flow regime is characterised by a Reynolds number, whereas a Damköhler number is chosen for the reaction zone.
- The mixture fraction field is retained to describe the turbulent mixing, and a Damköhler number characterises the flame.
- A velocity ratio (turbulence intensity to premixed laminar flame speed) and a length ratio (integral scale to premixed laminar flame thickness) may be constructed to delineate between regimes. Other length scales have also been used, for instance, thicknesses of profiles in mixture fraction space.

In a non-premixed turbulent flame, the reaction zones develop within a mean mixing zone whose thickness (l_z) is of the order of the turbulent integral length scale (l), that is:

$$l_z \approx l \approx \frac{k^{\frac{3}{2}}}{\varepsilon} \tag{9.103}$$

The thickness of the mixing layer can be estimated as:

$$l_d \approx \left(\frac{D}{\widetilde{\chi}_{st}} \right)^{\frac{1}{2}} \tag{9.104}$$

where $\widetilde{\chi}_{st}$ denotes the conditional mean value of the scalar dissipation rate.

In situations where the transport of species and heat by velocity fluctuations is faster than the transfer in the diffusion flame, a departure from laminar flamelet assumption is expected. Also, when the Kolmogorov scale (η) is of the order of the flame thickness, the inner structure of the reaction zone is prone to be modified by the turbulence. Since diffusion flames depend strongly on the local flow conditions, we may write:

$$l_d \approx \alpha_1 \eta \tag{9.105}$$

$$\widetilde{\chi}_{st} \approx \frac{\alpha_2}{\tau_\eta} \tag{9.106}$$

subject to the constraint that $\alpha_1 \geq 1$ and $\alpha_2 \leq 1$. Using the above scaling and $\sqrt{Re} = \tau_t/\tau_\eta$, the Damköhler number can be recast as:

$$Da = \frac{\tau_t}{\tau_c} = \frac{\tau_t}{\tau_\eta}\frac{\tau_\eta}{\tau_c} \approx \frac{\tau_t}{\tau_\eta}\frac{\alpha_2}{\widetilde{\chi}_{st}\tau_c} \approx \alpha_2 \sqrt{ReDa^*} \tag{9.107}$$

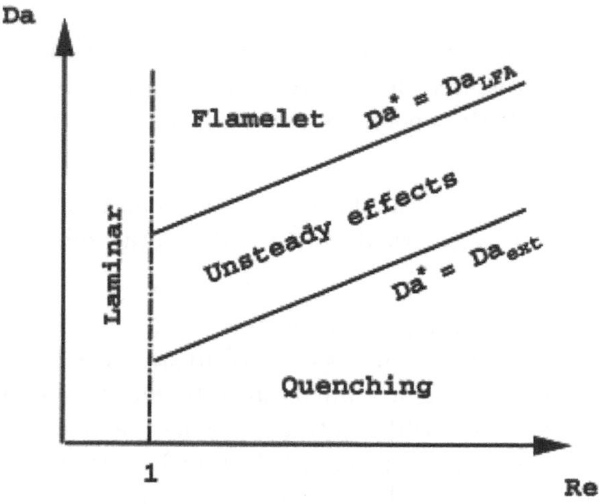

Figure 9.5 Regimes of non-premixed combustion. (From Veynante, D. and Vervisch, L., Turbulent combustion modelling, *Prog. Energ. Combust.*, **28**, 193–266, 2002. With Permission.)

Constant Damköhler numbers Da^* correspond to lines of slope of ½ in a log–log (Da, Re) diagram. When the chemistry is sufficiently fast (large Da values), the flame is expected to have a laminar flame structure. This condition may be simply expressed as $Da^* > Da_{LFA}$. On the other hand, for large chemical times (i.e., when $Da^* < Da_{ext}$), extinction occurs. Laminar flames are encountered for low Reynolds numbers ($Re < 1$). These results can be summarised in the regime diagram shown Figure 9.5.

9.5.2 Modelling Approaches

Extensive work has been carried out on numerical modelling of non-premixed combustion systems, mainly based on the assumption of infinitely fast chemistry with respect to mixing and molecular diffusion. There exist strong motivations for improving non-premixed and partially premixed turbulent combustion modelling. Some of these motivations are outlined below:

1. Development of new combustion technologies for aircraft engines, and more generally for gas turbines, operating in the non-premixed regime. These applications require accurate determination of position in the flow where combustion starts and the control of pollutant emissions.
2. Many practical systems include liquid injection of the fuel, followed by non-premixed and partially premixed combustion.

3. Even in burners operating in the premixed regime, the premixing of the reactants is not always complete at the molecular level, and some partial premixing may be observed. Sometimes, partial premixing is even desirable to limit pollutant emissions such as in stratified charge engines.

The major hypotheses usually made to formulate models for non-premixed turbulent combustion may be broadly organised into the following three major groups:

- Assumption of infinitely fast chemistry.
- Finite rate chemistry assuming a local diffusive–reactive budget similar to the one observed in laminar flames (flamelet assumption).
- Finite rate chemistry with treatment of molecular and heat transport separated from chemical reaction. Diffusion is addressed using turbulent micro-mixing modelling, while chemical sources are dealt with in an exact and closed form.

9.5.2.1 Eddy Dissipation Model

The eddy dissipation model is one of the oldest models used in modelling non-premixed combustion. This model is based on the assumption of infinitely fast chemistry where fuel and oxidiser cannot co-exist; as such, this model is strictly valid only for the flamelet regime of combustion [17].

Considering the generic combustion mechanism, that is:

$$\mathit{Fuel} + \boldsymbol{s}\mathit{Oxidiser} \rightarrow (\mathbf{1}+\boldsymbol{s})\,\mathit{Products}$$

the mean reaction rate of the fuel is expressed using the eddy dissipation model as follows:

$$\bar{\dot{w}}_F = -A\bar{\rho}\frac{\varepsilon}{k}min\left(Y_F, \frac{Y_o}{s}, B\frac{Y_p}{1+s}\right) \tag{9.108}$$

In this expression, A and B are two model parameters which require adjustments per the problem configuration.

Note that this model considers only the mixing time scale (k/ε) and neglects the chemical reaction time scale (τ_c) although there are some variants of this model where both the time scales are taken into account. Also, this model predicts an unphysically high reaction rate in regions with high shear, such as near a wall. This is a very simple model, however; it is easy to implement and is computationally very cheap. This is why this model is preferred in industrial calculations.

9.5.2.2 Presumed Probability Density Function (PDF) Approach

This approach simply presumes the shape of the probability density functions of the mixture fraction using a β function. This is expressed mathematically as [1]:

$$\tilde{P}\left(Z^{*}:x,t\right)=\frac{\left(Z^{*}\right)^{a-1}\left(1-Z^{*}\right)^{b-1}}{\int_{0}^{1}\left(Z^{*}\right)^{a-1}\left(1-Z^{*}\right)^{b-1}dZ^{*}} \tag{9.109}$$

The mean of the mixture fraction ($\tilde{Z}$) and its variance ($\tilde{Z}''^2$) can be expressed using this presumed PDF as follows:

$$\tilde{Z}=\int_{0}^{1}Z^{*}\tilde{P}\left(Z^{*};x,t\right)dZ^{*} \tag{9.110}$$

$$\widetilde{Z''^{2}}=\int_{0}^{1}\left(Z^{*}-\tilde{Z}\right)\tilde{P}\left(Z^{*};x,t\right)dZ^{*} \tag{9.111}$$

The integrals in the above expressions are determined using the following relation:

$$\int_{0}^{1}\left(Z^{*}\right)^{n}\tilde{P}\left(Z^{*};x,t\right)dZ^{*}=\frac{a(a+1)\ldots\left(a+(n-1)\right)}{(a+b)(a+b+1)\ldots\left(a+b+(n-1)\right)} \tag{9.112}$$

The quantities a and b are determined as:

$$a=\tilde{Z}\left(\frac{\tilde{Z}\left(1-\tilde{Z}\right)}{\widetilde{Z''^{2}}}-1\right)\geq 0 \tag{9.113}$$

$$b=a\left(\frac{1}{\tilde{Z}}-1\right)\geq 0 \tag{9.114}$$

It is evident from these expressions that additional transport equations need to be solved for determining $\tilde{Z}$ and $\widetilde{Z''^{2}}$.

The mean thermo-chemical quantities for a flame, that is, the temperature of the mixture and the mixture fraction of the fuel and the oxidiser, may be obtained using the presumed PDF approach assuming infinitely fast chemistry.

These quantities can be represented as functions of the mixture fraction, and the corresponding mean quantities are expressed as:

$$\tilde{Y}_F = \int_0^1 Y_F\left(Z^*\right)\tilde{P}\left(Z^*;x,t\right)dz \tag{9.115}$$

$$\tilde{Y}_O = \int_0^1 Y_O\left(Z^*\right)\tilde{P}\left(Z^*;x,t\right)dz \tag{9.116}$$

$$\tilde{T} = \int_0^1 T\left(Z^*\right)\tilde{P}\left(Z^*;x,t\right)dz \tag{9.117}$$

The mean thermo-chemical quantities, thus obtained, quantify the mean flame structure. Note that the relations are exact only when the chemical time is infinitely small. Therefore, for very large Damköhler numbers, the assumption of infinitely fast chemistry is an accurate description of a turbulent non-premixed flame. However, multi-step chemistry is required in other situations, and to handle such situations, infinitely fast chemistry may be replaced by a chemical equilibrium condition [18].

9.5.2.3 Flamelet Modelling

Experiments on flames in jets and direct numerical simulations suggest that there exist situations in burners where the chemistry is fast but not infinitely fast [1]. For a given state of mixing in the turbulent flow, flamelet models are derived, assuming that the local balance between diffusion and reaction is similar to the one found in a prototype laminar flame.

The two control parameters of planar and steady laminar strained flames are used for determining the flamelet model – the mixture fraction (Z) and its scalar dissipation rate (χ). In a turbulent flow, these two quantities fluctuate in space and time, but when the joint PDF is known, the mean properties of the flame may be calculated as follows [1]:

$$\tilde{Y}_i = \int_{Z^*}\int_{\chi^*} Y_i^{SLFM}\left(Z^*, \chi^*\right)\tilde{P}\left(Z^*, \chi^*;x,t\right)d\chi^* dZ^* \tag{9.118}$$

In the above expression, $Y_i^{SLFM}\left(Z^*, \chi^*\right)$ is the local flame structure in mixture fraction space and $\tilde{P}\left(Z^*, \chi^*;x,t\right)$ captures the statistics of fuel/air mixing.

The superscript SLFM refers to the steady laminar flamelet model. This model may be viewed as a direct improvement of the infinitely fast chemistry assumption, since it uses the same formalism, but with an additional parameter in the form of scalar dissipation rate (χ), thereby including the effects of finite rate chemistry.

It is evident from the above discussion that two quantities need to be determined for obtaining the mean flame properties $Y_i^{SLFM}\left(Z^*,\chi^*\right)$ and $\tilde{P}\left(Z^*,\chi^*;x,t\right)$. $Y_i^{SLFM}\left(Z^*,\chi^*\right)$ may be determined from the solutions of counter-flow diffusion flames [19]. Assuming thin quasi-one-dimensional structures convected and stretched by the turbulent fluid motions, the equations for the species and temperature can be expressed by neglecting higher-order terms as [1]:

$$\frac{\partial Y_i}{\partial t}=\dot{\omega}_i+\left(\frac{\chi}{Le_i}\right)\frac{\partial^2 Y_i}{\partial Z^2} \tag{9.119}$$

$$\frac{\partial T}{\partial t}=-\sum_{n=1}^{N}\frac{h_n\dot{\omega}_n}{C_p}+\chi\frac{\partial^2 T}{\partial Z^2} \tag{9.120}$$

Under the steady flamelet assumption, the time-derivatives in the above equations can be omitted, leading to the following expression:

$$\dot{\omega}_i=-\left(\frac{\chi}{Le_i}\right)\frac{\partial^2 Y_i}{\partial Z^2} \tag{9.121}$$

The solution of this equation for given concentrations and temperatures, and various values of χ, provides a flamelet library which is then used for obtaining the flamelet model.

The quantity $\tilde{P}\left(Z^*,\chi^*;x,t\right)$ captures the effects of turbulent fuel-air mixing. Most flamelet models assume that the mixture fraction and the scalar dissipation rate are two independent parameters such that [1]:

$$\tilde{P}\left(Z^*,\chi^*;x,t\right)=\tilde{P}\left(Z^*;x,t\right)\tilde{P}\left(\chi^*;x,t\right) \tag{9.122}$$

A β function is presumed for $\tilde{P}\left(Z^*;x,t\right)$, while a log-normal distribution is presumed in the case of $\tilde{P}\left(\chi^*;x,t\right)$. The expression for β –PDF is similar to that in the presumed PDF modelling approach. The log-normal distribution function is expressed as:

$$\tilde{P}\left(\chi^*;x,t\right)=\frac{1}{\chi^*\sigma(x,t)\sqrt{2\pi}}exp\left(-\frac{1}{2\sigma^2(x,t)}\left(\ln\chi^*-\gamma(x,t)\right)^2\right) \tag{9.123}$$

The quantities σ and γ are provided by the first and second moments of χ^*. These are expressed as:

$$\tilde{\chi} = exp\left(\gamma + \frac{\sigma^2}{2}\right) \tag{9.124}$$

$$\widetilde{\chi''^2} = \tilde{\chi}^2\left[exp\left(\sigma^2\right) - 1\right] \tag{9.125}$$

The mean reaction rate is expressed, using the above descriptions, as:

$$\overline{\dot{w}_F} = -\left(\overline{\rho\chi^*\middle|Z = Z^*}\right)\tilde{P}\left(\chi^*;x,t\right)\frac{Y_{FO}}{\left(1-\chi^*\right)} \tag{9.126}$$

9.5.2.4 Flame Surface Density Approach

The mean reaction rate is determined using the flame surface density as follows:

$$\overline{\dot{\omega}_i} = \dot{\Omega}_i\Sigma \tag{9.127}$$

Here, Σ represents the flame surface density, while $\dot{\Omega}_i$ denotes the local burning rate. The flame surface density is defined as [11]:

$$\Sigma = \overline{|\nabla Z|\delta\left(Z - Z^*\right)} = \overline{|\nabla Z|\left(Z = Z^*\right)}\overline{P}\left(Z^*\right) \tag{9.128}$$

The local burning rate can be expressed as:

$$\dot{\Omega}_i = \left\langle \frac{\dot{\omega}_i}{|\nabla Z|} \right\rangle_s \approx \int_0^1 \frac{\dot{\omega}_i}{|\nabla Z|} dZ = \int_{-\infty}^{\infty} \dot{\omega}_i d\xi \tag{9.129}$$

where ξ is the coordinate along the normal to the flame front.

9.5.2.5 Conditional Moment Closure Model

The main idea in conditional moment closure (CMC) modelling is to focus on particular states in combustion. Only conditional moments are considered in this modelling approach. The main advantage of this approach is that it can be applied to all combustion regimes and is relatively easy to implement. However, it is much more computationally expensive than flamelet based methods.

The CMC approach solves a balance equation for Q_i which can be expressed as [20]:

$$\left(\overline{\rho|Z=Z^*}\right)\frac{\partial Q_i}{\partial t}+\left(\overline{\rho u_l|Z=Z^*}\right)\frac{\partial Q_i}{\partial x_i}=\left(\overline{\rho\chi|Z=Z^*}\right)\frac{\partial^2 Q_i}{\partial Z^{*2}}+\left(\overline{\dot{\omega}_l|Z=Z^*}\right) \quad (9.130)$$

The accuracy of this modelling approach depends on the modelling of the conditional moment term $(\overline{\rho\chi|Z=Z^*})$, which requires an accurate model of χ. In addition, modelling is required for the conditional mean velocity $(\overline{u_l|Z=Z^*})$.

9.5.2.6 PDF Approach

The objective of PDF modelling is to relax all hypotheses concerning the shape of PDFs, such as that assumed in the presumed PDF approach. Once a methodology has been developed to calculate PDFs, it is possible to construct turbulent combustion closures in which all the values taken by species and temperature in the mixture fraction space may be accounted for.

The fine-grained PDF of a scalar (c) for one-time and one-point measurement is given as [1]:

$$f(v;x,t)=\delta(c-v) \quad (9.132)$$

The transport equation for this fine-grained PDF may be expressed as:

$$\frac{\partial(\rho f)}{\partial t}+\frac{\partial(\rho u_i f)}{\partial x_i}=-\frac{\partial}{\partial v}\left[\left\{\rho\left(\frac{\partial c}{\partial t}+u_i\frac{\partial c}{\partial x_i}\right)\right\}\delta(c-v)\right]$$
$$=-\frac{\partial}{\partial v}\left[\left\{\dot{\omega}+\frac{\partial}{\partial x_i}\left(\rho D\frac{\partial c}{\partial x_i}\right)\right\}\delta(c-v)\right] \quad (9.133)$$

On averaging this transport equation, we obtain:

$$\frac{\partial\left[\rho_v P(v)\right]}{\partial t}+\frac{\partial\left(\overline{\rho u_l}P(v)\right)}{\partial x_i}=-\frac{\partial}{\partial v}\left[\overline{\left\{\dot{\omega}+\frac{\partial}{\partial x_l}\left(\rho D\frac{\partial c}{\partial x_l}\right)\right\}}\Bigg|_v P(v)\right] \quad (9.134)$$

It is evident from the above equation that the reaction rate ($\dot{\omega}$) is inherently closed since:

$$\bar{\dot{\omega}}=\int_{v_{min}}^{v_{max}}\dot{\omega}P(v)dv \quad (9.135)$$

However, the molecular diffusion rate is not directly closed and needs to be further modelled. This is expressed as follows:

$$\frac{\partial}{\partial v}\left[\overline{\left\{\frac{\partial}{\partial x_l}\left(\rho D\frac{\partial c}{\partial x_l}\right)\right\}}\Bigg|_v P(v)\right]=-\frac{\partial}{\partial x_i}\left(\rho D\frac{\partial P(v)}{\partial x_i}\right)+\frac{\partial^2}{\partial v^2}\left[\overline{\left\{\rho D\frac{\partial c}{\partial x_l}\frac{\partial c}{\partial x_l}\right\}}\Bigg|_v P(v)\right] \tag{9.136}$$

In the case of non-premixed combustion, the scalar usually adopted for obtaining the PDF is the mixture fraction (Z). One may solve the balance equation for the joint PDF of the thermo-chemical variables, species, temperature, velocity, and so on. However, models must be proposed for capturing the micro-mixing and pressure fluctuations as well as the viscous effects.

9.6 TURBULENT PARTIALLY PREMIXED COMBUSTION

Flames in most practical applications cannot be described as purely premixed or non-premixed. Rather, a non-uniform mixing of the fuel and oxidiser may take place. This is referred to in the literature as partially premixed combustion, as opposed to premixed (uniform mixing) and non-premixed (no mixing) combustions. Partially premixed combustion may be encountered in many practical situations – gas-fired domestic burners, industrial furnaces, Bunsen burners, turbulent combustion and spray flames. Flame stabilisation in burners using burnt gases is another situation when partial premixing may take place.

Modelling of partially premixed turbulent combustion is usually achieved by combining the modelling approaches to premixed and non-premixed combustion. The presumed PDF approach, strained flamelet approach and the BML approach are some of the approaches generally utilised in partially premixed combustion modelling [1].

The premixed reaction-progress variable, c, determines the position of the flame front. Behind the flame front ($c = 1$), the mixture is assumed to be burnt, and the equilibrium or laminar flamelet mixture fraction solution is used. Ahead of the flame front ($c = 0$), the species mass fractions, temperature, and density are calculated from the mixed but unburnt mixture fraction. Within the flame ($0 < c < 1$), a linear combination of the unburnt and burnt mixtures is used.

The primary difficulty is faced in the definition of the reaction progress variable; that is, temperature and fuel mass fractions of unburnt and burnt gases can no longer be assumed to remain constant since these vary depending on the local equivalence ratio. This presents a difficulty in obtaining a closure of the transport equation of progress variable. The second difficulty is that the modelling should also take into account the variation of the mixture fraction. As such, the challenge becomes to properly model the joint probability density function of the mixture fraction and the reaction progress variable [1].

Example 9.2

Calculate the Damkohler number for the following situation:

Integral length = ⅕ of the domain length
Laminar flame thickness = 10^{-5} m
Mean velocity = 10 m/s
Turbulent intensity = 5%
Laminar burning velocity = 0.5 m/s
Domain length = 1 m

Solution

The characteristic flow time can be estimated from the above data to be:

$$\tau_{flow} = \frac{\text{Integral Length Scale}}{\text{Turbulent velocity}}$$

The turbulent velocity (v_{rms}) is calculated as:

$$\begin{aligned} v_{rms} &= \text{Turbulent Intensity} \times \text{Mean Velocity} \\ &= 0.05 \times 10 \\ &= 0.5 \text{ m / s} \end{aligned}$$

The characteristic flow time thus becomes:

$$\begin{aligned} \tau_{flow} &= \frac{\text{Integral Length Scale}}{\text{Turbulent velocity}} \\ &= \frac{\frac{1}{5} \times 1}{0.5} \\ &= 0.4\ s \end{aligned}$$

Similarly, the chemical time scale can be estimated as follows:

$$\tau_{chem} = \frac{\text{Laminar Flame Thickness}}{\text{Laminar Burning Velocity}}$$

$$= \frac{10^{-5}}{0.5}$$

$$= 2\times10^{-5}\ s$$

Therefore, the Damkohler number becomes:

$$Da = \frac{\tau_{flow}}{\tau_{chem}}$$

$$= \frac{0.4}{2\times10^{-5}}$$

$$= 20000$$

EXERCISES

9.1 Consider a flow situation with kinematic viscosity of the fluid and dissipation rates of 10^{-6} m²/s and 200 m²/s³, respectively. Calculate the Kolmogorov length scale for this situation.

9.2 If the turbulent velocity in Exercise 9.1 is 10 m/s, determine the Taylor micro scale length.

9.3 Consider the decay of turbulence during the induction stroke of an engine following the introduction of intake jet into the cylinder. The initial integral length scale l is taken to be $D_{cyl}/6$ and the root-mean-square (rms) turbulent velocity fluctuation is taken to be $u' = 10S_P$ where S_P is the mean piston speed. The Reynolds number is given by $\mathrm{Re} = S_P D_{cyl}/\nu = 1000$ where the kinematic viscosity of the in-cylinder gas is given by ν. Under the above conditions, determine:

a. The dissipation rate of turbulent kinetic energy ε.
b. The timescale (turnover time or lifetime) of an eddy of integral length scale.
c. Estimate the Kolmogorov timescale.
d. Estimate the Kolmogorov length scale.
e. The time scale of an eddy of any size between the integral length scale and Kolmogorov length scale.
f. Assuming that each eddy forms a new eddy of half of its size after its own lifetime, how long does it take before Kolmogorov eddies are created?

9.4 Estimate the Damkohler number and the ratio of the Kolmogorov length scale to the laminar flame thickness for conditions prevailing in the combustor of a utility-class gas-turbine engine. What flame regime does this indicate? Assume a premixed fuel-air mixture where the unburned gas temperature is 600 K, the burned gas temperature is 2000 K, the pressure is 15 atm and the mean velocity is 100 m/s. Consider the equivalence ratio to be unity, the fuel properties to be that of isooctane and the combustor can diameter to be 0.3 m. Also, assume the relative turbulence intensity to be 10% and the integral scale to be one-tenth of the can diameter. What happens to the flame regime if the mean velocity is doubled, keeping all other parameters constant?

9.5 Calculate the characteristic chemical and flow times, and hence determine the magnitude of the Damkohler number for the following conditions for a propane-air mixture:

$$P = 4 \text{ atm}$$

$$V_{rms} = 2 \text{ m/s}$$

$$L_o = 5 \text{ mm}$$

$$T_u = 350 \text{ K}$$

$$\Phi = 0.6$$

Comment on whether the conditions represent fast or slow chemistry.

9.6 Derive the URANS equations using Reynolds' decomposition technique.

9.7 Derive the $k-\varepsilon$ model transport equations starting from the Navier–Stokes equations.

REFERENCES

1. Veynante, D. and Vervisch, L., Turbulent combustion modelling, *Progress in Energy and Combustion Science*, **28**, 193–266, 2002.
2. Pope, S. B., *Turbulent Flows*, Cambridge, UK: Cambridge University Press, 2000.
3. Favre, A. J. A., Formulation of the statistical equations of turbulent flows with variable density. In: T. B. Gatski, C. G. Speziale, and S. Sarkar (Eds.) *Studies in Turbulence*. New York: Springer, 1992.
4. Peters, N., *Turbulent Combustion*, Cambridge, UK: Cambridge University Press, 2000.
5. Borghi, R. and Destriau, M., *Combustion and Flames, Chemical and Physical Principles*, Paris: Editions Technip, 1998.
6. Spalding, D. B., *Mixing and Chemical Reaction in Steady Confined Turbulent Flames. 13th Symposium (International) on Combustion*, Pittsburg, CA, 1971.

7. Moss, J. B. and Bray, K. N. C., A unified statistical model of the premixed turbulent flame, *Acta Astronaut*, **4**, 291–319, 1977.
8. Bray, K. N. C., Champion, M. and Libby, P. A., The interaction between turbulence and chemistry in premixed turbulent flames. In: R. Borghy, S. N. Murphy (Eds.) *Turbulent Reacting Flows, Lecture Notes in Engineering*, vol. 40, Berlin, Germany: Springer, 1989.
9. Gouldin, F. C., Bray, K. N. C. and Chen, J. Y., Chemical closure model for fractal flamelets, *Combustionand Flame*, **77**, 241–259, 1989.
10. Cant, R. S., Pope, S. B. and Bray, K. N. C., Modelling of Flamelet Surface to Volume Ratio in Turbulent Premixed Combustion. 23*rd* Symposium (International) on Combustion, Pittsburg, CA 1990.
11. Marble, F. E. and Broadwell, J. E., *The Coherent Flame Model of Non-Premixed Turbulent Combustion*. Project Squid TRW-9-PU, Purdue University, 1977.
12. Candel, S. and Poinsot, T., Flame stretch and the balance equation for the flame area, *Combustion Science and Technology*, **70**, 1–15, 1990.
13. Meneveau, C. and Poinsot, T., Stretching and quenching of flamelets in premixed turbulent combustion, *Combustion and Flame*, **86**, 311–332, 1991.
14. Mantel, T. and Borghi, R., A new model of premixed wrinkled flame propagation based on scalar dissipation equation, *Combustion and Flame*, **96**, 443, 1994.
15. Cheng, W. K. and Diringer, J. A., Numerical modelling of SI engine combustion with a flame sheet model, International Congress and Exposition, Detroit, 1991.
16. Choi, C. R. and Huh, K. Y., Development of a coherent flamelet model for a spark ignited turbulent premixed flame in a closed vessel, *Combustion and Flame* 114 (1998) 336–348.
17. Magnussen, B. F. and Hjertager, B. H., *On the Mathematical Modelling of Turbulent Combustion with Special Emphasis on Soot Formation and Combustion*. 16th Symposium (International) on Combustion, Pittsburg, CA 1976.
18. Coupland, J. and Priddin, C. H., Modelling of flow and combustion in a production gas turbine combustor. In: F. Durst, B.E. Launder, J.L. Lumley, F.W. Schmidt, J.H. Whitelaw (Eds.) *Turbulent Shear Flows*, Berlin, Germany: Springer, 1987.
19. Sung, C. J., Liu, J. B. and Law, C. K., Structural response of counter-flow diffusion flames to strain rate variations, *Combustion and Flame*, **102**, 481–492, 1995.
20. Bilger, R. W., Conditional moment closure for turbulent reacting flow, *Physics of Fluids*, **5**, 327–334, 1993.
21. Turns, S. R., *An Introduction to Combustion: Concepts and Applications*, McGraw-Hill, 2000.

Chapter 10

Combustion of Solid Fuels and Surface Reactions

10.1 HETEROGENEOUS COMBUSTION

When the reacting species are in different physical states, the reaction is called a heterogeneous reaction. In case of gaseous fuel combustion, the reacting components are all in gaseous state. For liquid fuel burning, the fuel first vapourises and then reacts. While reacting, the species are in gaseous state. This type of combustion can be called homogeneous combustion. When solid fuel, like coal, is burning in air, the reacting components are in solid and gaseous states. This is called heterogeneous combustion.

The gas-solid reactions include five processes [1]:

1. Convective-diffusive transport of reactant molecules to the surface
2. Adsorption of reactant molecules at the surface
3. Elementary reaction steps involving various combinations of adsorbed molecules, surface and gas phase
4. Desorption of product molecules from the surface
5. Product molecule transport from surface to away

The processes of adsorption, reaction and desorption are complex in nature, and detailed discussion about them are beyond the scope of this book. Instead of elaborating on these processes, three rate laws can be adopted depending on the nature of adsorption of reactant and/or product molecules on the surface. Table 10.1 summarises the rate laws, where $k(T)$ is the rate coefficient, and $[A]$ and $[B]$ are the gas phase concentration of molecules A and B, just adjacent to the surface. The difference of reaction rate expressions compared to our earlier understanding should be noted here.

TABLE 10.1 RATE LAWS FOR HETEROGENEOUS REACTIONS

	Conditions	Reaction Rate
Case I	Reactant molecule A is weakly adsorbed.	$R = k(T)[A]$ (10.1)
Case II	Reactant molecule A is strongly adsorbed.	$R = k(T)$ (10.2)
Case III	Reactant molecule A is weakly adsorbed, and product molecule B is strongly adsorbed.	$R = k(T)[A]/[B]$ (10.3)

10.2 BURNING OF COAL

Burning of coal involves two major components. First is the thermal decomposition or pyrolysis or devolatilisation. This process occurs during the initial heating when the absorbed/adsorbed combustible gases (volatile matters [VMs]) are released from solid particles. The released gases then burn following a homogeneous combustion route. This process of VM release and burning increases the temperature. The rise in temperature changes the physical as well as chemical properties of the solid particle. The porous solid particle, after the release of VM, is composed mainly of carbon and ash. At this temperature the solid particle or char becomes plastic, then it rehardens [2]. The elevated temperature also provides the initial energy to char to burn. The second part is burning of char. The size of the coal particle and the nature of the pores dictate the burning rate. The devolatilisation is very fast, while char combustion is slow. In pulverised fuel, devolatilisation time is of the order of 0.1 s. The char burnout time is 1 s; for particles burning in fluidised-bed combustors, the corresponding orders are 10 and 1000 s. Therefore the burning of the char has a major effect on the volume of the combustion chamber required to attain a given heat release.

Here we shall discuss about the char burning mechanism and simple models of that. The gas phase reactions are already discussed earlier. The burning mechanism of char is quite complex. Detailed discussion about coal combustion may be found in references [2,3,4,5].

The char particle surface mainly reacts with oxygen, carbon dioxide and water through the following four global reactions:

$$C + O_2 \rightarrow CO_2 \tag{10.4}$$

$$2C + O_2 \rightarrow 2CO \tag{10.5}$$

$$C + CO_2 \rightarrow 2CO \tag{10.6}$$

$$C + H_2O \rightarrow CO + H_2 \tag{10.7}$$

These four global reactions occur at the surface of the char particle. The gaseous reactants reach the surface by some mechanism like advection or diffusion. The gaseous reactants also penetrate through the pores of the char particle. The reactions are occurring outside the char surface as well as inside the char surface due to the gas trapped in the pores. The main product of these surface reactions is carbon monoxide that subsequently oxidises to dioxide.

The presence of particle surface in a gaseous stream produces a boundary layer. Outside the boundary layer, the free stream exists. Oxygen diffuses from the free stream towards the particle. The oxygen concentration varies from surface to free stream. Carbon monoxide generated through the surface reactions diffuses towards the free stream. In the process, it combines with oxygen through the gas phase reaction within the boundary layer. The gas phase global reaction is:

$$CO+\frac{1}{2}O_2 \rightarrow CO_2 \tag{10.8}$$

The simple models of char burning depend on these global reactions and rely on some assumptions also. Two such models are discussed below.

10.2.1 One-Film Model

In this simplified model, we consider the burning of a spherical char particle under certain assumptions. The particle is considered a sphere. It burns in a quasi-steady way in a quiescent and infinite medium. It is assumed that the medium is composed of oxygen and an inert gas. The model considers only one reaction of carbon with oxygen to form carbon dioxide (Equation 10.4). Although carbon monoxide is the main product in coal surface reaction, that is not reflected in this model. The gas phase is considered with oxygen, carbon dioxide and an inert gas. Oxygen diffuses up to the surface to form carbon dioxide. Carbon dioxide then moves outward. It is assumed that the char surface is impermeable towards gaseous substances. The property values are uniform over the char particle. The Lewis number is assumed to be unity.

Temperature and species mass fraction distributions are shown in Figure 10.1. It is assumed that the temperature is uniform within the char particle. The temperature drops from the coal surface temperature, T_s, and reaches a temperature, $T_{s,\infty}$, asymptotically. The carbon dioxide mass fraction is highest at the coal surface [$Y_{CO2,s}$], where it is formed per the assumption of the model. As the ambient medium is assumed to be constituted of oxygen and inert gas, the carbon dioxide mass fraction should be zero at infinity. As oxygen is diffusing in from the far stream to the surface, at infinite distance from the surface we get maximum oxygen mass fraction [$Y_{O2,\infty}$]. This corresponds to the

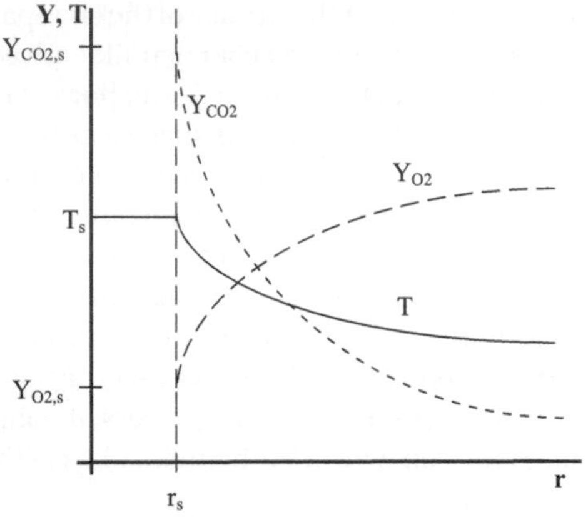

Figure 10.1 Temperature and species distribution in the one-film model.

mass fraction of the ambient medium. The oxygen mass fraction gradually decreases to its surface value, $Y_{O2,s}$.

Our objective is to find the burning rate of carbon. For that, we can use the mass and species conservation equations first. We can see the mass fluxes of three species – carbon, carbon dioxide and oxygen – in Figure 10.2. Carbon dioxide is moving outward while oxygen is moving in. The net mass flux, the difference between the mass fluxes of carbon dioxide and oxygen, is depicted as carbon mass flux. However, we can have carbon mass flux only at the surface. At other radii, strictly it is not carbon. So, we can write:

$$\dot{m}_C'' = \dot{m}_{CO2}'' - \dot{m}_{O2}'' \tag{10.9}$$

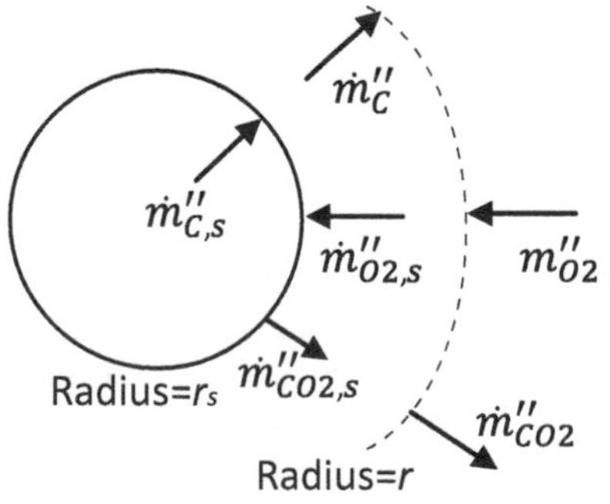

Figure 10.2 Mass fluxes for different species.

As mass flow rate is equal at each radius, we can write:

$$\dot{m}''_{C,s}\left(4\pi r_s^2\right)=\dot{m}''_{C}\left(4\pi r^2\right) \quad (10.10)$$

Form the chemical reaction, we can tell that:

$$1\,\text{kg of carbon}+\gamma\ \text{kg of oxygen}\rightarrow(1+\gamma)\text{kg of carbon dioxide}$$

So, it can be written that:

$$\dot{m}_C=\frac{1}{\gamma}\dot{m}_{O2}=\frac{1}{\gamma+1}\dot{m}_{CO2} \quad (10.11)$$

Now the mass flux of oxygen can be written as:

$$\dot{m}''_{O2}=Y_{O2}\left(\dot{m}''_{O2}+\dot{m}''_{CO2}\right)-\rho D\frac{dY_{O2}}{dr} \quad (10.12)$$

The mass flux terms can be substituted to carbon mass flux terms using Equation (10.11). The expression $\dot{m}_i=\dot{m}''_i\left(4\pi r^2\right)$ may be used along with Equation (10.11). After simplification, it can be written:

$$\dot{m}_C=\frac{4\pi r^2\rho D}{1+Y_{O2}/\gamma}\frac{dY_{O2}/\gamma}{dr} \quad (10.13)$$

The boundary conditions for oxygen mass fraction can be taken at char surface and infinity.

At

$$r=r_s, Y_{O2}=Y_{O2,s} \quad (10.14)$$

At

$$r=r_\infty, Y_{O2}=Y_{O2,\infty} \quad (10.15)$$

Separating the variables of Equation (10.13) and integrating between above two boundary limits, we find:

$$\dot{m}_C=4\pi r_s\rho D\left[ln\frac{1+\frac{Y_{O2,\infty}}{\gamma}}{1+\frac{Y_{O2,s}}{\gamma}}\right]=4\pi r_s\rho D\left[\ln\left(1+\frac{Y_{O2,\infty}-Y_{O2,s}}{\gamma+Y_{O2,s}}\right)\right] \quad (10.16)$$

Here, $\dot{m}_C$ means the burning rate or mass flow rate of carbon at the surface ($\dot{m}_{C,s}$). The oxygen mass fraction at the free stream is a known quantity. To find the carbon burning rate, we need to evaluate $Y_{O2,s}$. For that, we need to form another equation from the chemical kinetics.

Assuming that the reaction, $C+O_2 \rightarrow CO_2$, is a first-order equation of oxygen following the Equation (10.1), we can write the reaction rate as [4]:

$$\dot{m}''_{C,s} = k_c M_c \left[O_{2,s}\right] \tag{10.17}$$

where $\left[O_{2,s}\right]$ is the molar concentration of oxygen at the surface and k_c is the rate coefficient; k_c can be expressed in Arrhenius form. Converting the molar concentration into the mass fraction, we get:

$$\dot{m}_C = 4\pi r_s^2 k_c \frac{M_C M_{mix}}{M_{O2}} \frac{P}{R_u T_s} Y_{O2,s} = K_{kin} Y_{O2,s} \tag{10.18}$$

M_C, M_{mix} and M_{O2} are the molecular weights of carbon, mixture and oxygen, respectively. P denotes the pressure. T_s is the surface temperature and R_u is the universal gas constant. K_{kin} can be treated as the compact kinetic parameter that depends on particle radius, surface temperature and pressure. $Y_{O2,s}$ from Equation (10.18), can be substituted in Equation (10.16) and then solved to get the burning rate $\dot{m}_C$. However, the burning rate can be found using a resistance analogy rather than finding it from the above method. Equation (10.18) can be written as:

$$\dot{m}_C = \frac{Y_{O2,s} - 0}{1/K_{kin}} = \frac{\Delta Y(kin)}{R_{kin}} \tag{10.19}$$

which is in the form of potential difference. The oxygen mass fraction is acting as the potential here. The zero, added in the equation, can be treated as the oxygen mass fraction in the particle interior.

The oxygen mass fraction value is normally less than 0.233, for air. The value of γ is 2.66. So the expression $\frac{Y_{O2,\infty} - Y_{O2,s}}{\gamma + Y_{O2,s}}$ is a small value. With this rationale, we can expand the logarithmic term in Equation (10.16) into a series and we take the first term only. This will lead to:

$$\dot{m}_C = \frac{Y_{O2,\infty} - Y_{O2,s}}{\dfrac{\gamma + Y_{O2,s}}{4\pi r_s \rho D}} = \frac{\Delta Y(diff)}{R_{diff}} \tag{10.20}$$

Combining Equations (10.19) and (10.20), we can write:

$$\dot{m}_C = \frac{Y_{O2,\infty} - 0}{R_{kin} + R_{diff}} \tag{10.21}$$

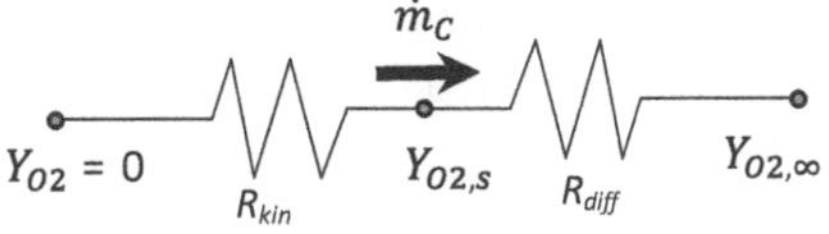

Figure 10.3 The resistance analogy for carbon particle burning.

Here, we note that burning of the carbon particle depends on two resistances. One is kinetic resistance, which depends on the chemical kinetics parameters. The second is diffusion resistance, which depends on the feature of oxygen diffusion. Note that the diffusion resistance depends on the mass fraction of oxygen at the particle surface. This makes the problem a non-linear one and the solution cannot be obtained in a straightforward way. The analogous circuit diagram is shown in Figure 10.3.

Now, we can evaluate the ratio of kinetic and diffusion resistances. If the ratio is of the order of 1, then both phenomena are controlling the burning mechanism. If the ratio is much smaller than unity, the burning is called diffusion controlled as the diffusion resistance dominates. In this case, diffusion is a slower mechanism compared to the reaction kinetics. At the surface, sufficient oxygen cannot be found (may be taken as 0). As soon as some oxygen diffuses to the surface level, it is consumed by the reaction. The burning rate eventually depends upon diffusion and hence the name. On the other hand, a ratio much higher than unity ensures kinetically controlled burning. In this case, a sufficient amount of oxygen is always present at the surface as diffusion is faster (may be taken as free stream oxygen). The reaction rate is actually controlling the burning rate. One can jot down the parameters and find the effect of different parametric values on the control of the burning rate.

To find the surface temperature, we need to solve the energy equation at the surface. We can assume that the energy transfer occurs at steady state. The amount of energy liberated at the surface is taken away towards the free stream by conduction in the gas layer at the surface and radiation. We can write:

$$\dot{m}_C \Delta h_C = -k_g 4\pi r_s^2 \left.\frac{dT}{dr}\right|_{r_s} + \varepsilon_s 4\pi r_s^2 \sigma \left[T_s^4 - T_a^4\right] \tag{10.22}$$

Δh_C is the heat of combustion for carbon–oxygen reaction. The gas phase thermal conductivity is k_g. The surface emissivity is ε_s, and σ is the Stefan-Boltzmann constant. The ambient temperature is denoted by T_a.

We need to find the temperature gradient at the surface, in the gas phase. The energy balance is considered in the gas phase; solving that, the temperature distribution is obtained (similar to the droplet evaporation model). The gradient can be written as:

$$\left.\frac{dT}{dr}\right|_{r_s} = \frac{Z\dot{m}_C}{r_s^2}\left[\frac{(T_\infty - T_s)\exp\left(-Z\dot{m}_C/r_s\right)}{1-\exp\left(-Z\dot{m}_C/r_s\right)}\right] \tag{10.23}$$

where

$$Z = \frac{c_{pg}}{4\pi k_g} \tag{10.24}$$

Equation (10.23) can be substituted in Equation (10.22) to obtain the following equation:

$$\dot{m}_C \Delta h_C = \dot{m}_C c_{pg}\left[\frac{\exp\left(-\dot{m}_C c_{pg}/4\pi k_g r_s\right)}{1-\exp\left(-\dot{m}_C c_{pg}/4\pi k_g r_s\right)}\right](T_s - T_\infty) + \varepsilon_s 4\pi r_s^2 \sigma\left[T_s^4 - T_a^4\right] \tag{10.25}$$

Due to the presence of the radiation term, the equation is non-linear. If we do not consider radiation (which should not be advisable as the burning temperature is high and radiation cannot be neglected at that high temperature), the equation becomes linear. $\dot{m}_C$ and T_s are the unknown terms in this equation. Equations (10.21) and (10.23) are to be solved simultaneously to find the unknown parameters.

Example 10.1

A 200-micron coal particle is burning in still air at 1 atm pressure. The particle temperature is 1800 K. The kinetic rate constant (k_c) is 13.9 m/s. Find the burning rate by assuming the one-film model. Take the mean molecular weight of the gases at the particle surface as 30kg/kmol. Discuss the combustion regime also. Assume that diffusion coefficient $D = 2.48\times10^{-4}\, m^2/s$.

Solution

The kinetic resistance can be calculated as:

$$R_{kin} = \frac{\gamma R_u T_s}{4\pi r_s^2 M_{mix} k_c P} = \frac{2.66(8315)1800}{4\pi\left(100\times10^{-6}\right)^2(30)(13.9)(100000)} = 7.6\times10^6\, s/kg$$

We assume that 1 atmospheric pressure = 1 bar. Density can be calculated from the ideal gas law as:

$$\rho = \frac{PM_{mix}}{R_u T_s} = \frac{(100000)(30)}{(8315)(1800)} = 0.2\,kg/m^3$$

The diffusion resistance can be found from:

$$R_{diff} = \frac{\gamma + Y_{O2,s}}{\rho D 4\pi r_s} = \frac{2.66 + 0}{(0.2)(2.48\times10^{-4})4\pi(100\times10^{-6})} = 4.26\times10^7\,s/kg$$

Here, we assumed zero oxygen at the burning surface. If we assume the same oxygen concentration as in the free stream, the oxygen mass fraction at the surface will be 0.233. Then,

$R_{diff} = 4.64\times10^7\,s/kg$. We see that, in both cases, the diffusion resistance is one order higher than the kinetic resistance. So, the burning in this condition is diffusion controlled.

Now we shall find the burning rate. As the burning is diffusion controlled, the burning surface is starved of oxygen. The zero oxygen mass fraction at the surface is a realistic assumption, and we should start our calculation with that only:

$$\dot{m}_C = \frac{Y_{O2,\infty}}{R_{diff}} = \frac{0.233}{4.26\times10^7} = 5.46\times10^{-9}\,kg/s$$

Now, $Y_{O2,s} = \dot{m}_C R_{kin} = (5.46\times10^{-9})(7.6\times10^6) = 0.041$.

So, there is an error. Now we can assume new value of oxygen mass fraction as 0.041 instead of zero and carry out the calculation until the error comes below the permissible error.

Note 1: We used the expression $\dot{m}_C = \left(\frac{Y_{O2,\infty}}{R_{diff}}\right)$ to find the burning rate. If we take $\dot{m}_C = \left(\frac{Y_{O2,\infty}}{R_{diff}} + R_{kin}\right)$, the rate of convergence can be faster.

Note 2: If we calculate the resistances for a particle diameter of 20 micron, we find both of those to be on the order of 10^8. In that condition, the ratio of resistances will be on the order of one. Further reduction in diameter will slide the burning to a kinetically controlled regime. If we consider burning of a coal particle, the particle diameter will gradually decrease. So the burning regime may shift during the process.

Example 10.2

Refer to Example 10.1 to estimate the surface temperature when the free stream temperature is 1100 K. Take Δh_C as $3.2765\times10^7\,J/kg$. Assume only convective heat transfer is present.

Solution

First, let us take $\dot{m}_C = 5.46\times10^{-9}\,kg/s$, as found in Example 10.1. We know that some error is there; still we can adopt the value. We can see the gas properties from the air property table. For this, we can assume the surface temperature is 1800K. We find $c_{pg} = 1286\frac{J}{kgK}$ and $k_g = 0.12\frac{W}{mK}$.

$$T_s = T_\infty + \frac{\Delta h_C}{c_{pg}}\frac{1-\exp\left(-\dot{m}_C c_{pg}\Big/4\pi k_g r_s\right)}{\exp\left(-\dot{m}_C c_{pg}\Big/4\pi k_g r_s\right)}$$

$$= 1100 + \frac{3.2765\times10^7}{1286}\left(\frac{1-\exp\left[\dfrac{-5.46\times10^{-9}\times1286}{4\pi(0.12)\left(100\times10^{-6}\right)}\right]}{\exp\left[\dfrac{-5.46\times10^{-9}\times1286}{4\pi(0.12)\left(100\times10^{-6}\right)}\right]}\right)$$

$$= 1100 + 1215.5$$

$$= 2315.5K$$

There is a large gap between the guess value of the surface temperature and the value we calculated. We can take up an iterative scheme to find the surface temperature. Now one can assume the temperature to be 2315.5 K and start the process from the resistance calculation and burning rate determination. Further calculations are not shown here. Readers can solve the example themselves.

Note 3: In this example, we neglected radiation. This is not at all realistic, as the temperature is very high. To take care of the radiation term, we can linearise it. The expression is of the following form: $A(T_s^4 - T_a^4)$. We can write it as $A\left(T_s^4 - T_a^4\right) = A\left(T_s^2 + T_a^2\right)\left(T_s + T_a\right)\left(T_s - T_a\right) = B\left(T_s - T_a\right)$. The parameter **B** is evaluated on the basis of the guess temperature and also solved iteratively.

Note 4: A possible iterative scheme for determining the burning rate and surface temperature follows:

1. Assume the temperature of the surface $(T_{s,G})$.
2. Find the kinetic rate constant (k_c) and other gas properties dependent on temperature.
3. Assume oxygen mass fraction at surface $(Y_{O2,s,G})$.
4. Calculate resistances.
5. Calculate burning rate $\dot{m}_C$.
6. Use the drop in the oxygen mass fraction across kinetic resistance to find $Y_{O2,s}$.

7. If $|Y_{O2,s,G} - Y_{O2,s}| \le \in 1$, proceed. Otherwise, $Y_{O2,s,G} = Y_{O2,s}$; go to step 3 ($\in 1$ is a predetermined permissible error limit for mass fraction).
8. Calculate the surface temperature on the basis of $\dot{m}_C$ found in Step 5.
9. If $|T_{s,G} - T_s| \le \in 2$, proceed. Otherwise, $T_{s,G} = T_s$; go to step 1 ($\in 2$ is a predetermined permissible error limit for temperature).
10. The burning rate ($\dot{m}_C$) and surface temperature (T_s) are determined.

10.2.2 Two-Film Model

The two-film model is more realistic compared to the one-film model, where carbon at the surface is first converted to carbon monoxide and the carbon monoxide is subsequently converted to carbon dioxide. Obviously, the two-film model is not as simple as the one-film model. But it cannot capture all the details of coal particle burning, which is a complex phenomenon.

Carbon at the surface reacts with carbon dioxide to produce carbon monoxide, as in Equation (10.6). Carbon dioxide is produced at flame sheet through the reaction of carbon monoxide with oxygen (Equation 10.8). Figure 10.4 shows the temperature and species distribution for a two-film model. Atmospheric oxygen is assumed not to penetrate up to the surface. So carbon from the surface cannot be converted to dioxide directly. The monoxide generated at the surface diffuses radially outward and combines with oxygen after reaching the flame sheet, up to where atmospheric oxygen can also diffuse. This reaction is assumed to be an infinitely fast one. Carbon monoxide and oxygen are consumed completely at the flame sheet, and thus carbon dioxide concentration is highest at the flame sheet. This carbon dioxide diffuses radially outward as well as inward. The inward-moving carbon dioxide reaches the surface to react with carbon. Temperature should also be highest at the flame sheet, where the combustion becomes complete.

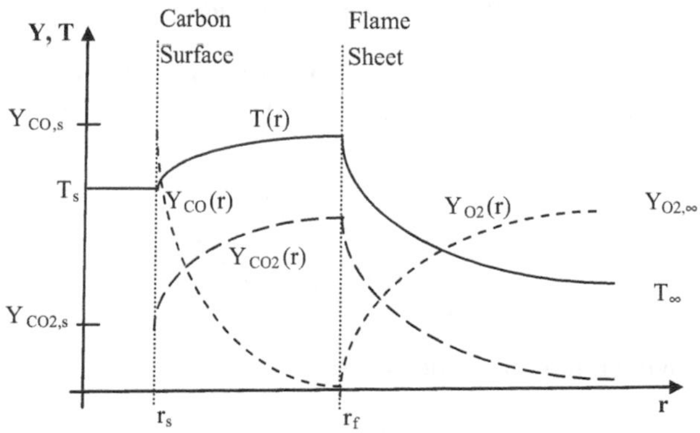

Figure 10.4 Temperature and species mass fraction distribution in the two-film model.

Mass balance can be carried out at the surface level and the flame level. Here, for the mass flow rate of carbon dioxide, we used two subscripts. Subscript *i* is used to denote the inward flow; the subscript *o* is used for the outward flow.

At the surface, we can obtain:

$$\dot{m}_C = \dot{m}_{CO} - \dot{m}_{CO2,i} \tag{10.26}$$

At the flame level, we can write:

$$\dot{m}_{CO} - \dot{m}_{CO2,i} = \dot{m}_{CO2,o} - \dot{m}_{O2} \tag{10.27}$$

or

$$\dot{m}_C = \dot{m}_{CO2,o} - \dot{m}_{O2} \tag{10.28}$$

Now, we can write the stoichiometric relations.

At the surface:

$$1\,kg\,\mathrm{C} + \gamma_s\,kg\,\mathrm{CO_2} = (\gamma_s + 1)\,kg\,\mathrm{CO} \tag{10.29}$$

At the flame:

$$1\,kg\,\mathrm{CO} + \gamma_f\,kg\;\mathrm{CO} = (\gamma_f + 1)\,kg\,\mathrm{CO_2} \tag{10.30}$$

We can find $\gamma_s = 3.66$ and $\gamma_f = \gamma_s - 1 = 2.66$. Now, we can write:

$$\dot{m}_{CO2,i} = \gamma_s \dot{m}_c \tag{10.31}$$

$$\dot{m}_{O2} = \gamma_f \dot{m}_c = (\gamma_s - 1)\dot{m}_c \tag{10.32}$$

$$\dot{m}_{CO2,o} = (\gamma_f + 1)\dot{m}_c = \gamma_s \dot{m}_c \tag{10.33}$$

The species equations can be taken up now. We shall take the carbon dioxide conservation equation for both the inner and outer zones. For the inner zone, we can write:

$$\dot{m}_c = \frac{4\pi r^2 \rho D}{\left(1 + {Y_{CO2}}/{\gamma_s}\right)} \frac{d\left({Y_{CO2}}/{\gamma_s}\right)}{dr} \tag{10.34}$$

with the following boundary conditions:

$$Y_{CO2}(r_s) = Y_{CO2,s} \tag{10.35}$$

$$Y_{CO2}(r_f) = Y_{CO2,f} \tag{10.36}$$

Similarly, for the outer zone we can write:

$$\dot{m}_c = \frac{-4\pi r^2 \rho D}{\left(1 - {Y_{CO2}}/{\gamma_s}\right)} \frac{d\left({Y_{CO2}}/{\gamma_s}\right)}{dr} \tag{10.37}$$

The boundary conditions are:

$$Y_{CO2}\left(r_f\right) = Y_{CO2,f} \tag{10.38}$$

$$Y_{CO2}\left(r \to \infty\right) = 0 \tag{10.39}$$

We can take the species conservation for the inert gas (like nitrogen). We get:

$$\dot{m}_c = \frac{4\pi r^2 \rho D}{Y_I} \frac{d(Y_I)}{dr} \tag{10.40}$$

The necessary boundary conditions are:

$$Y_I\left(r_f\right) = Y_{I,f} \tag{10.41}$$

$$Y_I\left(r \to \infty\right) = Y_{I,\infty} \tag{10.42}$$

Integration of Equations (10.34), (10.37) and (10.40) with their boundary conditions generates the following three equations:

$$\dot{m}_c - 4\pi \frac{r_f r_s}{r_f - r_s} \rho D ln \frac{1 + {Y_{CO2,f}}/{\gamma_s}}{1 + {Y_{CO2,s}}/{\gamma_s}} \tag{10.43}$$

$$\dot{m}_c = 4\pi r_f \rho D ln\left(1 - {Y_{CO2,f}}/{\gamma_s}\right) \tag{10.44}$$

$$Y_{I,f} = Y_{I,\infty} \exp\left[-\frac{\dot{m}_C}{4\pi r_f \rho D}\right] \tag{10.45}$$

These equations are algebraic in nature and involve five unknowns, $\dot{m}_c$, $Y_{CO2,f}$, $Y_{CO2,s}$, r_f and $Y_{I,f}$. At the flame, there remains only carbon dioxide and inert gas. We can eliminate $Y_{I,f}$ using the following relation:

$$Y_{I,f} = 1 - Y_{CO2,f} \tag{10.46}$$

To develop the closure equations, we need to invoke the surface kinetics. The reaction that occurs at the surface is first order in carbon dioxide concentration. The rate equation can be written as:

$$\dot{m}_c = 4\pi r_s^2 k_c \frac{M_{mix} M_C}{M_{CO2}} \frac{P}{R_u T_s} Y_{CO2,s} = k_{kin} Y_{CO2,s} \tag{10.47}$$

The rate coefficient k_c can be written in Arrhenius form [5].

From the above equations, we can arrive at the final form:

$$\dot{m}_c = 4\pi r_s \rho D ln(1+B) \tag{10.48}$$

where

$$B = \frac{2Y_{O2,\infty} - \left[\frac{\gamma_s - 1}{\gamma_s}\right] Y_{CO2,s}}{\gamma_s - 1 + \left[\frac{\gamma_s - 1}{\gamma_s}\right] Y_{CO2,s}} \tag{10.49}$$

As the value of B is less than 1, Equation (10.48) can be simplified, after linearisation, to:

$$\dot{m}_c = 4\pi r_s \rho D B \tag{10.50}$$

If sufficient carbon dioxide is not diffusing up to the surface, it will be a case of diffusion controlled burning. For that, we can assume $Y_{CO2,s} = 0$ and find the burning rate. Otherwise, the surface temperature needs to be estimated first. For that, we can adopt the similar process adopted in the case of the one-film model. Note that, in the case of diffusion controlled burning, an accurate estimation temperature calculation is required, because the mixture density is a function of temperature.

Example 10.3

Calculate the burning rate for Example 10.1 by using the two-film model. Consider diffusion controlled burning.

Solution

The mixture density is *0.2 kg/m³* as calculated in Example 10.1.

Diffusion coefficient $D = 2.48 \times 10^{-4}\ m^2/s$

For diffusion controlled burning in the two-film model, we assume $Y_{CO2,s} = 0$. Then:

$$B = \frac{2Y_{O2,\infty} - \left[\frac{\gamma_s - 1}{\gamma_s}\right] Y_{CO2,s}}{\gamma_s - 1 + \left[\frac{\gamma_s - 1}{\gamma_s}\right] Y_{CO2,s}} = \frac{2(0.233) - 0}{3.66 - 1 + 0} = 0.175$$

The burning rate can be calculated as:

$$\dot{m}_c = 4\pi r_s \rho DB = 4\pi\left(100\times10^{-6}\right)(0.2)\left(2.48\times10^{-4}\right)(0.175) = 10.9\times10^{-9}\ \mathbf{kg/s}$$

Note that the burning rate calculated in the one-film model was half of the value calculated here.

Example 10.4

A 100-micron diameter carbon particle is burning in atmospheric air. The surface temperature is 1800 K. Find the burning rate using the two-film model. Take the molecular weight of the mixture as 30 kg/kmol and the diffusion coefficient $D = 2.48\times10^{-4}\ \frac{m^2}{s}$.

Solution

Here, the conditions are similar to the previous example, but the particle diameter is smaller. We can take the gas density as 0.2 kg/m³. The kinetic rate constant for a carbon–carbon dioxide reaction, as suggested by Mon and Amundson [6], is:

$$k_c = 4.016\times10^8 \exp\left[\frac{-29790}{T_s}\right] = 4.016\times10^8 \exp\left[\frac{-29790}{1800}\right] = 26.075 m/s$$

At first, we shall assume a value of $Y_{CO2,s}$. Let us assume it is 0 for a diffusion controlled burning. We can calculate the following:

$$B = \frac{2Y_{O2,\infty} - \left[\frac{\gamma_s - 1}{\gamma_s}\right] Y_{CO2,s}}{\gamma_s - 1 + \left[\frac{\gamma_s - 1}{\gamma_s}\right] Y_{CO2,s}} = \frac{2(0.233) - \frac{3.66-1}{3.66} Y_{CO2,s}}{3.66 - 1 + \frac{3.66-1}{3.66} Y_{CO2,s}} = \frac{0.466 - (0.727) Y_{CO2,s}}{2.66 + (0.727) Y_{CO2,s}} \quad (10.51)$$

From Equation (10.47), we can write:

$$\dot{m}_c = 4\pi r_s \rho DB = 4\pi\left(50\times10^{-6}\right)(0.2)\left(2.48\times10^{-4}\right) B = 3.11\times10^{-8} B \quad (10.52)$$

We can find the value of $\dot{m}_c$. We calculate:

$$k_{kin} = 4\pi r_s^2 k_c \frac{M_{mix} M_C}{M_{CO2}} \frac{P}{R_u T_s} = 4\pi \left(50\times10^{-6}\right)^2 (26.075) \frac{(30)(12)}{44} \frac{10^5}{(8315)(1800)}$$

$$= 4.47\times10^{-8}$$

From Equation (10.48), we can write:

$$\dot{m}_C = 4.47\times10^{-8} Y_{CO2,s}$$

We can get:

$$Y_{CO2,s} = 0.224\times10^{8} \dot{m}_C \quad (10.53)$$

The algorithm for the solution is:

1. Assume a guess value of $Y_{CO2,s}$.
2. Calculate B from Equation (10.51).
3. Calculate $\dot{m}_C$ from Equation (10.52).
4. Calculate $Y_{CO2,s}$ from Equation (10.53).
5. If newly calculated $Y_{CO2,s}$ is close to guess value $Y_{CO2,s}$, adopt the solution as is. Otherwise, take the new$Y_{CO2,s}$ as guess value and go to step 2.

Two iterations are given in the table below. Still it is not converged. More iterations are required to complete the solution.

Iteration	$Y_{CO2,s}$ (Guess)	B	$\dot{m}_C$	$Y_{CO2,s}$ (New)
1	0	0.175	5.44×10^{-9}	0.122
2	0.122	0.137	4.26×10^{-9}	0.0955

10.3 COMBUSTION OF COAL

Historically coal is considered as the main source of energy. Coal was used as the fuel in heating boilers, transportation, power generation and others. To date, coal has the highest share as a fuel for energy generation.

Coal is analysed in two ways. One is known as ultimate analysis. In this method, the mass percentages of elemental components are the output. Another is known as proximate analysis. In this method, the mass percentage of char, VM, moisture, and ash are found. Char is the fixed carbon that is not combined chemically with any other element. VMs are the combustible gases (mainly different hydrocarbons) which are absorbed or adsorbed in the porous structure of coal. Moisture and ash are the two non-combustible components. Moisture takes heat from burning coal and evaporates. Ash is mainly mineral materials, non-combustible inorganic compounds. Sometimes sulfur is present in coal, although at a very low fraction. Sulfur can burn to produce energy. But the oxidation product, SO_2, produces acid, reacting with oxygen and water. For combustion requirements, proximate analysis is useful for combustion purposes.

Coals are classified on the basis of their carbon percentages. The highest carbon-containing coal is known as anthracite. Anthracite contains 86%–98% of fixed carbon. VM content is also low. This grade of coal took the longest to form and contains a very low percentage of ash. Bituminous coal is the next grade, which contains 70%–86% of fixed carbon with higher volatile content. This grade also contains a low, but relatively higher than anthracite, amount of ash. Subsequent grades are sub-bituminous and lignite, where the carbon percentage decreases and the ash content increases.

Coal combustion started with lump coal. The typical size of a lump was a few inches. During combustion of lump coal, the outer layer comes in contact with oxygen. After burning of the outer layer, a layer of ash is formed. The ash layer is porous, and oxygen diffuses through that and comes into contact with the next carbon layer. The amount of oxygen diffusion depends on the amount of ash formed and the porosity of the ash layer. However, oxygen cannot diffuse up to the core of the lump and always leaves some unburned coal. Anthracite is the best quality for lump coal burning, because it has the lowest amount of ash content and hence leaves less amount of unburned coal.

When lump coal combustion became obsolete, the industrial combustion of lump coal was substituted by combustion of pulverised coal. In this case, the coal is first crushed and then ground to micron-size dust. This dust coal is taken into the combustion chamber. The coal surface area has been increased and so the combustion is efficient. As there are small particles to burn, non-burning of coal is suppressed.

In pulverised fuel–fired furnaces, normally used in thermal power plants, coal dust is supplied through coal burners. Coal is usually conveyed pneumatically, with the help of a portion of air (known as primary air). Primary air is not sufficient to burn the fuel. The typical value of primary air is around 30% of the total air supplied. It is used for conveying as well as drying of coal dust.

For combustion of coal in a continuous feeding furnace, ignition plays a very important role. We should note here that anthracite ignition is normally at higher temperature. Anthracite burns with a short flame but for a long time. Coals, having more volatile content, burn with a longer flame. The volatility evolves from the coal during heating and burns surrounding the particle showing a longer flame [7]. Figure 10.5 shows pulverised coal particles covered with flames from VMs.

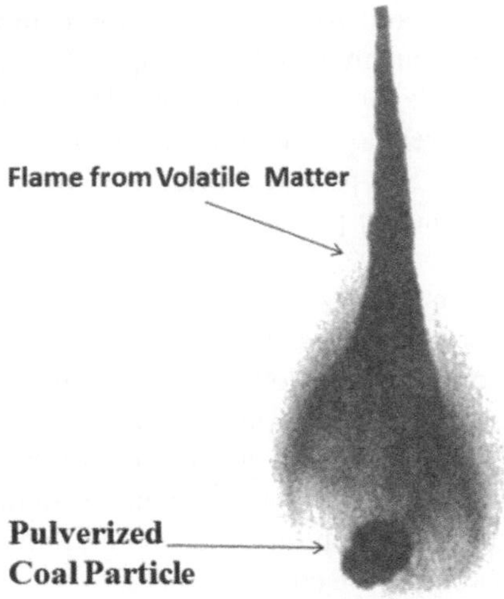

Figure 10.5 Shadowgraph of a burning coal particle covered with flame from VM.

Burning bituminous coal is easier for pulverised coal combustion as the released volatile matter heating catches flame quickly. Igniting char is much easier. Coal is ignited at 685°C for high volatile bituminous in a pulverised coal combustion chamber. The same is true for coke (without VM): 1090°C [8].

For coal combustion in a furnace, the residence time inside the combustion zone must be increased. The coal particles should be kept in the combustion zone for long enough for heat transfer and ignition. Coal particles enter with a velocity which takes them towards the downstream. The downstream zone is a comparatively low temperature zone, which disables the particle to burn, ultimately causing extinction of flame. Normally, downstream of a pulverised coal furnace is fitted with different heat transfer utilities. Combustion must occur at the proper place to ensure proper heat transfer in those utilities. A very common example is to use a corner firing arrangement. In this method, the coal burners are placed at four corners at a section of the furnace. Figure 10.6 shows the schematic arrangement. Four coal burners are mounted at four corners at a horizontal plane in the furnace. Usually pulverised coal furnaces are vertical in nature and coal is introduced at different vertical levels depending on the requirement. At one such vertical location, we can think of the horizontal plane where the burners are mounted. The direction of coal throw is tangential to an imaginary central circle. After entering the furnace, coal experiences a whirl motion, which increases its residence time as well as improves turbulence for better mixing and heat transfer. Modelling of such a furnace, used in boilers, is a vast topic and beyond the scope of this book. A detailed analysis and modelling may be found elsewhere [9].

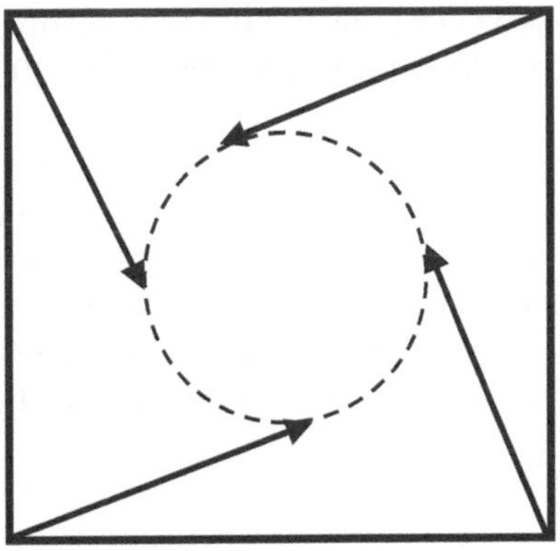

Figure 10.6 Corner fire arrangement of pulverized coal.

Although pulverised coal combustion is one of the most efficient methods of combustion, it suffers from loss of coal dust through leakages, causing a pollution hazard. Moreover, the pulverisation process uses a large amount of energy. Fluidised bed combustion is another method of combustion, which was developed in 1970s. In this method, coal is not pulverised but crushed to a size less than approximately 5 mm. Coal particles are supplied in a fluidised bed. The bed material is mainly limestone and other refractories. The coal particles usually experience a residence time of the order of 100–500 s and the combustion temperature is kept low by immersed water tubes in the bed to 1050K–1170K [10,11]. NO_x emission is low in such a method as combustion temperature is kept low. The SO_2 generated in combustion can be captured in the bed material through chemical reaction with limestone. Many types of coal can be burned successfully by this method.

10.4 COMBUSTION OF SOLID PROPELLANTS

A propellant is a generic name for chemicals used to propel projectiles from guns, rockets, missiles and other firearms. Usually the propellants are low-explosive materials. Sometimes those may include a high explosive. But in that case, the burning must be controlled (deflagration).

Solid propellants are mixtures of different heterogeneous components. The mixture comes as a composite material, which is known as composite solid propellant (CSP). The mixture contains primarily an oxidant in large proportion and a fuel-cum-binder. Usually ammonium perchlorate (AP) is used as the oxidant. The typical mass percentage may be 60%–84%. The fuel-cum-binder materials are generally hydroxyl-terminated polybutadiene (HTPB) or carboxyl-terminated polybutadiene (CTPB) [12,13]. This constitutes 12%–16% of the total mass of CSP. Along with the above, 2%–20% metallic fuel is also included.

Solid propellant combustion is a complex process. Some of the reactions are in condensed phase; some are in gas phase. But heterogeneous reactions are also there. They may vary with the composition of the propellant. It is not possible to describe the complete process for all propellants here. For a general understanding, we can take a combination of ammonium perchlorate-polybutadiene (AP-PB) as the propellant. The reactions can be divided into three steps:

1. AP decomposition
2. PB pyrolysis
3. Reaction of AP decomposition products and PB pyrolysis products

Decomposition of AP is a complex process. AP crystals undergo a dissociative sublimation initially. This produces:

$$NH_4^+ClO_4^-(s) \rightarrow NH_3(g) + HClO_4(g) \tag{10.54}$$

Perchloric acid is an explosive if comes in contact with oxygen. It is highly unstable and decomposes to:

$$HClO_4 \rightarrow ClO_3 + OH \tag{10.55}$$

$$ClO_3 \rightarrow ClO + O_2 \tag{10.56}$$

On absorption of energy, AP decomposing generates NH_3, OH, ClO and O_2 as shown in the above mechanism. Ammonia can be oxidised in the presence of the above species. A number of reactions can take place. Among those, the reaction of ammonia with OH is of importance, because of the energy released:

$$NH_3 + OH \leftrightarrow NH_2 + H_2O + \left(-15890 \frac{\text{kCal}}{\text{mole}}\right) \tag{10.57}$$

The details about the associated chemistry are discussed by Jacobs and Whitehead [14], Jacobs and Pearson [15], Coats and Nickerson [16].

PB is pyrolysed to generate 33 species, out of which, at 600K, the major products are ethylene and acetylene. As AP-PB surface temperature is on the order of 600K–700K [17], the pyrolysis products are considered at that temperature. The experimental study of Nagao and Hikata [18] showed that vinylcyclohexane, butadiene, ethylene and acetylene are the main pyrolysis products.

The structure of PB is $(-C-C=C-C-)_n$. The outermost single bonds are the weakest. On heating, these bonds break. Subsequently, the second single bonds are also broken, as they are weaker. The breakage of the outermost bonds produces the monomers. The subsequent bond breakage brings the following:

$$C_4H_6 \rightarrow C_2H_2 + 2CH_2 \rightarrow C_2H_2 + C_2H_4 \quad (10.58)$$

This is the reason why ethylene and acetylene are the major products of pyrolysis.

The next level is the oxidation of these two main products, which is called combustion. Acetylene and ethylene react with the decomposition products of AP.

Finely divided metals are added to almost all solid propellants. These particles provide a number of benefits. Normally they increase combustion characteristics and thus ballistic properties. During combustion, the metals are also oxidised to the corresponding metal oxides, liberating energy. Aluminium is the most popularly used metal fuel. Other metals can also be used, but one must be careful because of toxic exhausts. Table 10.2 shows different materials for solid propellant [19].

A number of very low-viscosity plasticisers are mixed with the propellant to increase the wettability. The plasticising materials improve mechanical strength, retard oxidation and inhibit brittleness of the propellant. Some curating agents are mixed with the propellant. Usually, isocyanate and diisocyanate groups are used as curating agents. As controlled release of energy is always required, some materials are used as the burn rate modifier. In most cases, they are used as a positive catalyst to improve the burning rate. In some cases, the burn rate must be supressed too.

TABLE 10.2 DIFFERENT SOLID PROPELLANTS

Metallic/Non-metallic Fuels	Oxidizers	Binders
Aluminium	Ammonium perchlorate	CTPB
Beryllium	Lithium perchlorate	Epoxydes
Magnesium	Potassium perchlorate	HTPB
Sodium	Ammonim nitrate	Nitrocellulose
Hydrocarbons	Potassium nitrate	Polybutadiene acrylic acid (PBAC)
Polymers	Sodium nitrate	Polybutadiene acrylonitrile (PBAN)
Plastics	Cyclotetramethylenetetranitramine-HMX	Asphalts
Rubber	Cyclotetramethylenetrinitramine-RDX	PVC

The steady linear regression of the burning surface is normally characterised as the burning rate of the propellant. The burning rate depends on a number of parameters. It can be written as:

$$\dot{m} = f\left(P, T_0, O_p, \varnothing, CF, U_g, G_p\right)$$

where P is the pressure, T_0 is the initial propellant temperature, O_p is the oxidiser particle size, $\varnothing$ is the oxidiser-fuel ratio, CF is the chemical formulation of the propellant, U_g is the transverse velocity of the combustion gases wetting the burning surface and G_p is the propellant grain shape factor [20]. From the above relation, we can identify the factors on which the burning rate depends. However, the burning rate is strongly dependent on the first two factors—pressure and initial temperature [21].

A number of empirical relations express the variation of burning rate with pressure. One may consult [20] to see the relations. The most popular relation is known as Vieille's law and is expressed as:

$$\dot{m} = ap^n \tag{10.59}$$

where $n = 0.77$ for AP combustion, and pressure ranges from 20–80 bar.

The burning rate increases with initial temperature. The effect of temperature is not as prominent as pressure. The variation with temperature can be expressed by changing Equation (10.59) as follows:

$$\dot{m} = Bp^n / \left(C - T_0\right) \tag{10.60}$$

where B and C are two constants, depending on the type of propellants [22].

Further discussion on solid propellant and metallic fuel combustion is beyond the scope of this book. Readers may consult [22,23] for additional information.

10.5 CATALYTIC REACTION

A catalyst is a chemical species that is involved in changing (increasing or decreasing) the rate of a reaction. Though the mechanism of each catalytic reaction is different, it can be said that catalysts form bonds with the reacting molecules and allow these to react to form a product. The product then detaches from the catalyst and leaves it unaltered so that it is available for the next reaction. In fact, we can describe most catalytic reactions as cyclic events in which the catalyst participates and is recovered in its original form at the end of the cycle.

To see how a catalyst affects the rate of a reaction, we must investigate its potential energy diagram. Essentially, a potential energy diagram is a graph of potential energy or Gibb's free energy of the reaction versus a reaction coordinate. Let us assume a spontaneous chemical reaction (ΔG is negative) between A and B to form product P, where C is the catalyst.

$$A + B \rightarrow P \tag{10.61}$$

All reactants must cross a specific activation energy barrier to form a transition state that will later on give the products. The lower the activation energy of a reaction, the higher is the ease with which a reaction proceeds and the higher is the kinetic rate of the reaction. Catalysts work by providing alternative mechanistic pathways involving a different transition state of lower energy. Thereby, the activation energy of the catalytic reaction is lower compared to the uncatalysed reaction, as shown in Figure 10.7.

Catalysis can be divided into two types on the basis of the phases of the participating species and catalyst:

1. Homogeneous catalysis
2. Heterogeneous catalysis

In a homogeneous catalytic reaction, the catalyst is in the same phase as the reactants. Typically, all the reactants and catalysts are either in one single liquid phase or gas phase. Most industrial homogeneous catalytic processes are carried out in solution phase. Ester hydrolysis involving general acid-base catalysts, polyethylene production with organometallic catalysts and enzyme catalysed processes are some of the important examples of industrial homogeneous catalytic processes.

Heterogeneous catalytic reactions are those in which the catalyst and reactant molecules are in different phases. Most often in these cases, the catalyst is

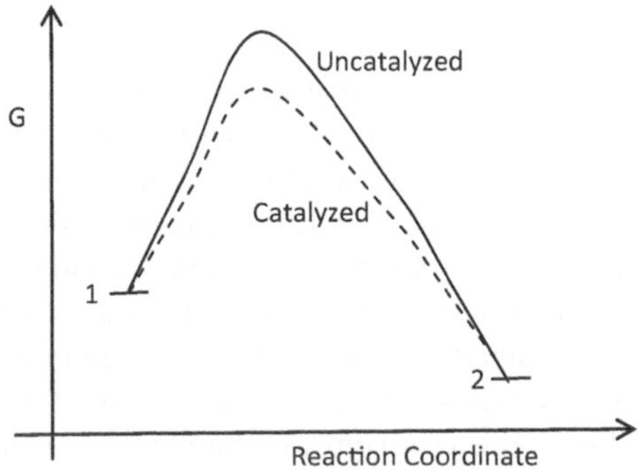

Figure 10.7 Energy requirement for catalysed and uncatalysed reactions.

solid, and it catalyses reactions of molecules in gas or solution phase. As solids—unless they are porous—are commonly impenetrable, catalytic reactions occur at the surface of the catalyst. Heterogeneous catalysis is enhanced by the increased surface area of the catalyst. To use the often-expensive materials (e.g., platinum) economically, catalysts are usually nanometer-sized particles (maximised surface area for certain amount of catalyst), supported on an inert, porous structure. Heterogeneous catalysts are the very basis of the chemical and petrochemical industry.

In general, it is believed that the entire surface of the solid catalyst is not responsible for catalysing any reaction. Only certain sites on the catalyst surface actually participate in the reaction, and these sites are called active sites on the catalysts. These sites may be the unsaturated atoms resulting from surface irregularities, or the atoms with chemical properties that enable the interaction with the adsorbed reactant atoms or molecules. Activity of the catalyst is directly proportional to the number of these active sites available on the surface and is often expressed in terms of turnover frequency. Turnover frequency is defined as the number of molecules reacting per active site per second at the condition of experiments. A solid catalytic reaction $A \rightarrow B$ goes through the following steps:

1. Transportation of reactant (A) from bulk fluid to pore mouth on the external surface of catalysts pellets
2. Diffusion of the reactant (A) from the pore mouth through the catalyst pores to the immediate vicinity of internal catalytic surface
3. Adsorption of reactant (A) onto the catalyst surface
4. Reaction of (A) on the catalyst surface producing product (B)
5. Desorption of the product (B) from the surface
6. Diffusion of the product (B) from interior part of the pores to the pore mouth on the external surface
7. Transfer of the product (B) from pore mouth on the external surface to the bulk fluid

Figure 10.8 shows the general pattern for catalytic combustion of hydrocarbons [24]. The pattern curve shows conversion of species variation with temperature. The process can be divided in four zones. At the beginning of zone I, the conversion starts. This temperature depends on the fuel and catalyst combination. As temperature increases, oxidation also increases, depending on the kinetics of the system, and this rise is almost linear. When the temperature is increased further, conversion increases in an exponential fashion. In this zone (zone II), heat generated by the reaction is greater than the supplied heat. The reaction becomes mass-transfer-controlled in zone III. Finally, in zone IV, almost no conversion exists because of depletion in reactants.

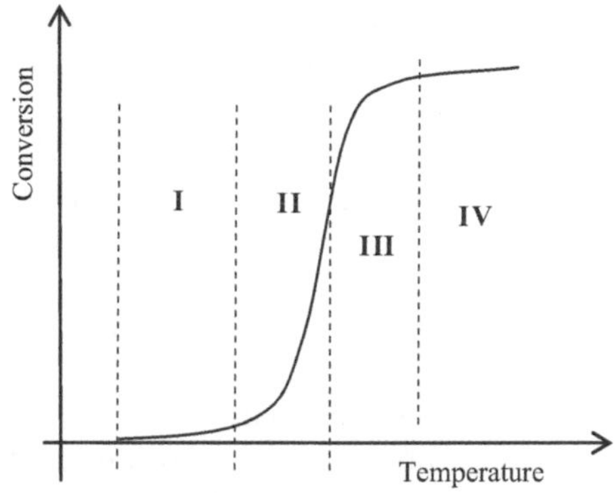

Figure 10.8 Catalytic conversion pattern of hydrocarbons.

The temperature at which mass transfer control becomes rate controlling is called the light off. In Figure 10.8, it is at the beginning of zone III. Note that kinetics of catalytic combustion are only relevant before light off. Once light off occurs, mass and heat transfer become dominant mechanisms for controlling the reaction. The combustor geometry, porosity of the catalyst/support are important in this region. The heat generation from combustion increases the temperature. Thus, the stability of catalysts at high temperatures is also of considerable interest.

Catalysts are used in combustion for complete oxidation of fuel at low temperature. This increases the efficiency of combustion. If fuels burn at lower temperature, the possibility of NO_x generation is less. Catalysts are also used for mitigating combustion-driven pollution, like the catalytic converter in the case of internal combustion (IC) engines. Normally platinum group metals are used as catalysts for gas phase oxidation of hydrocarbon fuels. The catalyst particles are impregnated on the surface of a support. The reactants are allowed to come in contact with the catalyst particles for effective combustion. A lot of research and many inventions include this area of catalytic combustors. Readers can consult [25–28] in the References section for additional information. Catalysts are also used in the case of solid propellant, as discussed in the previous section.

EXERCISES

10.1 Use the one-film model to determine the burning rate for carbon particles with radius 500, 50 and 5 microns burning in atmospheric air. Assume that the surface temperature is 1800 K, the kinetic rate constant $k_c = 3\times10^5 \exp\left(-\frac{17966}{T}\right)[m/s]$ and the mixture molecular weight is 29 kg/kmol. Also comment on the type of burning (kinetically controlled, diffusion controlled or intermediate) in each case. Assume a suitable diffusion coefficient.

10.2 Consider Exercise 10.1 with an unknown surface temperature. Find the burning rate as well as the surface temperature for the problem. Also comment on the type of burning. Assume that the ambient temperature is 300 K.

10.3 Develop the two-film model by filling up the steps omitted in the chapter.

10.4 A carbon particle of 100-micron diameter is burning in atmospheric air. Find the burning rate using the two-film model. Assume that the surface temperature is 2000 K. Assume the other parameters suitably.

10.5 Estimate the surface temperature for Exercise 10.4.

10.6 Estimate the burning rate (kg/s) of a carbon particle of 50-micron diameter in atmospheric air. Assume that the particle temperature is 1500 K. The kinetic rate constant $k_c = 1.9 m/s$ and the mean molecular weight of the gas at particle surface is 30 kg/kmol. Assume that the diffusion coefficient $D = 2.5\times10^{-4} \frac{m^2}{s}$. Which combustion regime prevails? If the burning rate were constant, independent of particle diameter, how long does it take to completely burn the particle? Assume that the density of coal is 2300 kg/m^3. Consider the one-film model and the two-film model.

10.7 Formulate the methodology to find the burning time of a coal particle using the (a) one-film model and the (b) two-film model. Consider variation in the burning rate with particle diameter and surface temperature.

REFERENCES

1. Mon, E. and Amundson, N. R., Diffusion and reaction in a stagnant boundary layer about a carbon particle, 2. An extension, *Industrial Engineering Chemistry Fundamentals*, **17**, 313–321, 1978.
2. McLean, W. J., Hardesty, D. R. and Pohl, J. H., Direct observations of devolatilizing pulverized coal particles in a combustion environment, *Symposium (International) on Combustion*, **18**, 1239–1248, 1981.
3. Fuertes, A. B., Hampsoumian E. and Williams, A., Direct measurement of ignition temperatures of pulverized coal particles, *Fuel*, **72**, 1287–1291, 1993.
4. Ranade, V. V. and Gupta, D. F., *Computational Modeling of Pulverized Coal Fired Boilers*, Boca Raton, FL: CRC Press, 2015.
5. Baranski, J. *Physical and Numerical Modeling of Flow Pattern and Combustion Process in Pulverized Fuel Fired Boiler*, PhD thesis, Sweden: Royal Institute of Technology, 2002.

6. Essenhigh, R. H., *Chemistry of Coal Utilization, 2nd volume.* Ed. Lowry, H. H., New York: Wiley, 1981.
7. Kubota, N., *Fundamentals of Solid Propellant Combustion*, New York: American Institute of Aeronautics and Astronautics, Incorporated, **90**, 1–52, 1984.
8. Kumar, N. R. R., *Fundamentals of Solid Propellant Combustion*, New York: American Institute of Aeronautics and Astronautics, Incorporated, **90**, 409–477, 1984.
9. Jacobs, P. W. M. and Whitehead, H. M., Decomposition and combustion of ammonium perchlorate, *Chemical Reviews*, **69**, 551–590, 1969.
10. Jacobs, P. W. M. and Pearson, G. S., Mechanism of the decomposition of ammonium perchlorate, *Combustion Flame*, **13**, 419–430, 1969.
11. Coats, D. E. and Nickerson, G. R., Analysis of a Solid Propellant Gas Generator using an Ammonium Perchlorate-Polybutadiene Composite Propellant, *Technical Report, AFRPL-TR-72-S8*, Air Force Rocket Propulsion Laboratory, United States Air Force, Edwards Air Base, 1972.
12. Sabadell, A. J., Summerfield, M. and Wengrad, J., Measurement of temperature profiles through solid-propellant flames using fine thermocouples, *AIAA Journal*, **3**, 1580–1584, 1965.
13. Nagao, M. and Hikata, T., Fundamental studies on combustion of solid propellant part I. Pyrolysis of a polymeric fuel binder, *Kogyo Kayaku Kyokaihi*, **27**, 234–241, 1966.
14. Goncalves, R. F. B., Iha, K., Machado, F. B. C. and Rocco, J. A .F. F., Ammonium Perchlorate and Ammonium Perchlorate-Hydroxyl Terminated Polybutadiene Simulated Combustion, *Journal of Aerospace Technology and Management*, **4**, 33–39, 2012.
15. Kishore, K. and Sridhara, K., Solid PropellantChemistry Condensed Phase Behaviour of Ammonium Perchlorate-Based Solid Propellants, *DRDO Monographs/ Special Publications Series*, New Delhi: Defence Scientific Information & Documentation Centre, 1999.
16. Waich, R. E., Jr., and Strauss, R. F., Fundamentals of rocket propulsion, *Reinhold Space Technology Series*, New York: Reinhold Publishing Corporation, 1960.
17. Warren, F. A., *Rocket Propellants*, New York: Reinhold Publishing Corp., 1958.
18. Glassman, I. and Yetter, R. A., *Combustion*, Academic Press, 2008.
19. Lee, J. H. and Trimm, D. L., Catalytic combustion of methane, *Fuel Processing Technology*, **42**, 339–359, 1995.
20. Lywood, W. J., Fowles, M. and Shipley, D. G., Catalytic combustion process, *US Patent Number*, 5,228,847, 1993.
21. McCarty, J. G., Wong, V. L. and Wood, B. J., Catalytic combustion process, *US Patent Number*, 6,015,285, 2000.
22. Miller, J. B. and Malatpure, M., Pd catalysts for total oxidation of methane: Support effects, *Applied Catalysis A: General*, **495**, 54–62, 2015.
23. Zou, X., Rui, Z., Song, S. and Ji, H., Enhanced methane combustion performance over NiAl2O4-interface promoted Pd/γ-Al2O3, *Journal of Catalysis*, **338**, 192–201, 2016.

Chapter 11

Combustion Emission

11.1 INTRODUCTION

At present, combustion system designers need to be careful about pollutant emissions from their designed systems. Early concerns, in the 1950s, about pollutants were based on the visible particulate emissions in smoke released by industrial processes. After setting up the standard acceptable limits, mitigation could be possible in next 30 years. Subsequently, in the 1960s, concerns developed on the pollutants emitted by the transportation sector, the large shareholder of which is internal combustion engines.

The major pollutants may be divided broadly into three categories. The first includes pollutant gases that contain emissions of CO, SO_2, SO_3, NOx, unburned and partially burned hydrocarbons, and so on. Greenhouse gases may be considered the second category. This group contains those gases which absorb and reemit thermal radiation. By this action, greenhouse gases increase global temperature. The main components are CO_2, CH_4, N_2O, H_2O, O_3, and so on. The third group, particulate matter, consists mainly of soot, fly ash, metal fumes, and other aerosols. A discussion on the impact of combustion on the atmosphere may be found in Prather and Logan [1].

Pollutants may be classified as primary and secondary types also. Primary pollutants are generated directly from the source. Secondary pollutants are formed via reactions of primary pollutants in atmosphere. Both types affect our health as well as environment.

11.2 EMISSION OF POLLUTANT GASES

11.2.1 Oxides of Nitrogen

Oxides of nitrogen that form during combustion can be of three types: nitrous oxide (N_2O), nitric oxide (NO) and nitrogen dioxide (NO_2). Generally, NO and NO_2 together and then are called NO_x. To form NO from nitrogen, a very high temperature is required. At high temperature, the nitrogen molecules dissociate to reactive nitrogen atoms. In some devices like gas turbine combustors, NO_2 is formed. NO, formed during combustion, reacts with atmospheric oxygen, after it is exhausted from combustion system. This reaction forms NO_2 as secondary pollutant. NO_2 is a hazardous gas for plants and animals. It generates photochemical smog. NO_2 further reacts with atmospheric water and oxygen to form nitric acid, as follows:

$$4NO_2 + 2H_2O + O_2 \rightarrow 4HNO_3 \tag{11.1}$$

This nitric acid, when it precipitates through rain, reduces the pH value of rain to as low as 2 from its normal value, which is around 6. This causes damage to soil and structures.

NO released at the upper atmosphere, mainly from aviation exhausts, reacts with ozone to form NO_2. This converts ozone to oxygen, creating depletion in the earth's ozone layer. NO_2 formed in atmosphere can break down into NO again and produces nascent oxygen:

$$4NO_2 \rightarrow NO + O \tag{11.2}$$

Nascent oxygen reacts with atmospheric oxygen to form ozone. Ozone in the lower atmosphere is a constituent of photochemical smog. The lower level ozone causes health hazards for animals too.

During combustion, NO can be formed from two sources of nitrogen. The main source is definitely air. The second source is nitrogen present in fuel.

At very high temperatures, the thermal mechanism of NO production is active. The thermal or Zeldovich mechanism [2] consists of two chain reactions:

$$O + N_2 \leftrightarrow NO + N \tag{11.3}$$

$$N + O_2 \leftrightarrow NO + O \tag{11.4}$$

And it is extended by another reaction:

$$N+OH \leftrightarrow NO+H \tag{11.5}$$

The above three reactions together are called the extended Zeldovich mechanism. The thermal mechanism is unimportant below 1800 K as activation energy for the first reaction is very high (319,050 kJ/kmol). This mechanism is prevalent over a wide range of the equivalence ratio, but the mechanism is inherently slow [3].

The N_2O intermediate mechanism is important for fuel-lean conditions. So the temperature is considerably low during this mechanism:

$$O+N_2+M \leftrightarrow N_2O+M \tag{11.6}$$

$$H+N_2O \leftrightarrow NO+NH \tag{11.7}$$

$$O+N_2O \leftrightarrow NO+NO \tag{11.8}$$

Fenimore observed formation of NO in the flame zone for premixed flames [4]. This formation is more rapid compared to thermal NO_x. He named it prompt NO_x. This formation starts with reaction of hydrocarbon radicals with nitrogen molecules. There are six reactions; of which the last four chain reactions prevail at higher equivalence ratio (around 1.2). The mechanism reactions are shown below:

$$CH+N_2 \leftrightarrow HCN+N \tag{11.9}$$

$$C+N_2 \leftrightarrow CN+N \tag{11.10}$$

$$HCN+O \leftrightarrow NCO+H \tag{11.11}$$

$$NCO+H \leftrightarrow CO+NH \tag{11.12}$$

$$NH+H \leftrightarrow N+H_2 \tag{11.13}$$

$$N+OH \leftrightarrow NO+H \tag{11.14}$$

When the equivalence ratio is more than 1.2, the prompt mechanism no longer remains rapid and, due to opening of other routes, it inhibits NO production [5]. NO formation is inhibited when the Zeldovich mechanism couples with prompt mechanism [6].

Nitrogen contained in fuels, like coal, is converted to hydrogen cyanide (HCN) during combustion. HCN can subsequently generate the NO that can be involved in the above mechanism.

11.2.2 Carbon Monoxide

Carbon monoxide (CO) is a toxic gas found in the exhaust of a combustion system. Exposure to carbon monoxide for a small duration causes headache and nausea. When the exposure is longer, it may lead to unconsciousness, even death. The amount of exposure depends on a combination of CO concentration and time of inhalation.

Carbon monoxide is an indicator of incomplete combustion and is normally produced in the fuel-rich combustion zone. In the case of premixed combustion, CO can be produced due to inhomogeneity in mixing. Inhomogeneous mixture leads to a dearth of oxygen at any part of the fuel air mixture and thus produces a rich premixed flame. In such situations, incomplete combustion of carbon produces CO. However, the tendency of CO production is low in the case of premixed combustion. But in non-premixed combustion, it is difficult to avoid CO production. Here, the fuel and oxygen mixes in the combustion area, and improper mixing is a probable situation. This creates CO during combustion. Liquid and solid fuel combustions are more prone to CO production. In the case of liquid fuel, if vapourisation (which is a function of fuel volatility) and subsequent mixing is not fast enough, CO can be produced. In the case of coal combustion, the ash layer hinders oxygen diffusion to inner carbon layer. This also leads to CO generation. In the case of partially premixed flame, the rich premixed zone is always susceptible to CO production. A supply of excess oxygen may not always be a remedial step for these cases.

Quenching of the flame in certain zones in a combustion chamber is also responsible for CO production. As an example, we can take an internal combustion (IC) engine cylinder. The cylinder body is cooled using flowing coolant in the cylinder jacket. Thus, the cylinder wall temperature is always maintained at a lower level compared to flame temperature. The flame is quenched at the vicinity of the cylinder wall. Similarly, there are some crevice volumes inside the combustion chamber. Flame cannot propagate into those if its size is below the quenching length. These types of situations of flame extinction lead to incomplete combustion and CO production.

Carbon monoxide can be produced from dissociation of carbon dioxide at high temperature. This high temperature may exist in the flame zone, like the flame of spark ignition engines. With a decrease in temperature, this CO can be oxidised to CO_2 with time (required for equilibrium). The reaction is:

$$CO + OH \rightarrow CO + H \tag{11.15}$$

If sufficient time is not provided for the reaction to occur, CO is liberated through the exhaust.

11.2.3 Unburned Hydrocarbon Gases

Unburned hydrocarbons (UHCs) are considered gaseous pollutants for the atmosphere. These are also an indicator for incomplete combustion, like CO. If we study the elementary reaction steps of any hydrocarbon fuel, we find various hydrocarbons are generated as intermediate products. If sufficient reactants are not present to complete the reactions or the amount is insufficient for reaction completion, UHCs may result. The basic reasons for CO generation, discussed above, are true for UHC generation also. UHCs are produced mainly from rich mixtures as well as from flame quenching.

11.2.4 Oxides of Sulfur

Oxides of sulfur normally result from combustion of sulfur-containing fuels. Sulfur is found in coal or fuel oil in different percentages. It can be present either in its elemental form or as a compound. Sulfur combines with oxidising agents very fast compared to other reactions. Sulfur is first oxidised to sulfur dioxide (SO_2). The formation time of SO_2 is comparable with H_2O formation and much faster than the formation of carbon dioxide [7]. SO_2 subsequently combines with oxygen to form sulfur trioxide (SO_3) either in the combustion device or in the atmosphere after coming out as exhaust. These two oxides together can be called SO_x. Both oxides are considered environmental pollutants and are hazardous for animal health due to coughing and choking.

SO_3 can react with water to form sulfuric acid:

$$SO_3 + H_2O \rightarrow H_2SO_4 \tag{11.16}$$

Sulfuric acid formed in atmosphere causes acid rain. A severe corrosion problem occurs due to acid formation in low-temperature heat recovery devices of thermal power plants, like economisers and air-preheaters. SO_3 generally forms in coal combustion if the fuel contains sulfur. At the end of the flue gas path, if the temperature falls below the dew point, SO_3 precipitates in the form of H_2SO_4. This precipitation damages the heat transfer devices as well as the induced draft fan by acid corrosion.

11.3 EMISSION OF GREENHOUSE GASES

Among the greenhouse gases, CO_2 and H_2O are natural combustion products for hydrocarbon fuel, which is the major fuel of present use. Around 70% of CO_2 is generated by the energy and transport sector. Table 11.1 shows the

TABLE 11.1 GLOBAL CO_2 EMISSIONS FROM DIFFERENT COMBUSTION SOURCES IN 2017

Source of Emission	Emission (Million Tonnes)
Fuel combustion	32294.2
Coal combustion	14512.7
Oil combustion	11169.1
Natural gas combustion	6437.0

Source: CO_2 Emissions from Fuel Combustion Highlights, *International Energy Agency*, www.iea.org, 2017.

amount of CO_2 emission from different fuel sources. Different abatement issues are taken up to reduce the atmospheric CO_2. Technology changes are also in vogue.

Nitrous oxide (N_2O) is another greenhouse gas that is a combustion product. Amine radical reacts with NO to produce nitrous oxide (Reaction 11.7). Amine radical may be generated from ammonia (NH_3) or hydrogen cyanide (HCN) during combustion.

At low temperatures (<1500 K) nitrogen may be converted to nitrous oxide reacting with nascent oxygen (Reaction 11.6).

N_2O is commonly known as laughing gas and is considered an atmospheric pollutant. It is also responsible for stratospheric ozone layer depletion. Fluidised bed combustion is more prone to N_2O production. N_2O may be converted to NO when it reacts with nascent oxygen (Reaction 11.8).

Methane is treated as another greenhouse gas, and it also comes from the combustion route. Methane can be generated as an intermediate product for hydrocarbon combustion as the smallest single bond component. Subsequent oxidation of methane results in consumption of methane during combustion. Incompleteness in combustion may generate methane as an exhaust component. Methane emission is very common in the case of bio-mass burning or low-temperature burning of coal.

11.4 EMISSION OF PARTICULATE MATTER

Soot is the most vulnerable contributor in particulate matter emission from flame. The formation mechanism of soot is discussed in Chapter 7. Soot generation is present in non-premixed and partially premixed flames. If the same can be consumed completely within the flame through oxidation, it is desirable. The practical combustion devices are mostly non-premixed/partially premixed in nature. Moreover, in high-speed devices like IC engines or gas turbines, the soot particles generated may not get sufficient time to reside in the combustion chamber for oxidation. Coal-fired furnaces also exhaust soot as a combustion product.

Soot particles are carcinogenic and cause bronchial trouble. They are considered an atmospheric pollutant. As suspended particles, they contribute towards smog formation. Soot particles produced in gas turbine combustion chambers may hamper the efficient operation of turbine also.

Coal-fired power plants are sources of another particulate matter known as fly ash. The pulverised coal-fired systems are the major generators of fly ash. Here the coal is fed in very small particulate form (almost powdered) and the particles burn in suspended form. After burning of char material, ash generated from one such particle is also very small. The typical size of a fly ash particle varies from 0.5 micron to 300 microns. These particles are carried with the flue gas and discharged into the atmosphere through the chimney. These particles are also harmful for the environment. If released into the atmosphere, they normally precipitate down almost immediately to cover the whole area with ash. Inhalation of these particles is harmful for animals and humans. These particles cover the leaves of the plants and hinder photosynthesis.

Some combustion-generated nanoparticles are also not very friendly for our environment [8]. However, their measurement is not easy, and any environmental measure has not yet been initiated for them. We should understand the presence of these type of particulate emissions from combustion devices also.

11.5 ABATEMENT OF EMISSION

Energy generation from combustion produces certain undesirable components. These components cause damage to our atmosphere, habitats and overall environment. With the growing need of energy, we are searching for new non-conventional sources. But we have not yet ended our dependence on conventional sources, mainly fossil fuels. The technology of fossil fuel combustion is more developed and better understood compared to others. Existing devices cannot be scrapped overnight unless we find comparable new technology.

In this context, we are now also focused towards maintaining the environment through control of undesired emissions. A number of statutory bodies in the world form regulations about the environment. These regulatory authorities are forming the limits for pollutant emissions, keeping in view the present environmental position and threat on one hand and the available technologies for abatement on the other. These authorities are tightening these norms to minimise harmful emissions.

European countries follow the well-known standard, the EURO norm, for emissions control in the automobile sector. Table 11.2 shows the norm for

TABLE 11.2 ALLOWABLE LIMITS FOR DIFFERENT POLLUTANTS FOR LIGHT DUTY VEHICLES IN VARIOUS EURO LEVELS

Levels	Effective Implementation Date	Fuel Type	Allowable Emission Limits				
			CO	HC	NO_x	PM	Particulate Number (PN)
EURO I	July 1, 1992	Petrol	2.72 g/km	0.97 g/km			
		Diesel	2.72 g/km	0.97 g/km		0.14 g/km	
EURO II	January 1, 1996	Petrol	2.2 g/km	0.5 g/km			
		Diesel	1.0 g/km	0.7 g/km		0.08 g/km	
EURO III	January 1, 2000	Petrol	2.3 g/km	0.20 g/km	0.15 g/km		
		Diesel	0.66 g/km	0.56 g/km		0.05 g/km	
EURO IV	January 1, 2006	Petrol	1.0 g/km	0.10 g/km	0.08 g/km		
		Diesel	0.50 g/km	0.30 g/km (NO_x alone: 0.25 g/km)		0.025 g/km	
EURO V	September 1, 2009	Petrol	1.0 g/km	0.10 g/km	0.06 g/km	0.005 g/km (DI only)	
		Diesel	0.50 g/km	0.23 g/km (NO_x alone: 0.18 g/km)		0.005 g/km	6.0×10^{11} /km
EURO VI	September 1, 2014	Petrol	1.0 g/km	0.1 g/km	0.06 g/km	0.005 (DI only)	6.0×10^{11} /km
		Diesel	0.50 g/km	0.17 (NO_x alone: 0.08g/km)		0.005	6.0×10^{11} /km

light-duty vehicles. The table shows how the control regulation became stricter from EURO I to EURO VI. Many other countries are following norms in line with EURO. The objective of stricter norms is to take a step towards being green tomorrow by invoking up-to-date technology and removing obsolete devices with time.

Now we shall discuss a few control methodologies for emissions control from combustion.

11.5.1 Control of NO_x Emission

Thermal power plants are a major source of NO_x emission. As the fuel used in thermal power plants, coal or fuel oil, contains some nitrogen, fuel NO_x as well as thermal, prompt or intermediate NO_x can be emitted by these plants. Control of the emissions can be thought of in two ways. The first is controlling the combustion; the second is capture after combustion.

During combustion, keeping the flame temperature lower may help in NO_x reduction. Supply of less oxygen in high-temperature zone can be one method of NO_x reduction. Another method is not to allow the flue gas to stay a long time in the oxidising region. Keep in mind, however, that these methods may lead to loss of combustion efficiency as well as incomplete combustion.

Pulverised fuel fired boilers normally use 20%–25% excess air. This amount is determined to maximise combustion efficiency and also to minimise energy loss. If we supply less excess air, NO production can be reduced, but this comes with reduced boiler efficiency.

Another common technique is the burner out of service method. In this method, one or more burners at each burner level can be selected to supply air only. Necessary fuel can be supplied through other burners. This method generates fuel lean zones within the furnace to reduce NO_x formation.

In some boilers, the over fire port is provided at an elevation above the burner planes. Ten to twenty per cent of the air is supplied through this over fire port. This air is termed over fire air (OFA). This air reduces the temperature of the flue gas after the combustion zone to reduce NO_x production. The velocity of OFA is a tricky parameter. This has an influence on the mixing of this air with the flue gas. Also this velocity determines the heat transfer from flue gas to heat transfer surfaces in the secondary path of the flue gas, like convective superheater, reheater, and so on.

The fuel supplied can also be divided into two components. The major part (80%–90%) of coal is supplied through the regular burners. There can be a level of burners at a certain elevation above the regular burner location. This method is called the coal reburn technique. The temperature in the original combustion zone is reduced by this method, due to fuel lean combustion. The NO generated in the original zone gets a chance to be reduced while it is passing through the reburn zone. An alternative fuel can also be supplied as reburn fuel.

A low NO_x burner is another development to reduce emission. In this burner, a provision for tertiary air is created. Coal along with primary air enter the furnace through the central nozzle. Secondary air flows through an annular passage, surrounding the central nozzle. A swirler is normally provided in this passage. There is one outermost annular passage which is designed for tertiary air. In this method, air mixes with fuel at different locations, controlling the equivalence ratio. The temperature also varies with lean and rich mixture.

Exhaust gas can be recirculated in the combustion zone to reduce the temperature of combustion. This leads to less NO_x formation during combustion. In fluidised bed combustion, submerged water tubes in the fluidised bed also reduce the furnace temperature at the level of 1150 K–1200 K. NO_x emission can be substantially reduced in fluidised bed combustion.

Post-combustion capture of NO can be done by using reagent like urea (CH_4N_2O) or ammonia (NH_3). These reagents can selectively react with NO to reduce to nitrogen. Normally, the reagents, in a low percentage, are dissolved in water and injected into the flue gas path. The reactions likely to occur are:

$$4NO+4NH_3+O_2 \rightarrow 4N_2+6H_2O \tag{11.17}$$

$$4NO+2CH_4N_2O+O_2 \rightarrow 4N_2+4H_2O+2CO_2 \tag{11.18}$$

The above reactions occur within a very narrow temperature range of 1100 K–1350 K. Note that at temperature above 1400 K, ammonia is converted to NO rather than reducing it. This makes the reagent injection location selection very critical. However, in the presence of titanium oxide or vanadium oxide as a catalyst, the above reactions occur at 600 K–700 K. Platinum and palladium can also play a similar catalytic role. The catalyst section is typically placed before the air-preheater in the flue gas path, and the reagent solution is vaporised and injected into the flue gas.

Internal combustion engines, both SI and CI, are emitters of NOx. IC engine fuels normally carry hardly any sulfur. Thus, fuel NO_x may be negligible here. SI engine and CI engine operation regimes and combustion modes are grossly different. But thermal and prompt NO_x can be expected from the combustion chamber. A detailed discussion may be found in Heywood [9].

Control of NO_x during combustion cannot be done effectively. Post-combustion capture can be the method of emission control. NO_x generated in IC engines can be reduced by using a catalytic converter. A catalytic converter is a chamber containing small honeycomb flow passages. The chamber diameter is larger than the exhaust manifold diameter to reduce the flow velocity of the gas. This honeycomb section is normally made of ceramic materials. Catalyst particles are impregnated on the surface of the flow passage. The exhaust gas, while flowing through the small flow passages, comes into contact with the catalyst particles. For NO_x reduction, rhodium is used as the catalyst material. In the presence of the catalysts, the following reactions occur to reduce the NO_x produced:

$$2NO + 2CO \rightarrow N_2 + 2CO_2 \tag{11.19}$$

$$2NO + 5CO + 3H_2O \rightarrow 2NH_3 + 5CO_2 \tag{11.20}$$

$$2NO + CO \rightarrow N_2O + CO_2 \tag{11.21}$$

$$2NO + 2H_2 \rightarrow N_2 + 2H_2O \tag{11.22}$$

$$NO + 5H_2 \rightarrow 2NH_3 + 2H_2O \tag{11.23}$$

$$2NO + H_2 \rightarrow N_2O + H_2O \tag{11.24}$$

When the temperature is about 700 K, it reduces 95% of the NO_x. The performance decreases with decreases in temperature.

New concepts for combustion technology have been introduced to the gas turbine industry to reduce NO_x. This includes lean premixed (LPM) combustion (or lean–premixed pre-vaporised [LPP] combustion when liquid fuels are employed), rich-burn quick-quench lean-burn (RQL) combustion, catalytic combustion and selective catalytic reduction (SCR) [10,11]. Among these, the RQL techniques are hampered by soot formation and incomplete mixing between fuel-rich combustion products and air. Catalytic combustion suffers from challenges associated with cost, safety and durability. In SCR, a chemical is added to the exhaust gas to convert harmful NO_x to compounds with less pollutants, but the major drawbacks are size and cost. Lean-premixed combustion appears to be the most promising technology recently because it offers a practical solution to reducing the emissions of nitrogen oxides (NO_x) and it is simple to implement in practical combustors. In LPM combustion, the temperature of the flame is significantly reduced due to excess air present in the combustion zone which consequently eliminates production of thermal NO_x. In addition, the lean operating condition elevates low maintenance requirements because the lower combustor exhaust temperature increases the lifetime of the turbine blades and other mechanical components [12].

However, in the LPM approach, the combustors are operated at lower equivalence ratios, so the operability range between the flashback and blowout regimes of such LPM combustors are much narrower compared to conventional (diffusion) combustors [13]. Hence, while lean, premixed operation allows for reduction in NO_x production, the overall stability of the combustion process is reduced.

11.5.2 Control of SO_x Emission

Thermal power plants are the main sources of SO_x emissions. Coal and fuel oil contain organic and inorganic sulfur. Natural gas normally carries no sulfur. One way to reduce sulfur is by using low-sulfur fuel. Sulfur-containing coal can be used in furnace after coal beneficiation. Inorganic sulfur, usually in the form of pyritic or sulfate form, can be removed to a great extent by beneficiation. Organic sulfur cannot be removed by this means. Combustion control to reduce sulfur is also not a solution as sulfur dioxide formation is a very fast process.

Post-combustion capture of SO_x can be done by injecting dry limestone ($CaCO_3$) directly into flue gas. Limestone absorbs sulfur dioxide in the presence of oxygen, producing calcium sulfate:

$$2CaCO_3 + 2SO_2 + O_2 \rightarrow 2CaSO_4 + 2CO_2 \tag{11.25}$$

The product can be separated in a particulate separation device. This technique is known as dry sorbent injection.

Wet gas desulfurisation can be considered another post-combustion capture method. Here, a slurry tower is made for wet scrubbing of the flue gas. Slurry containing water and finely ground limestone is sprayed from the top of the tower. Flue gas is passed in the opposite direction and interacts with the slurry. SO_2 is converted first to sulfurous acid (H_2SO_3). The acid subsequently reacts with limestone:

$$CaCO_3 + H_2SO_3 \rightarrow CaSO_3 + CO_2 + H_2O \tag{11.26}$$

$CaSO_3$ generated in this process is further oxidised at the bottom of the chamber to $CaSO_4$ and collected at the bottom as hydrated crystal ($CaSO_4$, $2H_2O$) or gypsum. The scrubber is placed at the base of the stack of the power plant. Corrosion of low-temperature heat-transfer devices cannot be prevented by this method.

Another method employs a spray of very fine limestone slurry before the electrostatic precipitator (ESP). The sulphite and sulphates formed by the reaction of limestone and sulfur dioxide can be arrested in ESP. This method is called dry gas desulfurisation.

In fluidised bed combustion, concurrent removal of sulphur dioxide is possible. Desulphurisation is accomplished by the addition of limestone directly to the bed together with the crushed coal. Limestone absorbs sulphur dioxide with the help of oxygen from the excess air.

The calcium sulphate produced in this process is a dry waste product that is either regenerated or disposed off. Reductions in sulphur dioxide up to 90% have been achieved in fluidised bed pilot plants.

11.5.3 Control of CO and UHC Emission

Carbon monoxide combines with oxygen to form carbon dioxide at about 850 K–950 K. UHCs also combine with oxygen at similar temperatures. In thermal power plants, CO and UHC produced in the furnace have sufficient time, temperature and oxygen from excess air to be converted to CO_2 and water in the subsequent passage. However, CO and UHC produced in IC engines do not have sufficient residence time inside the combustion chamber for subsequent oxidation. The same feature can also be noted for gas turbines.

In IC engines, CO and UHC can be oxidised as a post-combustion reaction. A thermal converter can be one such device, where the exhaust gas velocity is reduced to increase residence time. The exhaust gas is taken in this converter at temperatures of 850 K–950 K. To use this thermal converter effectively, the exhaust cannot be cooled below a certain limit while coming out of the engine cylinder. This in turn reduces the device's efficiency. However, the oxidation reactions can occur at 550 K–600 K in the presence of platinum or palladium catalyst. The catalytic converter, discussed earlier, can be used for these conversions. The platinum and/or palladium particles can be kept in the catalytic converter's ceramic section. This can effectively reduce CO and hydrocarbon emissions without hampering the engine efficiency due to higher exhaust temperatures from the engine. The catalytic converters, which oxidise CO and UHC, are called two-way catalytic converters. If NO_x is reduced in the converter along with oxidation of CO and UHC, the converter is termed a three-way catalytic converter.

Lean premixed combustion also reduces CO and UHC emissions along with NO_x emissions from gas turbine combustors.

11.5.4 Control of Carbon Dioxide Emission

To control CO_2 emission, carbon capture and storage are employed to reduce carbon loading to our atmosphere. CO_2 is considered the major contributor to greenhouse gases. Control of CO_2 emissions thus reduces global warming due to greenhouse gas emissions. However, the methods evolved are not really cost effective yet for the transport sector and small-scale power generation units.

Fossil fuels can be gasified to synthetic gas (syngas). In this process, the fuel can be gasified by substoichiometric oxygen or by steam. The syngas contains hydrogen, carbon monoxide, carbon dioxide, water, methane and other hydrocarbons along with tar. Tar can be separated from the syngas by cooling. The syngas then reacts with water to perform water gas shift reaction in a reactor. The reaction oxidises carbon monoxide and produces CO_2 and hydrogen:

$$CO + H_2O \leftrightarrow CO_2 + H \tag{11.27}$$

The product is first dried to remove water. CO_2 can then be separated from hydrogen by pressure swing adsorption on a suitable adsorbing material, like zeolite. Subsequently it can be removed by a desorption process and stored.

The post-combustion CO_2 capture pivots on the technology of separating CO_2 from other components of the product gases, mainly from nitrogen, the major constituent of combustion product. One effective method of separation is absorption of CO_2 in mono-ethanol amine. In this method, a 25%–30% aqueous solution of mono-ethanol amine is used as absorbent. The combustion product is first cooled below 40°C and supplied to a chamber containing amine solution. The product gas enters into the amine chamber from the bottom and moves towards the top. During this passage, the CO_2 is absorbed the amine. The product without CO_2 escapes from the chamber from the top. Before leaving the chamber, the gas is passed through water to take away the amine vapour from the product gas. The CO_2-rich amine is taken out from the absorption chamber and supplied to a stripper unit. In the stripper unit, amine is again regenerated by increasing its temperature to 100°C–150°C with steam. The steam carries the CO_2 from the amine and the regenerated amine is recirculated to the absorption chamber. The steam and CO_2 mixture is taken out from the stripper and supplied to a condenser. In the condenser, CO_2 is separated from the steam and taken to a storage unit, where it is compressed and stored.

If oxygen is supplied to the combustion chamber, the nitrogen loading on the product of combustion will be absent. In another method (oxyfuel combustion), oxygen is supplied for combustion instead of air. Oxygen can be separated from air first by cryogenic separation. The combustion product, coming out of oxyfuel combustion, consists of CO_2, H_2O and excess oxygen, if any. The product can be cooled via condensation to separate the water component. CO_2 along with excess oxygen can be taken to the storage unit.

Oxyfuel combustion technology can also use oxides of metals, like iron, nickel, copper and so on, as the oxidiser. Fine particles of metal oxides are supplied to the combustion chamber with fuel. Fuel can react with these particles, keeping the particles in fluidisation to improve mixing. Reduced particles in the process can be separated and taken to another oxidation chamber for regeneration. In this method, particles become unsuitable after a number of cycles of operation. Moreover, it is possible that particle-laden flue gas also remains.

The stored CO_2 cannot be released into the atmosphere. It can be transported for permanent storage to geological sites. Normally the gas is compressed and put into evacuated mines or deep underground for permanent storage.

11.5.5 Control of Particulate Emission

Fly ash can be considered the major particulate coming out of a thermal power plant. A small percentage of ash generated in pulverized fuel-fired boiler furnaces is deposited at the furnace bottom as bottom ash. As the coal particles are very small in size, ash particles generated after combustion are also too small to precipitate at the bottom. The remaining large portion (as high as 80%–90%) of the ash generated is carried with the flue gas towards the stack. It is deposited on various surfaces in the boiler, reducing the heat transfer efficiencies. The soot generated during combustion also takes a similar path as fly ash. A periodic cleaning of those surfaces is necessary. Normally soot blowers are employed for periodic cleaning of the surfaces.

ESP is a common device to take the particulate matter from the flue gas. In this device, two electrodes are present. They are normally called the emitter and the collector. A very high voltage is applied across these two electrodes to charge the particulate matters in the flue gas. The charged particles are then attracted towards the collector and collected there. The collected ash is sometimes deposited in the hopper, placed below the collector electrode, by shaking the collector electrode. The collection efficiency of ESP can be as high as 99.9%.

Fabric filters are also employed for particulate separation. The flue gas is allowed to flow through small-diameter fabric bags. While passing through these bags, the particulate matter is separated and stored in the bags. The bags can be cleaned with air blast if necessary. Normally these filters are not suitable for heavy duty boilers. The collection efficiency of these filters can be as high as 99.5%.

Soot particles from gas turbines are normally generated in the primary zone. The soot formed in the primary zone is oxidised in the subsequent zones.

Soot is a major emission in IC engines also. It comes mainly from CI engines because of the non-premixed nature of combustion. Metal wire mesh or ceramic mats are used to capture soot particles. But deposition of soot particles blocks the flow passage and increases the back pressure of the engine, which reduces the engine efficiency. Normally the deposited soot particles are burned by the electric heater with the help of excess oxygen in the exhaust gas. Normally the traps are placed close to the cylinder to take advantage of high exhaust temperature for soot oxidation.

11.6 EMISSION QUANTIFICATION

Pollution emissions are normally quantified in terms of concentration, which can be done either by volume or by mass. On a mass basis, the concentration is expressed as a percentage. The volume or molar basis concentration is usually expressed in parts per million (ppm) or parts per billion (ppb). Normally the emission quantity is very small and thus the measure is expressed in such terms.

The emission index (EI) can be specified for a species, and it is expressed as a ratio of mass of that species to the mass of fuel burned. So it is defined as the mass of emission of *i*th species generated from unit mass of fuel burning:

$$EI_i = \frac{m_{i,emitted}}{m_{F,burnt}} \tag{11.28}$$

The quantity is a dimensionless one. The value will be very low, so it is often specified as g/kg like units. It is a useful parameter to compare devices operating on the same fuel.

If we know the concentration of the combustion product, the EI can be found. EI can be expressed as:

$$EI_i = \left(\frac{X_i}{X_{CO2} + X_{CO} + X_{UHC}}\right)\left(\frac{mM_i}{M_F}\right) \tag{11.29}$$

X is the mole fraction, M is the molecular weight and m is the number of moles of carbon in one mole of fuel. In the denominator, one has to take the concentration of all products that contain carbon (as carbon is coming from fuel).

Emission is quantified either on a dry basis or on a wet basis. On a dry basis, water is not taken as a component of the combustion product. On a wet basis, water is taken into consideration.

When a general hydrocarbon fuel is oxidised with excess oxygen or incomplete combustion takes place, oxygen is present in the product. The generic reaction of combustion of a hydrocarbon fuel (C_mH_n) can be written as:

$$C_mH_n + aO_2 + 3.76aN_2 \rightarrow mCO_2 + \left(\frac{n}{2}\right)H_2O + bO_2 + 3.76aN_2 + pollutants \tag{11.30}$$

If n_i is the number of moles of pollutant that is generated from one mole of hydrocarbon fuel, the concentration on dry basis can be written as:

$$X_{i,dry} = \frac{n_i}{m + b + 3.76a} \tag{11.31}$$

On a wet basis, we can write:

$$X_{i,wet} = \frac{n_i}{m + \frac{n}{2} + b + 3.76a} \tag{11.32}$$

So we can find a relation between dry basis and wet basis concentration through the oxygen balance:

$$X_{i,dry} = X_{i,wet}\left[1+\frac{\frac{n}{2}}{4.76a-\frac{n}{4}}\right] \tag{11.33}$$

If the oxygen concentration on a dry basis in the product is measured, we can write:

$$X_{O2,dry} = \frac{b}{m+b+3.76a} = \frac{a-m-\frac{n}{4}}{4.76a-\frac{n}{4}} \tag{11.34}$$

Rearranging we get:

$$a = -\frac{m+\left(1-X_{O2,dry}\right)\left(\frac{n}{4}\right)}{\left(1-4.76X_{O2,dry}\right)} \tag{11.35}$$

The concentration measurement has a drawback due to variation in the amount of oxygen in the flue gas due to excess oxygen. The concentration of pollutant can be given on the basis of a defined oxygen level in the exhaust gas. When the oxygen in the product is changed, the value of pollutant concentration should be corrected per the existing oxygen concentration in the product (measured oxygen level):

$$X_{\text{defined O2}} = X_{\text{measured O2}}\left(\frac{\text{Number of moles in the exhaust at measured oxygen level}}{\text{Number of moles in the exhaust at defined oxygen level}}\right) \tag{11.36}$$

The number of moles in the exhaust can be calculated from the chemical equation once the oxygen supplied can be found from Equation (11.35).

Example 11.1

The following measurements are taken during an exhaust test of an IC engine: $CO_2 = 12.5\%$, $CO = 0.15\%$, $O_2 = 2.4\%$, C_6H_{14}(equivalent) = 365 ppm, NO = 80 ppm. All measurements are taken on a dry basis. The fuel of the engine is iso-octane. Find the emission index of CO and UHC (which is expressed as hexane equivalent).

Solution

Molecular weight of CO = 28
Molecular weight of iso-octane = 114
Molecular weight of hexane = 86
$m = 8$
We can use Equation (11.29):

$$
\begin{aligned}
EI_{\text{UHC}} &= \left(\frac{X_{\text{UHC}}}{X_{\text{CO2}} + X_{\text{CO}} + X_{\text{UHC}}}\right)\left(\frac{mM_{\text{UHC}}}{M_F}\right) \\
&= \left(\frac{365\times10^{-6}}{0.125+0.0015+365\times10^{-6}}\right)\left(\frac{8\times86}{114}\right) \\
&= 0.0174 = 17.4\,g/kg
\end{aligned}
$$

$$
\begin{aligned}
EI_{\text{CO}} &= \left(\frac{X_{\text{CO}}}{X_{\text{CO2}} + X_{\text{CO}} + X_{\text{UHC}}}\right)\left(\frac{mM_{\text{CO}}}{M_F}\right) \\
&= \left(\frac{0.0015}{0.125+0.0015+365\times10^{-6}}\right)\left(\frac{8\times28}{114}\right) \\
&= 0.0232 = 23.2\,g/kg
\end{aligned}
$$

Example 11.2

The NO emission in the exhaust is measured as 20 ppm (by volume), with an oxygen concentration of 13% by volume on a wet basis. Find the NO concentration corrected to 3% oxygen level.

Solution

We can consider Equation (11.30) to be the chemical equation. From species balance we get, $b = a - 2$.

The percentage of oxygen in the exhaust on a wet basis is 13%. So:

$$\frac{b}{1+2+b+3.76a} = 0.13$$

Solving we get:
$b = 3.588$
$a = 5.588$
From equation (11.36), we can write:

$$X_{\text{NO, 3\%}} = X_{\text{NO, 13\%}} \times \frac{N_{exhaust,\ wet,\ 13\%}}{N_{exhaust,\ wet,\ 3\%}}$$

$$N_{exhaust,\,wet,\,13\%} = 27.68$$

For a 3% oxygen level:

$b = 0.368$

$a = 2.368$

$$N_{exhaust,\ wet,\ 3\%} = 12.267$$

So $X_{NO,\ 3\%} = 45.13\ ppm$

EXERCISES

11.1 Derive Equation (11.33).

11.2 Find the emission index of NO in Example 11.2.

11.3 The boiler furnace fired with natural gas has an NO concentration of 46 ppm and an oxygen concentration of 4.3% (both by volume) in the exhaust gas on a dry basis. Determine (a) the percentage of excess air, (b) the NO concentration corrected to 3% oxygen level, and (c) the EI of NO.

REFERENCES

1. Prather, M. J. and Logan, J. A., Combustion's impact on global atmosphere, *Symposium (International) on Combustion*, **25**, 1513–1527, 1994.
2. Zeldovich, J., The oxidation of nitrogen in combustion and explosions, *Acta Physicochimica*, **21**, 577–628, 1946.
3. Turns, S. R., *An Introduction to Combustion: Concepts and Applications*, New York: Tata McGraw-Hill, 2012.
4. Fenimore, C. P., Formation of nitric oxide in premixed hydrocarbon flames. *Symposium (International) on Combustion*, **13**, 373–380, 1971.
5. Miller, J. A. and Bowman, C. T., Mechanicsm and modeling of nitrogen chemistry in combustion, *Progress in Energy and Combustion Science*, **15**, 287–338, 1989.
6. Bowman, C. T., Control of combustion generated nitrogen oxide emissions: Technology driven by regulation, *Symposium (International) on Combustion*, **24**, 859–878, 1992.
7. Harris, B. W., Conversion of sulfur dioxide to sulfur trioxide in gas turbine exhaust, *Journal of Engineering for Gas Turbines and Power*, **112**, 585–589, 1990.
8. Paul, B., Datta, A., Datta, A. and Saha, A., Occurrence and characterization of carbon nanoparticles below the soot laden zone of a partially premixed flame, *Combustion Flame*, **156**, 2319–2327, 2009.
9. Heywood, J., *Internal Combustion Engine Fundamentals*, New York: McGraw-Hill, 2018.

10. Lefebvre, A. H., The role of fuel preparation in low-emission combustion, *Journal of Engineering for Gas Turbines and Power*, **117**, 617–654, 1995.
11. Correa, S. M., Power generation and aero-propulsion gas turbines: From combustion science to combustion technology, *Proceedings of the Combustion Institute*, **28**, 1793–1807, 1998.
12. Lefebvre, A. H., *Gas Turbine Combustion*, Ann Arbor, MI: Edwards Brothers, 1999.
13. Glassman, I., *Combustion*, 3rd ed., San Diego, CA: Academic Press, 1996.
14. CO_2 Emissions from Fuel Combustion Highlights, *International Energy Agency*, 2017, www.iea.org.

Chapter 12

Combustion Diagnostics

12.1 OVERVIEW

Combustion is the backbone of energy systems. A demand is always there to make the system more and more efficient. To achieve this, combustion researchers carried out an enormous number of experimentations, in both the laboratory and in industry. Always the focus was to ensure an accomplished burning with controllability, which requires a reliable measurement system. The hostile environment of a combustion chamber, due to the high temperature and flame, is also noteworthy. Measurement in that environment requires some special attention. To achieve this, a robust measurement system is required. The common parameters to measure in combustion are temperature, pressure, velocity, concentration and so on.

Temperature is measured by thermometer or thermocouple. Velocity is measured using anemometer or pitot tube. But these methods employ a probe that resides at the location and measures. The probe is provided with sufficient time before measurement to be in equilibrium with the system. But the intrusion of the probe in the flame environment distorts the flow field as well as the flame. The probe can also act like a flame holder. In some cases, the probe conducts energy out from the flame, which is a feature for the thermocouple wires in particular. Moreover, the physical size of the probe affects the spatial accuracy and resolution of the measurement. The temporal resolution of the measurement depends on the response time of the probe, that is, the time required to come into equilibrium.

To overcome the flame distortion due to probe insertion, non-intrusive measurement techniques, with no intrusive probe, are preferred. In these methods, optical techniques are mainly used. Nowadays, with the advancement of laser, optical diagnostics is better developed. In this method, flame distortion can be avoided. Actual shape and size can be reflected in the measurement. The optical methods mostly do not require any response time, so it is possible to gather data very fast. Measurement is possible for flame transience and short duration phenomenon. The measurement volume is very small. The spatial resolution of non-intrusive methods is also better than corresponding intrusive methods. With the introduction of lasers, a new horizon in non-intrusive combustion diagnostics has opened up [1].

Combustion diagnostics is a very vast and specialised area. It is difficult to discuss the methods in details within the scope of this book. However, this chapter attempts to outline briefly a few combustion diagnostic methods.

12.2 FLOW FIELD MEASUREMENT

For flow velocity measurement, the pitot tube is the traditional instrument. The measurement principle is simple. In combustion, the pitot tube is not very popular for its large intrusion inside the flow area. The high temperature of the environment restricts the use of the pitot tube. But if the area is large enough and the probe material can withstand temperature, this traditional instrument can be used to measure velocity within the combustion zone.

As an intrusive technique, hot wire anemometry is popular for measuring flow velocity in a combustion environment. Using this probe, we can measure the mean velocity and fluctuation also. An anemometer probe is a very fine wire (4–10 micron diameter) held within two prongs (Figure 12.1), which is heated from an external source. Heat transfer from this wire probe to its surroundings is calibrated to measure the velocity. The flow should be cooler than the probe wire. In combustion that becomes a difficulty. However, one can measure the cold flow also in the combustion device.

The probe wire is made of platinum or tungsten. When a current is passed through the probe, it produces Joule heating. Heat is lost from the wire by conduction (through the holder), convection and radiation. At equilibrium, heat generated will be equal to the heat loss. We can assume that natural convection heat loss is negligible compared to the forced convection. If we neglect the conduction and radiation loss, we get:

$$Q = iR = (\pi dl)h(T_w - T_\infty) \tag{12.1}$$

where, d and l are the diameter and length of the wire, respectively; h is the heat transfer coefficient; T_w and T_∞ are the wire and ambient temperatures, respectively; and i and R are the current and resistance, respectively. The resistance is a function of wire temperature and h can be written in terms of flow velocity. Equating those, we can find the flow velocity.

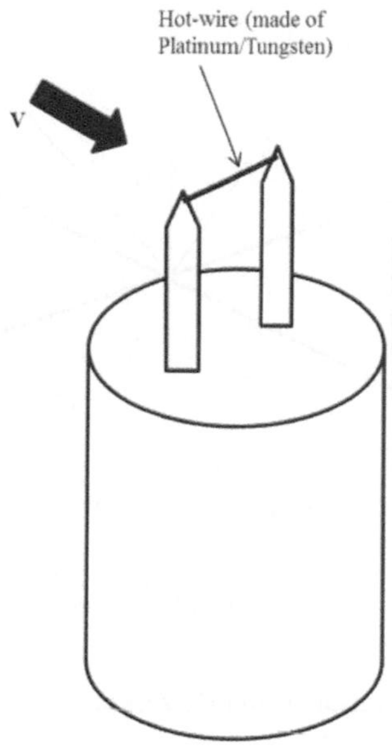

Figure 12.1 Anemometer probe.

While measuring, either current or temperature is kept constant. The voltage drop across the probe is measured and is calibrated to find the velocity. If we want to get the flow direction, we need to put another sensor in the flow domain. To find the flow fluctuations, the voltage fluctuation can be used. For further reading on this subject, the book by Bruun [2] can be consulted.

Laser Doppler velocimetery (LDV) is a non-intrusive method of velocity measurement. This method can measure all components of instantaneous velocity of the flow field. The laser Doppler velocimeter sends a monochromatic laser beam toward the target and collects the scattered radiation. According to the Doppler effect, the change in wavelength of the scattered radiation is a function of the targeted object's relative velocity. Thus, the velocity of the object can be obtained by measuring the change in wavelength of the reflected laser light, which is done by forming an interference fringe pattern. The flow is normally seeded with small, neutrally buoyant particles that can flow with the fluid. These particles are illuminated with laser and scatter light. The frequency of the illuminating laser is known. The scattered light detection is done by a photomultiplier tube (PMT) or a similar instrument. This generates a current in proportion to absorbed photon energy and amplifies that current. The difference between the incident and scattered light frequencies is called the Doppler shift. By analyzing the Doppler-equivalent frequency of the laser light scattered (intensity modulations within the crossed-beam probe volume) by

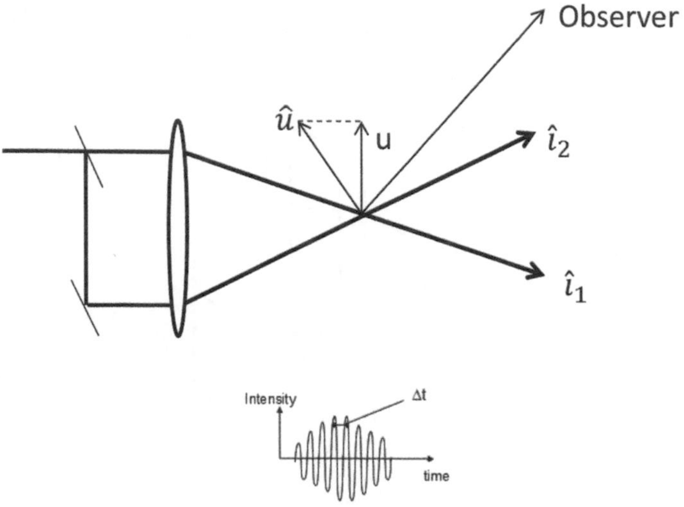

Figure 12.2 Schematic diagram of an LDV setup.

the seeded particles within the flow, the local velocity of the fluid can be determined. Yeh and Cummins [3] first introduced the measuring principle to the scientific community (Figure 12.2).

In an LDV setup, usually an incident laser beam (of frequency f) is split into two parts: $\hat{\imath}_1$ and $\hat{\imath}_2$. This operation is done by optical beam splitters and mirrors. The two incident rays are then converged at a single point of focus for measurement. The two rays illuminate the same area. As the path length is different for two rays, an interference pattern is generated at the illuminated area. One collector is kept to get scattered light from both rays. This scattered light will have two different frequencies, f_1 and f_2, due to two Doppler shift frequencies. We can write:

$$f_1 = f + f_{D,1} = f + \hat{u}\ (\hat{r} - \hat{\imath}_1)/L \tag{12.2}$$

$$f_2 = f + f_{D,2} = f + \hat{u}\ (\hat{r} - \hat{\imath}_2)/L \tag{12.3}$$

where $\hat{r}$ is the unit vector in the direction of scattering, $\hat{u}$ is the particle as well as fluid velocity, $f_{D,1}$ and $f_{D,2}$ are the two Doppler shifts, and L is the wave length of the source. The detector detects a burst of signal with a frequency (f_1-f_2):

$$f_1 - f_2 = \frac{\hat{u}\left(\hat{\imath}_1 - \hat{\imath}_2\right)}{L} = 2usin\varnothing / L \tag{12.4}$$

where u is the component of the velocity in the direction of $(\hat{\imath}_1 - \hat{\imath}_2)$, and $\varnothing$ is the angle between the two rays. The velocity can be calculated using the above equation.

Using a single monochromatic light, we can find the velocity in one direction. For determining the velocity in other axes, different frequency sources can be used. The measurement for three velocity components can be done simultaneously.

Another very popular measuring technique for velocity is particle image velocimetry (PIV). PIV is not a point-by-point measurement method like LDV. Rather, PIV is a full-frame measurement system. This is a non-intrusive laser-based method.

The laser coming from the source is taken to a suitable optics to generate a light sheet. Usually the light coming from a point source is passed through a cylindrical plano-convex lens to get a laser sheet. The light sheet is now passed through the test section. It illuminates a plane in the test section. Observation is made in the perpendicular direction of the illuminated test plane. The information is taken to a camera. Figure 12.3 shows the arrangement schematically.

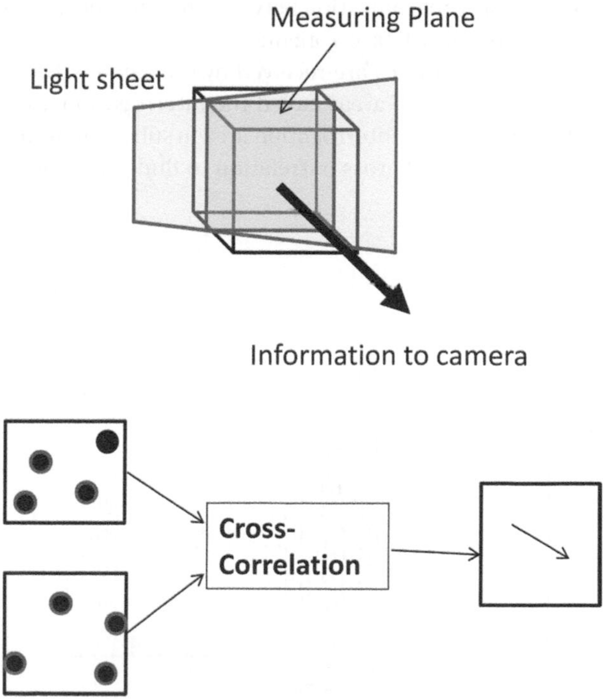

Figure 12.3 Measurement using PIV.

The fluid flowing in the test section is seeded with light and neutral particles that can move with flow velocity. When the test section is illuminated, we can see the motion of the particles. There are a number of seeding particles. They are selected on the basis of the flow and fluid types. For combustion, titanium dioxide fine particles (2–5 microns in diameter) are normally used. The camera takes two subsequent snaps. If we track one particle, we can identify the displacement of that particle during this time. If the time duration is known, velocity can be found knowing the two positions.

Here we should understand that the time gap between two snapshots must be very small. And there is more than one similar particle. It is difficult to track one particle. For full-frame measurement, where we get information about the full area at a time, there is no benefit to tracking a single particle. But we understand the physics of measurement now. Normally, the laser light is of sufficient high power and it is pulsed. The camera shutter is opening two times in a very short interval. The time interval depends on the velocity to measure. The larger the velocity, the lower should be the interval. When the camera's shutter is open, the particles are in motion. So there is a possibility of getting a blur image. This can be avoided by pulsing the laser for a short while the camera's shutter is open. There is requirement of synchronisation between the camera and the laser. If the image is blurred, the result will be erroneous.

When two subsequent frames are received by the camera, the image area is divided into a number of small areas called the interrogation area. Figure 12.3 shows information of one such interrogation area in subsequent time. These two images are taken to statistical cross correlation to find the velocity. Figure 12.4

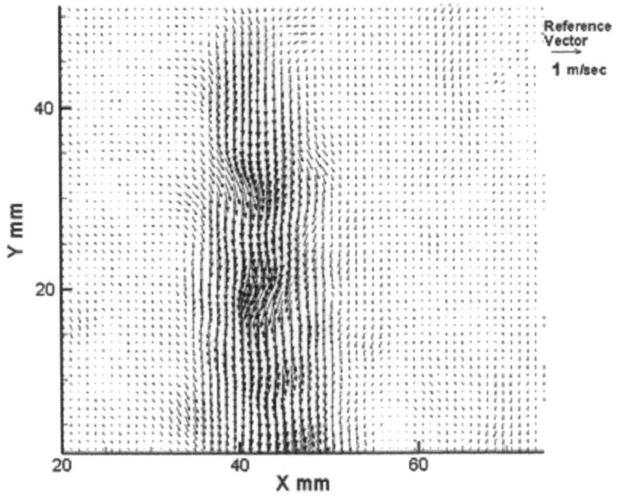

Figure 12.4 Velocity vectors for a turbulent jet obtained from PIV.

shows a velocity vector diagram of a turbulent jet from PIV. The theory of cross correlation is beyond the scope of this book, but Keane and Adrian [4] can be recommended for those interested in further study.

The above method of PIV measures a two-dimensional (2D) flow-field only. Stereoscopic PIV can be used for getting a three-dimensional (3D) velocity. Here also a plane of the test section is illuminated. The information is captured by two cameras simultaneously. Both the cameras make an angle with the perpendicular direction of the illuminated plane. Previously, with one camera in the perpendicular direction, we got the component of displacement on the illuminated plane only. However, stereoscopic camera placement can find the actual displacement of the particle and subsequently the 3D velocity.

12.3 TEMPERATURE MEASUREMENT

Thermocouples are the traditional temperature measuring instrument. When the junction points of two dissimilar wires are kept at different temperatures, a voltage is generated known as the Seebeck effect. This principle is used to measure temperature. The electromotive force generated can be calibrated to find the temperature of one junction if the other junction temperature is known. In combustion measurement, for high temperature, normally Pt%–Pt13%Rh (R-type) and Pt-Pt10%Rh (S-type) are used. The measurement method is intrusive in nature. So minimum wire diameter helps in not disturbing the flow pattern. However, 100-micron wires also create a 250- to 400-micron junction size. This offers a resistance to flow when placed inside the flame. The sensitivity of the R-type thermocouple is not very high. It is 7 μV/°C and thus it is very difficult to get a reading to the extent of one decimal place. The junction or bid formed to measure temperature has a response time of the order of 250 ms. So it is difficult to measure anything at a faster rate.

The temperature measured here is the surface temperature of the junction, not the gas temperature at the flame. The two temperatures may differ due to radiative and conductive heat transfer. Sometimes, there is a reaction at the thermocouple junction. As Pt is a catalyst material, it may happen that on Pt-containing thermocouple junctions, more reaction happens. That also makes a difference in temperature.

The thermocouple, when inserted in the flame, radiates a considerable amount of energy. The higher the temperature to measure, the higher will be the loss. The measured temperature must be corrected. If we consider the thermocouple junction as a lump body and take care of its radiation and convection, the corrected temperature can be found as:

$$T_{j,c} = T_j + \frac{\sigma\varepsilon}{h}\left(T_j^4 - T_\infty^4\right) \tag{12.5}$$

where $T_{j,c}$ is the corrected temperature; T_j is the uncorrected junction temperature; T_∞ is the surrounding ambient temperature; σ and ε are Stefan-Boltzmann constant and emissivity of the junction, respectively; and h is the heat transfer coefficient.

Conduction heat loss through the thermocouple error can also be estimated and corrected. But for small diameter wires, this loss can be neglected.

The thermocouple can sag due to thermal expansion and measure the temperature of a different location. So, care must be taken during measurement to tighten the sag of the thermocouple wire.

In non-premixed or partially premixed flames, the use of the thermocouple can be difficult due to soot deposition at the junction.

Measurement of flame temperature is one of the most important and challenging tasks in combustion diagnostics. Two-colour pyrometry is considered a simple, non-intrusive and inexpensive method for flame temperature method. This method is based on the detection of thermal radiation of an object at two different wavelengths.

The radiation emitted from a heat source is a strong function of temperature. So measurement of wavelength provides a hint of the temperature of the object. However, radiation of the body also depends on emissivity of the surface of the object. Two-colour pyrometry is an emissivity independent temperature measurement method. It considers the emitted radiation of the body at two close wavelengths to eliminate the wavelength-dependent influence of the surface. It is found that the temperature of the object is proportional to the ratio of the two spectral radiances.

Two-colour pyrometry was invented by Hottel and Broughton in 1932. They used this method to determine the true temperature of a luminous gas flame of a furnace. This method was explored and developed for the last one hundred years in different applications like soot measurement, combustion engines, coal burners and industrial furnaces and gasifiers.

The radiation from a black body is considered a continuous spectra. The intensity of the radiation is a strong function of wavelength and temperature. The intensity of the emitted radiation at wavelength λ and temperature T is expressed by the Planck's law:

$$\Phi_\lambda(T) = \frac{C_1}{\lambda^5}\left(e^{\frac{C_2}{\lambda T}} - 1\right)^{-1} \tag{12.6}$$

where $C_1 = 2hc^2$ and $C_2 = hc/k$, h is Planck's constant, c is speed of light in vacuum and k is Boltzmann's constant. For $\lambda \ll \frac{C_2}{T}$ and T < 3000 K, the above expression can be reduced to:

$$\left(e^{\frac{C_2}{\lambda T}} - 1\right)^{-1} \approx e^{-\frac{C_2}{\lambda T}} \tag{12.7}$$

So:

$$\Phi_\lambda(T) = \frac{C_1}{\lambda^5} e^{-\frac{C_2}{\lambda T}}, \tag{12.8}$$

This is referred as Wien radiant law (Figure 12.5).

The above expressions are valid for a black body. The natural heat sources behave like a gray body for a specific range of wavelength. Emissivity ε_λ at the given wavelength is considered for practical surfaces. Thus, the equation modifies to the following:

$$\Phi_\lambda\,(T) = \varepsilon_\lambda \frac{C_1}{\lambda^5} e^{-\frac{C_2}{\lambda T}} \tag{12.9}$$

If the spectral radiations of a body at temperature T at wavelength λ_1 and λ_2 are denoted by Φ_λ and $\Phi_{\lambda 2}$, then:

$$\Phi_{\lambda 1}\,(T) = \varepsilon_{\lambda 1} \frac{C_1}{\lambda_1^{\,5}} e^{-\frac{C_2}{\lambda_1 T}} \tag{12.10}$$

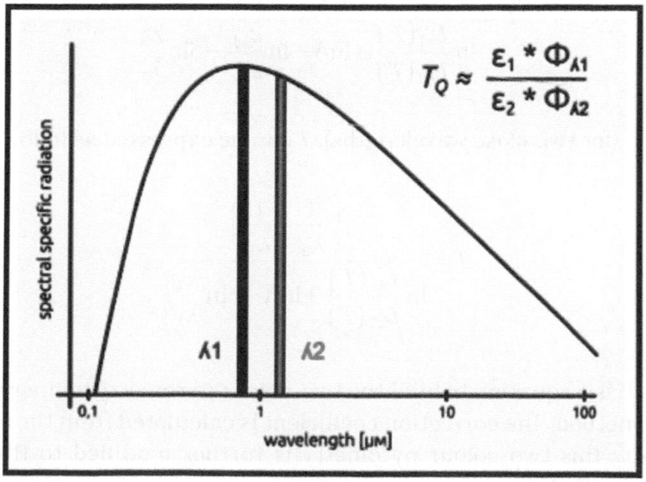

Figure 12.5 Measurement of the radiation at two wavelengths and determination of the temperature from the ratio of the intensity of the radiation (working principle of two-colour pyrometers).

$$\Phi_{\lambda 2}(T) = \varepsilon_{\lambda 2} \frac{C_1}{\lambda_2^{\,5}} e^{-\frac{C_2}{\lambda_2 T}} \tag{12.11}$$

The ratio of spectral radiations is:

$$\frac{\Phi_{\lambda 1}(T)}{\Phi_{\lambda 2}(T)} = \frac{\varepsilon_{\lambda 1} \frac{C_1}{\lambda_1^{\,5}} e^{-\frac{C_2}{\lambda_1 T}}}{\varepsilon_{\lambda 2} \frac{C_1}{\lambda_2^{\,5}} e^{-\frac{C_2}{\lambda_2 T}}} \tag{12.12}$$

This ratio is proportional to the ratio of signal detected at detectors at wavelength λ_1 and λ_2. These signals can be represented by the symbol $L_{\lambda 1}(T)$ and $L_{\lambda 2}(T)$, respectively, as follows:

$$\frac{\Phi_{\lambda 1}(T)}{\Phi_{\lambda 2}(T)} = A \frac{L_{\lambda 1}(T)}{L_{\lambda 2}(T)} \tag{12.13}$$

where A is the correction coefficient.

Now the temperature can be derived from the above equations as follows:

$$T = \frac{C_2\left(\frac{1}{\lambda_2} - \frac{1}{\lambda_1}\right)}{\ln\frac{L_{\lambda 1}(T)}{L_{\lambda 2}(T)} + \ln A - \ln\frac{\varepsilon_{\lambda 1}}{\varepsilon_{\lambda 2}} - 5\ln\frac{\lambda_2}{\lambda_1}} \tag{12.14}$$

For $\varepsilon_{\lambda 1} \approx \varepsilon_{\lambda 2}$ (for two close wavelengths), T can be expressed as follows:

$$T = \frac{C_2\left(\frac{1}{\lambda_2} - \frac{1}{\lambda_1}\right)}{\ln\frac{L_{\lambda 1}(T)}{L_{\lambda 2}(T)} + \ln A - 5\ln\frac{\lambda_2}{\lambda_1}} \tag{12.15}$$

This is the basic equation behind the two-colour pyrometry temperature measurement method. The correction coefficient is calculated from the calibration experiment. This two-colour pyrometry is further modified to three-colour pyrometry for better accuracy.

Schlieren imaging is used in combustion diagnostics to identify the flame structure. The principle depends on variation of the refractive index of the medium. Light rays are shifted due to refraction in a medium. If the refractive

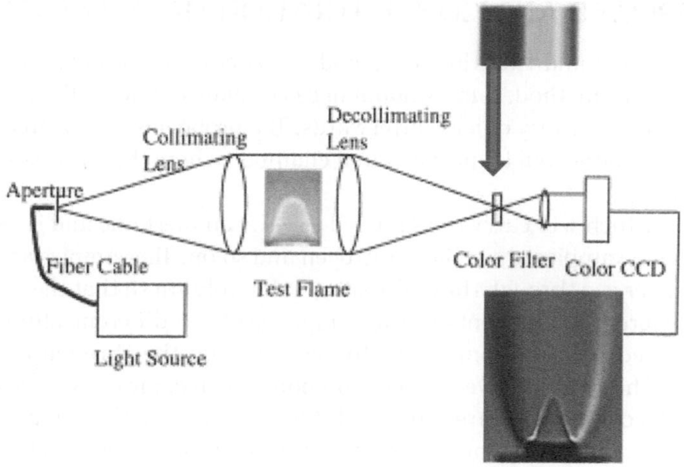

Figure 12.6 Rainbow Schlieren deflectometry setup.

index varies at different places, the refractive shift will be different. The main reason for change in the refractive index is density variation. Again, density variation can be caused by temperature variation or concentration variation. If the refractive shift can be quantified, it can be calibrated with density variation and in turn with temperature variation or concentration variation (Figure 12.6).

We shall discuss here the measuring principle of rainbow Schlieren deflectometry. A white light is first passed through suitable optics to obtain a parallel light beam. This beam of white light is passed through the test section. If there is no density variation in the test section, the parallel light rays move straight, fall on decollimating lens and converge on a point. A colour filter is placed at the convergence plane. Light is allowed to pass through this colour filter and finally falls on the Charged Coupled Device/Complementary Metal-Oxide-Semiconductor (CCD/CMOS) sensor of the camera. Now, if density variation is discovered, the light rays will shift depending on density values. Now the rays will fall on different colours depending on the amount of shift. This shift can be calibrated to find the density variation and thus the temperature [5].

The shift of rays due to refractive index (density) variation is utilised in another measurement technique called holographic interferometry. Here also, the arrangement is similar to the Schlieren setup. The collimated beam of light is passing through the test section. Then it is converged to a point. The light passes through the cold test section and then through the hot setup, while combustion is ongoing. Due the refractive shift, the path length will be different. This will make a phase shift if we consider interference of the two beams. This interferometric hologram pattern can reveal the shift of the rays. The fringe patterns can be calibrated to find the density difference and subsequently the temperature [6].

12.4 SPECIES AND CONCENTRATION MEASUREMENT

The well-known and traditional method is gas chromatograph, an intrusive method. In this method, some amount of gas is taken out of the flame and then must be quenched very quickly afterwards. The probe has a very small orifice. The probe is connected to an evacuated chamber, where the gas is stored after collection.

The gas which will carry the sampled gases is an inert one and is called the carrier gas. Examples are helium, nitrogen and so on. The samples are carried by the carrier gas through the chromatographic column so that they can react better. Different components of the sample gas have different affinities. The time required for each component to pass through the chromatograph column is not the same. A lower affinity component can come out very fast. In the process, the components are separated. After separation, the components are carried to a detector by the carrier gas. A number of essential detectors can be used, for example, the flame ionisation detector, the thermal conductivity detector and so on. The detectors generate electric signals depending on the amount of component present. The signals can be mapped with the concentration. The resultant curve is called a chromatogram. A chromatogram is shown in Figure 12.7. Different components are marked as 1, 2, and so on.

Raman invented a technique where scattering of energy was found for inelastic collision of light photon and a molecule [7]. The molecule can shift its energy level due to this collision. If the molecule energy level goes to a different rotational level but within the same vibrational level, it is called rotational scattering. In this case, the energy level change is very small and difficult to detect. When the molecule changes vibrational level, called vibration scattering, the energy change is considerable and can be measured (Figure 12.8).

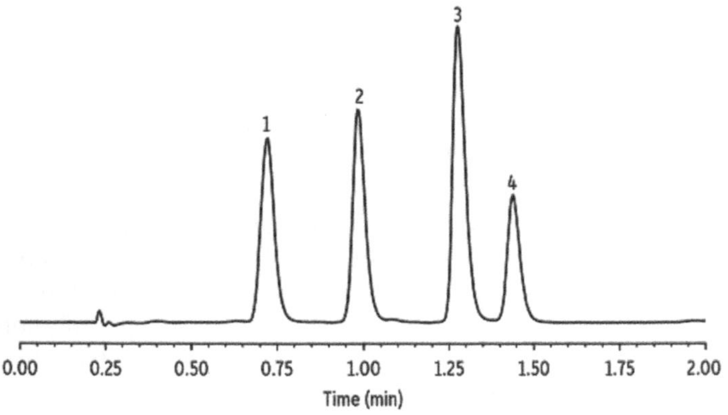

Figure 12.7 A typical chromatogram.

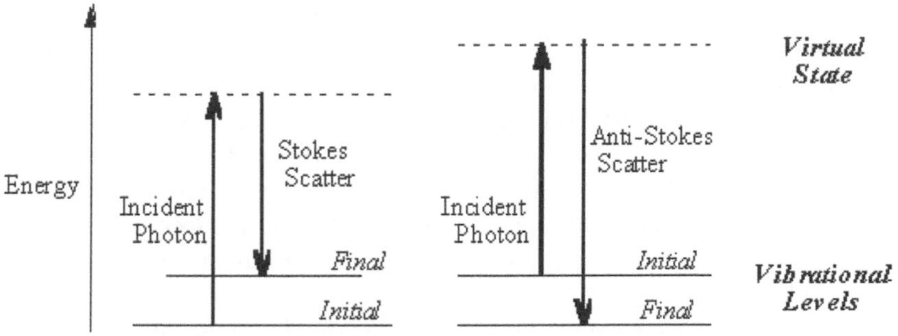

Figure 12.8 Stokes and anti-Stokes Raman scattering.

During collision, vibrational scattering can be of two types. When the molecule gains energy, it finally goes to a higher vibrational energy level. This is called Stokes Raman. In the opposite case, the molecule loses energy, and the molecule finally comes to a lower energy level. This type of scattering is known as anti-Stokes Raman scattering.

The amount of scattering can be measured and quantified. Each molecule has its characteristic energy states. From the Raman signal, we can identify the concentration of different components. Raman scattering is suitable for constituents having more than 0.1% concentration. Many species can be detected simultaneously with this non-intrusive method.

Temperature can also be measured in a number of ways from the spectral distribution. The accuracy of the measurement, depending on measuring period, is 2%–5%.

Coherent anti-Stokes Raman spectroscopy is an advanced method for measuring temperature and concentration of the species. The method became popular after the work by Regnier and Taran [8]. The process enables us to measure temperature and concentration simultaneously. This can measure concentration for the species with a concentration of more than 0.1%. The method is robust and is suitable in industrial applications. The process takes place through third-order non-linear susceptibility. In this method, a molecule is taken to an excited state using a pump beam. Then it is brought back to a higher energy level compared to its initial level. Subsequently, it is taken to an excited state by a second beam, and it releases signal beam to come to its initial energy level. Temperature and concentration are related to this signal and hence can be detected.

Raman scattering can detect concentrations at large percentages. Laser-induced fluorescence (LIF) can be used to detect very small concentrations, even at the sub-ppm level. When a molecule emits light spontaneously due to its excited state, it is called fluorescence. The duration of fluorescence is very small

(10^{-5}–$10^{-4°}$ s), but much higher than scattering. The molecules or radicals of a flame can be energised by a laser. If the fluorescence is emitted at the same frequency, it is called resonant fluorescence. It is difficult to detect the signal or differentiate the signal resonant one. So there should be a frequency shift in emission.

Depending on the species to measure, excitation laser frequency can be tuned. The species to be detected must have a known emission and absorption spectra. From those, suitable excitation and detection frequencies can be identified. A laser sheet can excite a 2D plane for LIF. This is known as planar LIF (PLIF).

Determination of concentration after getting the signal includes a number of considerations. Proper calibration is necessary for determination of the values. LIF can be used for temperature measurement also.

12.5 PRESSURE MEASUREMENT

Pressure measurement of flame is normally done by pressure sensors. For combustion chambers of nearly atmospheric pressure, like boiler furnaces, the pressure can be measured by a manometer or Bourdon pressure gauge. However, these sensors are not suitable for accurate dynamic measurement. Piezoelectric or piezoresistive pressure sensors can be used for taking dynamic measurement. The major problem in measurement is the high temperature, which is not suitable for the sensors. To overcome this, a special cooling arrangement is required for the sensors.

12.6 SOOT MEASUREMENT

Thermocouples can be used to measure the soot produced in a flame. This method is an intrusive one. The measuring principle is based on thermophoretic movement of soot particles towards the thermocouple junction and the temperature history of the thermocouple. The method is known as thermocouple particle densitometry (TPD) [9,10]. The thermocouple junction is to be placed at the location where the soot volume fraction is to be measured. For a non-sooty flame, the temperature measured will come to a steady state value. In case of a sooty flame, a steady state cannot be achieved due to continuous deposition of soot on the thermocouple junction by thermophoresis. The soot mass flux due to thermophoresis is:

$$m'' = \frac{D_T Nuf \rho_p}{2d}\left[1-\left(\frac{T_j}{T_g}\right)^2\right] \quad (12.16)$$

where D_T is the thermophoretic diffusion coefficient, f is the soot volume fraction, ρ_p is the density of the soot particles, T_g is the gas temperature and T_j is the junction temperature.

The junction temperature changes with time. Initially, soot deposition increases the radiation loss, because soot emissivity is more than the junction emissivity. So the temperature decreases. Susequently, when the junction is completely covered with soot, temperature decreases because of increase in diameter.

After some mathematical manipulation, we arrive at the following relations [11]:

$$\frac{dG}{dt} = \beta f = m \tag{12.17}$$

$$G = \frac{1}{4}\left(\frac{T_g}{T_j}\right)^8 - \frac{1}{6}\left(\frac{T_g}{T_j}\right)^6 \tag{12.18}$$

In variable diameter stage, β is constant. So m will be a constant, and f is a constant because it is soot volume fraction at a particular location. Now, we can plot G with time. The slope of the curve will be known to us. Then f can be found assuming the value of β.

Light extinction is an easy non-intrusive method for detection of soot volume fraction. Monochromatic light can be passed through a flame, and the intensity is detected at the opposite end. When the light is passing without a test section, the attenuation of the light intensity is negligible and the intensity falling on the sensor can be detected. While passing through the test section (a flame in this case), attenuation will be more as some amount will be scattered by the particulate of the flame. Here one needs to assume that the scattering of light is from the soot only. Now, the transmittivity of flame can be determined by taking the ratio with and without test section intensities. This can be done in a point-by-point measurement or by a full field measurement. A attenuated intensity image is shown in Figure 12.9.

The ratio of intensities can be expressed as follows:

$$\frac{I}{I_0} = \exp\left(-\frac{k_{ext}}{L}\int f ds\right) \tag{12.19}$$

where I and I_0 are the intensities with and without test section. k_{ext} is the extinction coefficient, L is the wavelength, f is the soot volume fraction and s is the path length.

The extinction coefficient can be suitably chosen or determined to get the value of the soot volume fraction [12]. The method is suitable for dynamic

Figure 12.9 Light intensity attenuation in flame.

measurement also. The spatial and temporal resolution depends on that of the detection device.

When soot particles are heated, they emit light. Laser-induced incandescence (LII) is a technique based on detection of the emitted light. Radiation is absorbed by soot particles very effectively. The particles behave in a way almost similar to a black body. This is the reason why soot is very common as a pigment. More or less everything black has soot-based additives, for instance, black plastic material or toner for laser printers. A material that absorbs effectively also emits effectively. This is the reason for its incandescence.

The soot particles are exposed to a short energetic laser pulse in order to heat them to temperatures around 3500°C. The high temperature is necessary for increased light emission. The absolute strength of this emission is related to the soot volume fraction. The emission is to be calibrated for quantitative soot volume fraction measurements. LII has a high temporal and spatial resolution.

The LII signal typically is approximately 10 ns long. The duration of the excitation pulse is less than that. The reason for this is that the heated soot particles retain their original temperature by heat conduction to the surrounding gas. The heat conduction process depends on the area of the particles whereas the absorption of laser radiation is volume-dependent. This area-to-volume relationship makes it possible to derive size information from the particles by studying the decay time of the LII signal. A detailed description can be found in the work of Santoro and Shaddix [13].

12.7 DROPLET AND SPRAY MEASUREMENT

Spray properties are important parameters for the measurement for liquid fuels. Spray cone angle and breakup lengths are readily required for spray characterisation. These two properties can be identified by image processing of the spray (Figure 12.10).

Spray cone angle is the divergence angle with which the spray is coming out of a nozzle. When the image is obtained, two lines can be drawn tangent to the extreme lines of spray on either side. We can measure the angle between these two lines, which is the spray cone angle.

The breakup length is the minimum distance from the nozzle mouth, where the jet is first broken. This length can also be found quite accurately from image processing.

The spatial distribution of spray at any distance from the nozzle mouth can be measured using a patternator. A number of small-area-measuring tubes are kept together to make a patternator. The amount of liquid deposited in each tube can generate a spatial distribution of liquid. Pattaernation can be done by manual measurement. It is sometimes done digitally through image processing also.

Droplet size and its number distribution are other very important parameters for spray and combustion analysis and modeling. This can be done using

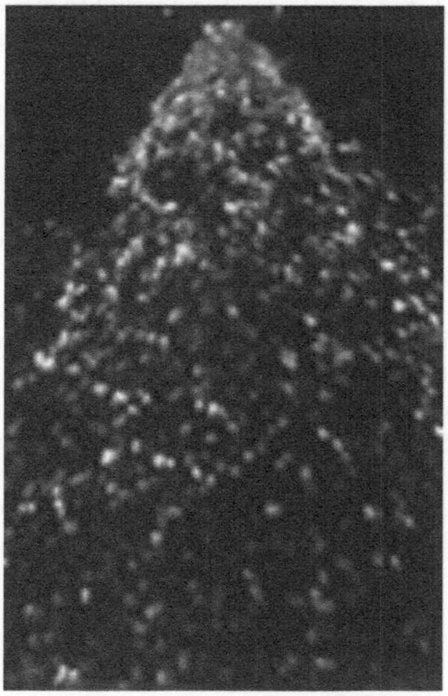

Figure 12.10 Image of a spray.

phase Doppler particle analyser (PDPA). It works on the principle of LDV. The Doppler shift is calibrated with the droplet size to obtain the droplet size distribution.

REFERENCES

1. Eckbreth, A. C., *Laser Diagnostics for Combustion Temperature and Species*, Royal Tunbridge Wells, UK: Gordon and Breach Publisher, 1996.
2. Bruun, H. H., *Hot Wire Anemometry—Principles and Signal Analysis*, Oxford, UK: Oxford University Press, 1995.
3. Yeh, Y. and Cummins, H. Z., Localized fluid flow measurements with an He-Ne laser spectrometer, *Applied Physics Letters*, **4**, 176–178, 1964.
4. Keane, R. D. and Adrian, R. J., Theory of cross-correlation analysis of PIV images, *Applied Scientific Research*, **49**, 191–215, 1992.
5. Xiao, X., Puri, I. K. and Agarwal, A. K., Temperature measurements in steady axisymmetric partially premixed flames by use of rainbow schlieren deflectometry, *Applied Optics*, **41**, 1922–1928, 2002.
6. Xiao, X., Choi, C. K and Puri, I. K., Temperature measurements in steady two-dimensional partially premixed flames using laser interpherometric holography, *Combustion Flame*, **120**, 318–332, 2000.
7. Raman, C. V. and Krishnan, K. S., A new type of secondary radiation, *Nature*, **121**, 501–502, 1928.
8. Regnier, P. P. and Taran, J. P. E., On the possibility of measuring gas concentrations by stimulated Anki-Stokes scattering., *Applied Physics Letters*, **23**, 240–242, 1973.
9. Eisner, A. D. and Rosner, D. E., Experimental studies of soot particle thermophoresis in non-isothermal combustion gases using thermocouple response technique, *Combustion Flame*, **61**, 153–166, 1985.
10. McEnally, C. S., Koylu, U. O., Pferfferle, L. and Rosner, D. E., Soot volume fraction and temperature measurements in laminar non-premixed flames using thermocouples, *Combustion Flame*, **109**, 701–720, 1997.
11. Sahu, K. B., Datta, A., Sen, S. and Sarkar, A., Concentration distributions of intermediate hydrocarbons and soot in propane-air triple flames, *Journal of the Energy Institute*, **82**, 185–196, 2009.
12. Arana, C. P., Pontoni, M., Sen, S. and Puri, I. K., Field measurements of soot volume fractions in laminar partially premixed coflow ethylene/air flames, *Combustion Flames*, **134**, 362–372, 2004.
13. Santoro, R. J. and Shaddix, C. R., Laser-induced incandescence. In: K.K, Hoinghaus, J.B. Jeffries (Eds.) *Applied Combustion Diagnostics*, Taylor & Francis Group, 252–286, 2002.

Appendix: Thermodynamic Properties

Tables A.1–A.3 were prepared using curve-fit coefficients available in CHEMKIN Database [Kee, R.J., Rupley, F.M. and Miller, J.A., "The Chemkin Thermodynamic Data Base," Sandia Report, SAND87-8215B, March 1991].

Table A.4 was obtained from CHEMKIN Database [Kee, R.J., Rupley, F.M. and Miller, J.A., "The Chemkin Thermodynamic Data Base," Sandia Report, SAND87-8215B, March 1991].

TABLE A.1 SENSIBLE ENTHALPY CHANGE [$\Delta h(T) = h(T) - h(298.15)$kJ / kmol] FOR DIFFERENT GASES

Temperature (K)	CO $\Delta h_f(\frac{kJ}{mol}) = -110541$	CO_2 $\Delta h_f(\frac{kJ}{mol}) = -393546$	H_2O $\Delta h_f(\frac{kJ}{mol}) = -241845$ $h_{fg}(\frac{kJ}{mol}) = 44010$	O_2 $\Delta h_f(\frac{kJ}{mol}) = 0$	N_2 $\Delta h_f(\frac{kJ}{mol}) = 0$	H_2 $\Delta h_f(\frac{kJ}{mol}) = 0$
298.15	0	0	0	0	0	0
300	54	69	62	55	54	54
400	2980	4004	3459	3031	2974	2954
500	5943	8301	6947	6097	5920	5875
600	8955	12899	10528	9254	8905	8807
700	12029	17748	14209	12502	11942	11749
800	15175	22809	18005	15837	15045	14700
900	18400	28046	21930	19249	18222	17668
1000	21696	33424	25992	22721	21468	20664
1100	25046	38909	30190	26231	24769	23703
1200	28440	44487	34516	29774	28117	26788
1300	31874	50147	38962	33349	31509	29918
1400	35344	55880	43519	36954	34938	33090
1500	38847	61679	48179	40588	38403	36305
1600	42379	67536	52937	44251	41898	39561
1700	45936	73443	57784	47941	45421	42856
1800	49517	79396	62714	51658	48969	46190
1900	53118	85389	67722	55400	52539	49560
2000	56737	91416	72802	59166	56128	52966

(Continued)

TABLE A.1 (*Continued*) SENSIBLE ENTHALPY CHANGE [$\Delta h(T) = h(T) - h(298.15)$kJ / kmol] FOR DIFFERENT GASES

Temperature (K)	CO $\Delta h_f(\frac{kJ}{mol}) = -110541$	CO_2 $\Delta h_f(\frac{kJ}{mol}) = -393546$	H_2O $\Delta h_f(\frac{kJ}{mol}) = -241845$ $h_{fg}(\frac{kJ}{mol}) = 44010$	O_2 $\Delta h_f(\frac{kJ}{mol}) = 0$	N_2 $\Delta h_f(\frac{kJ}{mol}) = 0$	H_2 $\Delta h_f(\frac{kJ}{mol}) = 0$
2100	60371	97473	77949	62957	59735	56406
2200	64020	103558	83157	66770	63358	59880
2300	67682	109665	88422	70606	66994	63386
2400	71354	115793	93740	74464	70642	66922
2500	75036	121939	99107	78342	74302	70489
2600	78727	128102	104520	82242	77971	74084
2700	82426	134278	109974	86160	81648	77706
2800	86132	140468	115467	90099	85334	81355
2900	89844	146671	120995	94056	89027	85030
3000	93563	152884	126557	98031	92726	88729
3100	97287	159109	132150	102025	96432	92452
3200	101017	165344	137771	106035	100143	96197
3300	104751	171589	143419	110063	103860	99964
3400	108491	177845	149092	114107	107583	103753
3500	112236	184112	154788	118168	111310	107561
3600	115985	190389	160506	122244	115043	111390
3700	119740	196676	166245	126336	118780	115238
3800	123500	202975	172003	130442	122523	119104
3900	127264	209284	177779	134564	126270	122989
4000	131033	215604	183573	138699	130022	126892

TABLE A.2 ENTHALPY OF FORMATION (Δh_f^*) FOR DIFFERENT GASES IN KJ/KMOL

Temperature (K)	CO	CO_2	H_2O	O_2	N_2	H_2
298.15	−110541	−393546	−241845	0	0	0
300	−110530	−393547	−241864	0	0	0
400	−110121	−393617	−242856	0	0	0
500	−110017	−393712	−243821	0	0	0
600	−110156	−393844	−244752	0	0	0
700	−110477	−394013	−245636	0	0	0
800	−110924	−394213	−246459	0	0	0
900	−111450	−394433	−247207	0	0	0
1000	−112022	−394659	−247877	0	0	0
1100	−112619	−394875	−248473	0	0	0
1200	−113240	−395083	−249003	0	0	0
1300	−113881	−395287	−249475	0	0	0
1400	−114543	−395488	−249893	0	0	0
1500	−115225	−395691	−250265	0	0	0
1600	−115925	−395897	−250595	0	0	0
1700	−116644	−396110	−250888	0	0	0
1800	−117380	−396332	−251149	0	0	0
1900	−118132	−396564	−251382	0	0	0
2000	−118902	−396808	−251592	0	0	0
2100	−119687	−397065	−251781	0	0	0
2200	−120488	−397338	−251953	0	0	0
2300	−121305	−397626	−252111	0	0	0
2400	−122137	−397931	−252259	0	0	0
2500	−122984	−398253	−252397	0	0	0
2600	−123847	−398594	−252530	0	0	0
2700	−124724	−398952	−252658	0	0	0
2800	−125616	−399329	−252783	0	0	0
2900	−126523	−399725	−252907	0	0	0
3000	−127446	−400140	−253032	0	0	0
3100	−128383	−400573	−253159	0	0	0

(Continued)

TABLE A.2 (*Continued*) ENTHALPY OF FORMATION (Δh_f^*) FOR DIFFERENT GASES IN KJ/KMOL

Temperature (K)	CO	CO_2	H_2O	O_2	N_2	H_2
3200	−129335	−401025	−253288	0	0	0
3300	−130303	−401495	−253421	0	0	0
3400	−131285	−401983	−253559	0	0	0
3500	−132283	−402489	−253702	0	0	0
3600	−133295	−403013	−253850	0	0	0
3700	−134323	−403553	−254006	0	0	0
3800	−135366	−404110	−254167	0	0	0
3900	−136424	−404684	−254336	0	0	0
4000	−137497	−405273	−254513	0	0	0

TABLE A.3 GIBBS FUNCTION OF FORMATION (Δh_f^*) FOR DIFFERENT GASES IN KJ/KMOL

Temperature (K)	CO	CO_2	H_2O	O_2	N_2	H_2
298.15	−137163	−394428	−228607	0	0	0
300	−137328	−394433	−228525	0	0	0
400	−146332	−394718	−223929	0	0	0
500	−155403	−394983	−219084	0	0	0
600	−164470	−395226	−214049	0	0	0
700	−173499	−395443	−208861	0	0	0
800	−182473	−395635	−203550	0	0	0
900	−191386	−395799	−198141	0	0	0
1000	−200238	−395939	−192652	0	0	0
1100	−209030	−396056	−187087	0	0	0
1200	−217768	−396155	−181482	0	0	0
1300	−226453	−396236	−175836	0	0	0
1400	−235087	−396301	−170155	0	0	0
1500	−243674	−396352	−164446	0	0	0
1600	−252214	−396389	−158714	0	0	0
1700	−260711	−396414	−152962	0	0	0

(*Continued*)

TABLE A.3 (*Continued*) GIBBS FUNCTION OF FORMATION (Δh_f^*) FOR DIFFERENT GASES IN KJ/KMOL

Temperature (K)	CO	CO_2	H_2O	O_2	N_2	H_2
1800	−269164	−396425	−147194	0	0	0
1900	−277576	−396424	−141412	0	0	0
2000	−285948	−396410	−135618	0	0	0
2100	−294281	−396384	−129815	0	0	0
2200	−302576	−396346	−124003	0	0	0
2300	−310835	−396294	−118183	0	0	0
2400	−319057	−396230	−112357	0	0	0
2500	−327245	−396061	−106524	0	0	0
2600	−335399	−396061	−100687	0	0	0
2700	−343519	−395957	−94844	0	0	0
2800	−351606	−395840	−88996	0	0	0
2900	−359661	−395708	−83144	0	0	0
3000	−367684	−395562	−77288	0	0	0
3100	−375677	−395403	−71428	0	0	0
3200	−383639	−395229	−65563	0	0	0
3300	−391571	−395041	−59695	0	0	0
3400	−399474	−394838	−53822	0	0	0
3500	−407347	−394620	−47945	0	0	0
3600	−415192	−394388	−42064	0	0	0
3700	−423008	−394141	−36179	0	0	0
3800	−430796	−393879	−30289	0	0	0
3900	−438557	−393602	−24395	0	0	0
4000	−446291	−393311	−18497	0	0	0

TABLE A.4 ENTHALPY OF FORMATION $\left(\Delta h_f^*\right)$ AT REFERENCE TEMPERATURE (298.15 K) FOR COMMON FUELS IN KJ/KMOL

Fuel	Enthalpy of Formation
Methane, CH_4	–74831
Ethane, C_2H_6	–84667
Propane, C_3H_8	–103847
n-Butane, C_4H_{10}	–124733
n-Heptane, C_7H_{16}	–187820
n-Octane, C_8H_{18}	–208447
n-Decane, $C_{10}H_{22}$	–249659
Ethene, C_2H_4	52283
Ethyne, C_2H_2	226748

Index

Note: Page numbers in italic and bold refer to figures and tables respectively.